EPICYCLIC DRIVE TRAINS

EPICYCLIC DRIVE TRAINS
Analysis, Synthesis, and Applications

by HERBERT W. MÜLLER

Werner G. Mannhardt, translator
John H. Glover, technical editor

WAYNE STATE UNIVERSITY PRESS
Detroit

Library of Congress Cataloging in Publication Data

Müller, Herbert W., 1914–
  Epicyclic drive trains.

  Translation of: Die Umlaufgetriebe.
  Bibliography: p.
  Includes index.
  1. Gearing, Planetary. I. Mannhardt, Werner G. II. Glover,
John H. III. Title.
TJ202.M8313   621.8'33   81-11422
ISBN 0-8143-1663-8   AACR2

ISBN-13 : 978-0-8143-1663-4   ISBN-10 : 0-8143-1663-8

CONTENTS

TABLES

Preface to the American Edition

Since its publication in 1971, this book has found increasing acceptance at German universities and technical schools as well as in pertinent industries. I felt encouraged, therefore, to present its translation to English-speaking engineers. The publication of this edition is due to the painstaking and careful work of the translator, Mr. Werner G. Mannhardt, whose logical train of thought is of particular value to the student in the clarification of sometimes complex problems. He has been decisively supported by the technical editor, Mr. John H. Glover, who contributed valuable advice based on his own practical experience with epicyclic drive trains, and who also reviewed the manuscript with great care. My special appreciation is due these two gentlemen for their painstaking efforts.

I also want to thank the publishing house of Springer, for permitting this English translation, and the Wayne State University Press, for careful editing and excellent cooperation.

Herbert W. Müller
Summer, 1980

From the Preface to the German Edition

In view of the numerous publications about epicyclic drive trains it may seem superfluous to write a book about this subject. However, a thorough study of the field revealed that previous authors mostly limited themselves to especially interesting subproblems, or to the analysis of particular, executed transmissions. A comprehensive treatment of revolving drive trains, which could serve as a reliable guide for the design engineer and at the same time as a textbook, was not available. The present book is an attempt to meet this need.

It was my particular objective to arrive at a clear and concise description of the multitude of revolving drive trains in terms common to all, grouping transmissions of the same type. Unified methods of analysis, orderly arrangement of material, and definitions in part novel were introduced to represent in a generally valid form the characteristic parameters which aid in the selection and design of the best drive train.

Herbert W. Müller
January, 1971

Translator's Note

The American edition of the present book is a completely revised version of the German original. While it closely follows the established train of thought, some sections have been completely rewritten to reflect recent developments. In many cases the derivations of formulas have been extended and intermediate steps have been retained to help the reader follow difficult arguments.

The fact that some of the discussed drive trains had not been described previously in English-language publications presented a major difficulty in preparing this translation, because it made it necessary to coin a number of new terms and assign names and symbols to some drive-train components. Great care has been taken to select simple and self-explanatory terms which do not inadvertently conflict with already established terminology.

Epicyclic drive trains play an increasingly important role in many industries. Each three-shaft drive train possesses six different usable speed-ratios, and compound transmissions can be constructed by coupling together two or more drive trains with one or two of their shafts. Thus, a large number of speed-ratios can be obtained with even simple compound transmissions, and selecting the most efficient and economical drive from this multitude is becoming increasingly important.

Besides being eminently practical by providing concise guide rules and worksheets, this book holds a fascination of its own by hinting that the analysis and synthesis of complex drive systems could be further facilitated by establishing a multivalued logic system analogous to the bivalent logic system upon which digital electronics is built. It is hoped that this book is not an end in itself but rather will spawn new developments. Accordingly, suggestions for improvement in concept, emphasis, and clarity are welcome.

I would like to take this opportunity to express my appreciation to Dr. Müller for his help in the preparation of this translation. Thanks are due to John H. Glover for his many suggestions and his help in defining new terms and symbols, as well as for his patience in combing through numerous drafts of the manuscript to identify inconsistencies and errors. Thanks are due also to Darle W. Dudley for his review and encouragement. Finally I want to thank my wife for her patience in typing the several drafts of the manuscript, and the Wayne State University Press for their help in bringing several years of work to a promising conclusion.

Werner G. Mannhardt

SYMBOLS

Note: other, less frequently used symbols are defined where they are first introduced.

F	Degree of freedom or degree of mobility of a drive train (used when static degree of freedom is not considered)
F_{kim}	Kinematic degree of freedom designating the number of independent shaft *speeds* which must be given, or can be freely chosen
F_{op}	Operating degree of freedom $F_{op} = F_{kim} + F_{stat} = N_{cs}$
F_{stat}	Static degree of freedom designating the number of independent shaft *torques* which must be given or can be freely chosen
i	Speed-ratio (transmission-ratio) of a constrained drive train

$$i_o = i_{12} = \frac{n_1}{n_2}$$

$$i_{o'} = i_{1'2'} = \frac{n_{1'}}{n_{2'}}$$

Basic speed-ratio (carrier shaft locked)

$$i_{1s} = \frac{n_1}{n_s}$$

$$i_{ac} = \frac{n_a}{n_c}$$

First subscript of this speed-ratio represents numerator shaft, second subscript denominator shaft; third shaft (*2* or *b*) is locked

$$i_{1\,II} = \frac{n_1}{n_{II}}$$

Series speed-ratio of a bicoupled transmission with a locked connected coupling shaft S

k	Speed-ratio between any two shafts of a revolving drive train with more than one degree of mobility, such as a planetary transmission with three rotating connected shafts

$k_{1s} = \dfrac{n_1}{n_s}$

First subscript of this speed-ratio represents numerator shaft, second subscript denominator shaft

$k_{ac} = \dfrac{n_a}{n_c}$

N_{cs} Number of connected shafts in a drive train

n Shaft speed (rpm)

P Power

P_L Power-loss

P_{Lt} Tooth-friction power-loss

P_{R1}, P_{R2} Rolling power of shaft *1*, or shaft *2*; in revolving hydrostatic drives, hydraulic power of shaft *1* or shaft *2*

P_{RI}, P_{RII} Analogous to P_{R1}, P_{R2}, series power of shaft *I* or shaft *II* of a bicoupled drive train

P_{s1}, P_{s2} Coupling-power of shaft *1* or shaft *2*

P_t Power transmitted in a gear mesh by pure rolling

$r1 = \pm 1$ Exponent with the sign of the rolling-power P_{R1} of shaft *1*

$rI = \pm 1$ Exponent with the sign of the series power P_{RI} of shaft *I*, analogous to $r1$

T Torque

I, II, III   Characterize the component transmissions of compound revolving drive trains; in constrained bicoupled drive trains, *I* represents main component transmission and *II* auxiliary component transmission

Φ Pressure angle

ϵ_0 Ratio of theoretical input power of auxiliary component transmission to external power of a constrained or variable bicoupled transmission (losses are neglected)

ϵ Ratio of actual input power of auxiliary component transmission to input power of a constrained or variable bicoupled transmission (losses are considered)

$\zeta = 1 - \eta$ Tooth-friction loss factor

η Efficiency

$\eta_o = \eta_{12} \approx \eta_{21}$ Basic efficiency (that is, efficiency of a revolving drive train with a locked carrier shaft)

η_{s1}
$\eta_{2<s}^1$ Subscripts closest to η represent input shafts; others output shafts
$\eta_{2>s}^1$

μ_t Mean tooth-friction coefficient

φ Angular position; adjustment ratio of a variable bicoupled drive train

φ' Adjustment ratio of auxiliary (or variable) component transmission of a variable bicoupled drive train

ω Angular velocity

MATHEMATICAL SYMBOLS

$\neq$ not equal to

$\approx$ approximately equal to

$<$ less than

$>$ greater than

$\leq$ less than or equal to

$\geq$ greater than or equal to

$\lessgtr$ less than, equal to, or greater than

$\ll$ much less than

$\gg$ much greater than

$\triangleq$ analogous to, or may be substituted for; example: $1 \triangleq I$, $2 \triangleq II$, and $s \triangleq S$. Like the $=$ sign, $\triangleq$ operates in both directions, that is, $1 \triangleq I$ implies that $I \triangleq 1$, etc.

$\sim$ proportional to

$n!$ factorial $n = n(n-1) \ldots 1$

$\binom{n}{r}$ combinations of n different things taken r at a time $= n!/(r!)(n-r)!$ Example: $\binom{3}{2} = 3!/(2!)(1!) = (3 \cdot 2 \cdot 1)/(2 \cdot 1 \cdot 1) = 3$

SYMBOLS WHICH MAY ALSO BE USED AS SUBSCRIPTS

$A, B, C, D, \ldots$	connected shafts of compound drive trains which consist of several component transmissions
$a, b, c, \ldots$	connected shafts of a simple revolving drive train whose internal structure is unknown or arbitrary
c, c'	symbol or subscript indicating a connected coupling shaft
F	free coupling shaft of a bicoupled transmission
f, f'	symbol or subscript indicating a free coupling shaft
in	subscript indicating an input
m, m'	symbol or subscript indicating a monoshaft
out	subscript indicating an output
p, p_1, p_2	planet, planet meshing with gear 1, planet meshing with gear 2
S	connected coupling shaft of a bicoupled transmission (analogous to the carrier shaft of a simple revolving drive train)
$s, s', s'', \ldots$	carrier (spider) or carrier shaft of a revolving drive train
tot	subscript indicating a total quantity such as power, efficiency, etc.
z	number of teeth of a gear (see ISO 701-1976)
I, II	monoshafts of a bicoupled transmission (the connected shafts of a bicoupled transmission which are analogous to shafts 1 and 2 of a simple revolving drive train)
$1, 2$ $1', 2'$ $1'', 2''$	connected shafts of a revolving drive train; they may be the central gears (shafts) in reverted "planetary" transmissions, corresponding to the connected shafts of the associated basic drive train or the central gear (shaft) and the planet gear (shaft) of open "planetary" transmissions
$3, 4, 5, \ldots$	gears or shafts of conventional drive trains (e.g., intermediate drive trains)
$', '', ''', \ldots$	superscript designations for second, third, etc., component transmissions of compound revolving drive trains

DEFINITIONS

adjustment range	difference between the highest and the lowest speeds or speed-ratios (at a given constant input speed) of a variable-speed drive
adjustment-ratio	quotient of speed-ratio limits of a variable-speed drive
auxiliary (component) transmission	component transmission II of a constrained bicoupled drive train, or the variable component transmission of a variable bicoupled drive train
basic bicoupled transmission	bicoupled transmission with a locked coupling shaft S
basic efficiency	efficiency of the basic transmission
basic transmission	revolving drive train with a locked carrier shaft s
bicoupled transmission; bicoupled drive train (see also "higher bicoupled transmission")	compound drive train which consists of two simple revolving drive trains internally coupled at two of their three shafts: the free-coupling shaft and the connected-coupling shaft
central gears	gears whose axes coincide with the central axis of an epicyclic gear train
circulating power	futile power which contributes nothing to the external output power, but flows in a closed path through both the main and the auxiliary component transmissions of some constrained bicoupled drive trains
connected shaft	input or output shaft which transmits an external power to or from the drive train

constrained bicoupled drive train	bicoupled drive train with a fixed monoshaft and, thereby, one degree of freedom
constrained drive train	drive train with only one degree of freedom
conventional transmission	drive train with spatially fixed axes such as simple reduction drives
coupling point	operating condition of a revolving drive train where all three connected shafts rotate with the same speed
coupling-power	power transmitted by a simple or compound revolving drive train while all three connected shafts rotate with the same speed
degree of mobility	degree of freedom of a drive train (that is, the number of independent shaft speeds)
difference shaft	shaft of a simple or compound revolving drive train whose absolute torque is equal to the difference of the absolute torques transmitted by the other two shafts
epicycle	circle generating an epicycloid or hypocycloid
epicyclic drive train	mechanical system of drive members, some of which move around the circumference(s) of one or more central drive members on an arm or carrier; thus epicyclic drives may have two or three connected shafts. Drive members moving on an epicycle have a motion compounded of a rotation about their own axes and a circular orbit or revolution about the central axis.
epicycloid	curve traced by a point of a circle rolling on the outside of a fixed circle (see "hypocycloid")
four-shaft bicoupled transmission	transmission type obtained when the free-coupling shaft of a bicoupled transmission becomes an additional connected shaft
free coupling shaft	shaft which couples two or more different drive trains but has no external connection
functionally-equivalent bicoupled transmission	simple bicoupled transmission with same power-flow and theoretical efficiency as a given reduced bicoupled transmission

futile power	real additional power comparable to the reactive power in electrical circuits; it contributes nothing to the transmitted external power. The circulating power present in some constrained bicoupled drive trains and variable bicoupled drive trains is a futile power.
higher bicoupled transmission	bicoupled transmission which consists of more than two simple three-shaft transmissions, each of which has at least two of its three shafts coupled to one or more of the other component transmissions
hypocycloid	curve traced by a point of a circle rolling on the inside of a fixed circle (see "epicycloid")
loss-symmetrical drive train	drive train which maintains the same efficiency when its input and output shafts are interchanged
main (component) transmission	component transmission I of a constrained bicoupled transmission; two of its three rotating shafts are the connected shafts of the compound transmission
minus train	revolving drive train with a negative basic speed-ratio
monoshaft	shaft of a compound transmission which is not internally coupled to another component transmission; monoshafts which are either input or output shafts of the compound transmissions are called "connected monoshafts"; if locked by a rigid connection with the housing, they are called "fixed monoshafts"
open planetary drive train (see also "epicyclic drive trains")	simple epicyclic drive train with a single central gear and a connected planet shaft
partial-locking	internal state of operation of a three-shaft positive-ratio drive train which is capable of self-locking but whose carrier shaft is not the sole output shaft

partial-power shaft	connected shaft of a revolving drive train which carries only a fraction of the total external power
planetary drive train	because the planets rotate about their own axes while executing an orbit about a central axis, the terms "planetary drive train" and "epicyclic drive train" often are used as synonyms. There exist other, nonmechanical drive trains with rotating shafts and drive members which revolve about the central axis but which do not simultaneously rotate about their own axes (e.g., the pistons in fig. 44). Where the discussion is extended to these lesser known drives, the term "revolving drive train" has been used.
plus train	revolving drive train with a positive basic speed-ratio
power division	operating condition in which the drive train operates with two or more output shafts but only a single input shaft (not to be confused with power-branching, defined in section 2)
power summation	operating condition in which the drive train operates with two or more input shafts but only a single output shaft
reduced bicoupled transmission	bicoupled transmission simplified by combining the two carriers, the two identical central gears, and the identical planets of its components into single elements
reverted epicyclic (planetary) transmission	epicyclic (planetary) transmission whose connected shafts are coaxial
revolving-carrier speed-ratio	two-shaft speed-ratio i of a simple revolving drive train which is obtained when the carrier (housing) rotates but one of the other two (external) shafts is locked
revolving drive train (see also "epicyclic drive train")	mechanical or hydrostatic drive train with the frame (housing, carrier) suspended on a stationary axis in such a way that it can revolve around it, together with all members of the drive train (epicyclic or planetary

drives, hydrostatic revolving drives [52], or revolving mechanisms [55])

rolling-power
that part of the total power transmitted by the pure rolling of gears or traction wheels; rolling speed is measured relative to the carrier only

self-locking
internal state of a mechanism locked by friction so that the output speed is zero regardless of the magnitude of the input torque; the efficiency is negative, which means that the mechanism can move only when additional external power is supplied to the output shaft; some two-shaft and three-shaft positive-ratio transmissions become self-locking when a carrier shaft is the sole output shaft

simple planetary transmission
planetary transmission with one carrier, at least one planet, and one or two central gears or traction wheels; in operation, the shafts of any two or three of these component members may be connected shafts

summation shaft
shaft of a simple or compound revolving drive train whose torque is equal and opposite to the sum of the torques of the other two connected shafts

superposition drive
drive train with more than one degree of mobility (that is, more than two connected shafts) which can be operated with power division or power summation, that is, with power superposition

total-power shaft
the connected shaft of a three-shaft transmission which transmits the total external power

variable bicoupled transmission
constrained bicoupled transmission with a variable auxiliary (component) transmission

variable transmission
conventional (or revolving) drive train with an infinitely variable (basic) speed-ratio

I. PROPERTIES OF EPICYCLIC DRIVE TRAINS

A. Differentiation from Conventional Drive Trains

Section 1. General Principles and Sign Conventions

Revolving drive trains, and their laws of motion, stem from the conventional drive trains with fixed axes of rotation. Therefore, we shall begin with a brief discussion of the fundamental principles governing the operation of conventional transmissions.

These principles are valid for all types of conventional drive trains such as gear trains, traction drives, belt drives, or hydraulic transmissions, as long as the influence of the slip (in friction-coupled mechanical drives) or the leakage losses (in hydrostatic drives) can be neglected.

A conventional transmission consists of a housing and two external shafts which are internally connected through gears. During operation the housing must be secured to a base. The most important characteristic of a transmission is its speed-ratio,

$$i = \frac{n_3}{n_4} = \text{const.} \tag{1}$$

which is determined by its internal construction. To conform with ISO Standard 701, i rather than m is used to denote the speed-ratio. The symbols n_3 and n_4 stand for the speeds of the external shafts which can be arbitrarily labeled *3* and *4* (fig. 1).

The two external shafts may have the same or an opposite direction of rotation. To account for this possibility, the direction of rotation of each external shaft must be indicated by a plus or minus sign. This leads to R 1, which describes a sign convention that holds true for any revolving drive train:

The speeds of all parallel shafts of a drive train which rotate   R 1
in the same direction are designated by the same sign.

In this context it is immaterial which of the two possible directions of rotation is defined as positive.

3

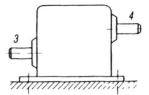

Fig. 1. Conventional transmission with fixed axes of rotation.

According to this sign convention, the speed-ratio

$i > 0$ (positive) when the two external shafts have the same directions of rotation

and

$i < 0$ (negative) when the two shafts have opposite directions of rotation.

A sign convention for the direction of rotation of nonparallel shafts is purposely omitted, since such a definition is without practical significance for the analysis of revolving drives.

A second characteristic of a transmission is the ratio of the external torques acting on its shafts. The directions of these torques also must be identified by appropriate signs. If a transmission shaft is *driven* by an external torque, it turns in the *same* direction in which the torque acts on it so that an *input shaft* is characterized by the fact that its torque and speed have the *same* sense of rotation. In mathematical terms this means that its torque and speed have the *same sign*.

If the *output shaft* of a transmission drives a power consuming machine (which could be replaced by a friction brake) it must overcome the "brake torque." However, this external torque acts *against* the direction of rotation, so that the external torque and speed of the output shaft have an opposite sense of rotation. In mathematical terms this means that its torque and speed have opposite signs.

This leads to R 2, which gives a sign convention for the torques acting on the external shafts:

> *Torque and speed have the same signs for an input shaft, but*   R 2
> *opposite signs for an output shaft.*

If the shaft power of a transmission is calculated according to this rule, the input power becomes positive and the output power negative. This is in agreement with the principle of conservation of energy, which states that for a system in the state of equilibrium, the sum of all input powers must equal zero. In mathematical terms a power output can be considered as a negative power input. Thus we can formulate a further guide rule:

An input power is always positive; an output power always negative. R 3

If two gear stages or two drive trains are connected through a common coupling shaft as shown in fig. 2, the speeds of the two coupled gears have

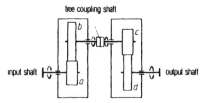

Fig. 2. Direction of the torques acting on the two component gear trains. The sign of the torque changes at the coupling shaft.

equal magnitudes and signs, so that $n_b = n_c$. The torques which act on these gears have equal magnitudes but opposite directions and, therefore, have opposite signs,

$$T_b = -T_c ,$$ (2)

as follows from the equilibrium conditions for the reaction torques $-T_b$ and T_c. However, eq. (2) is valid only for a "free coupling shaft," that is, a shaft whose torque is not altered by, for example, a gear mounted on it. This condition is expressed in the following guide rule:

The two equal torques which act on a free coupling shaft have R 4
opposite signs.

The sign conventions previously established are universally valid for all types of drive trains. Their consistent consideration will make it easier to understand and analyze even complicated revolving drives.

The ratio of the external shaft torques of a transmission as shown in fig. 1 can be determined without knowledge of its inner construction when only its transmission ratio i and its internal friction losses or its mechanical efficiency η are known. Since such a drive train can neither generate nor consume energy, the sum of all energy supplied per unit time (the sum of all input powers) must equal zero:

$$\Sigma P = P_3 + P_4 + P_L = 0 .$$ (3)

In this equation, the power P_3 or P_4 of the respective output shaft and the power loss P_L, which is converted into heat, are removed from the system and, therefore, have a negative sign.

The power loss P_L can be related to the input power P_{in} through the "loss factor" ζ where:

$$\zeta = -\frac{P_L}{P_{in}} \tag{4}$$

or through the "efficiency" η where:

$$\eta = -\frac{P_{out}}{P_{in}} = -\frac{-P_{in} - P_L}{P_{in}} = 1 + \frac{P_L}{P_{in}} = 1 - \zeta . \tag{5}$$

The efficiency is usually defined as the ratio between the absolute values of the output and the input power and, therefore, has a positive value. However, if the signs of the powers are chosen according to the previously given rules, then the ratio between the output and the input power becomes negative. Consequently it becomes necessary to introduce a minus sign in eq. (5), so that the efficiency becomes positive as usual. The same correction must be applied to the loss factor in eq. (4).

From eq. (5) it follows that:

$$P_{out} = -\eta P_{in} ,$$

or

$$T_{out} n_{out} = -\eta T_{in} n_{in} ,$$

so that the torque ratio becomes:

$$\frac{T_{out}}{T_{in}} = -\eta \cdot \frac{n_{in}}{n_{out}} . \tag{6}$$

If shaft *4* is the input shaft, the torque ratio becomes:

$$\left(\frac{T_3}{T_4}\right)_{P_3 < 0} = -\eta_{43} \cdot \frac{n_4}{n_3} = -\eta_{43} \cdot \frac{1}{i} , \tag{7}$$

where the sequence of the indices of η indicates the direction of the power-flow for which this efficiency is valid. Thus, η_{43} denotes the efficiency when shaft *4* is the input and shaft *3* the output. When shaft *3* is the input shaft, the torque ratio becomes:

$$\left(\frac{T_3}{T_4}\right)_{P_3 > 0} = -\frac{1}{\eta_{34}} \cdot \frac{n_4}{n_3} = -\frac{1}{\eta_{34} i} . \tag{8}$$

This shows that in conventional transmissions the torque ratio depends not only on the efficiency and the speed-ratio i but also on the direction of the power-flow.

In a transmission with several stages I, II etc., through which the power flows in series, power losses occur in every stage. These power losses add up to a total power loss P_L where

$$P_L = P_{LI} + P_{LII} + P_{LIII} + \cdots .$$

The overall efficiency η of such a multistage transmission can be calculated from the efficiencies η_I, η_{II}, η_{III}, . . . or from the loss factors ζ_I, ζ_{II}, ζ_{III}, . . . of the individual stages as follows:

$$\eta = \frac{-P_{out}}{P_{in}} = \frac{-P_{outI}}{P_{inI}} \cdot \frac{-P_{outII}}{P_{inII}} \cdot \frac{-P_{outIII}}{P_{inIII}} \cdots ,$$

where $-P_{outI} = P_{inII}$, $-P_{outII} = P_{inIII}$ etc. Thus,

$$\eta = \eta_I \eta_{II} \eta_{III} \cdots = (1 - \zeta_I)(1 - \zeta_{II})(1 - \zeta_{III}) \cdots , \tag{9}$$

or if ζ is much smaller than 1:

$$\eta \approx 1 - (\zeta_I + \zeta_{II} + \zeta_{III} + \cdots). \tag{10}$$

Section 2. Conventional and Revolving Drive Trains

To satisfy the equilibrium conditions of engineering mechanics, the sum of all torques which act on a drive mechanism must equal zero. Since, however, in a drive train of the type shown in fig. 1 the sum of the shaft torques can equal zero only when $i \approx +1$, a reaction torque T_R exerted by the base must normally act on the transmission housing, so that

$$T_3 + T_4 + T_R = 0 .$$

If the housing which thus constitutes an active support for the internal shafts is not secured on a base, but rather is allowed to rotate about a concentric axis, the drive train acquires a third shaft which likewise can transmit power to or from the system (figs. 3 and 4).

Consequently, a conventional drive train with fixed axes of rotation becomes a revolving drive train where all members which are actively supported by the rotating housing, or "carrier," revolve about the carrier axis when power is transmitted through the drive train.

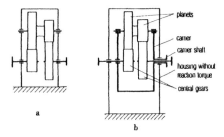

Fig. 3. Derivation of a reverted planetary gear train from a coaxial conventional transmission with a positive speed-ratio: *a*, conventional transmission with fixed axes of rotation; *b*, planetary transmission.

If the conventional transmission was a gear train or traction drive, then it becomes a revolving gear train or traction drive, and all of its internal shafts and gears, whose axes do not coincide with the central axis of rotation of the carrier, revolve around this central axis while rotating about their own axes at the same time, much like planets revolve around the sun. Drive trains of this type are therefore called "planetary drive trains" or "planetary transmissions." The revolving gears are called "planet gears" or "planets," the non-revolving central gears are called "sun gears" when their planet mesh is inside the orbit of the planets, or "annular (ring) gears" when their planet mesh is outside the orbit of the planets.

If the conventional drive train housing was a hydrostatic transmission, it becomes a revolving hydrostatic drive, as shown in fig. 44, whose originally stationary power-transmitting parts such as the pressure lines revolve in unison with the housing.

The simplest and most widely used types of revolving drive trains are the planetary transmissions. Therefore we shall present the fundamental principles which govern the operation of the revolving drive trains by using plan-

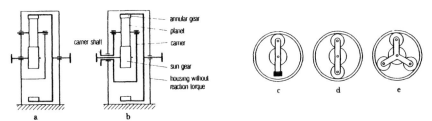

Fig. 4. Derivation of a planetary gear train from a coaxial conventional transmission with a negative speed-ratio: *a*, conventional transmission; *b*, planetary transmission; *c-e*, some possible arrangements of the planets.

etary transmissions as examples. As explained in section 33k, these basic principles which pertain to the speed-ratios, the torque ratios, the external and internal powers, as well as to the efficiencies and the capability to be self-locking, are also valid, without restriction, for hydrostatic or other revolving drives. The term "revolving drive train," therefore, is used throughout this text in discussions which have general validity, even when planetary transmissions are used as examples. The term "planetary drive train" is reserved for discussions which relate only to this particular design.

In actual planetary drive trains, the rotating housing structure is reduced to a "spider" or "carrier" which supports only the planet shafts. The total gear train is encased in a new housing which, unlike the housing of a conventional transmission, does not require a supporting torque from the base as long as the three external shafts can freely rotate (figs. 3, 4, and 6).

Frequently, however, one of the central gears is attached rigidly to the housing, and is thus prevented from rotating (fig. 5). The torque of the locked gear must then be counteracted by the housing from which it is transferred to the foundation. In the following text, transmissions of this type are denoted as "two-shaft transmissions." Likewise, revolving transmissions with three rotating shafts are called "three-shaft revolving drive trains," or simply "three-shaft transmissions."

If a conventional transmission with parallel shafts (fig. 6a) is converted into a planetary transmission, then only one of the two shafts can be arranged coaxially with the carrier and supported in the housing. The other shaft moves in a circular path around the carrier axis while simultaneously rotating about its own axis. Transmissions of this type are called "open-train planetary transmissions." To distinguish the open-train planetaries from the coaxial planetaries (figs. 3b and 4b) the latter are called "closed-train," or "reverted planetary transmissions," because the central gears and the planet(s), which are commonly supported at the carrier shaft, form a closed gear train 1-p_1-p_2-2, and the power-flow reverts back from the planet(s) to the central axis (figs. 19–28 and 32–42). In open planetary transmissions, the gears form an "open" gear train where the first gear, which is keyed to a connected shaft, is a central gear and the last gear, which is again keyed to a connected shaft, is a revolving planet, or vice versa. In a reverted planetary transmission, the planets usually are free-wheeling, intermediate gears, or "idlers."

It is, of course, possible to convert an open-train planetary (of the type shown in fig. 6b) into a closed-train planetary (of the type shown in fig. 7) by inserting a universal joint shaft between the planet shaft and the output shaft.

The revolving mass of a planet causes an imbalance which can be compensated by a counterweight (fig. 4c). In most cases, however, balancing is achieved through a symmetrical arrangement of several—usually three—

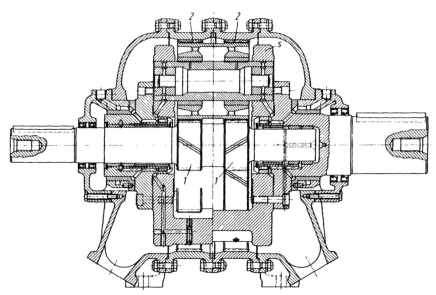

Fig. 5. Two-shaft planetary transmission with a locked central gear manufactured by Voith KG, Heidenheim. Series PH, for rated torques $T_s \approx 1.1 \cdot 10^2$ to $1.1 \cdot 10^6$ Nm and $i_o = -1.3$ to -9.

planets (figs. 4d and 4e). Thus, if the gear train is manufactured with sufficient accuracy, the tangential forces which act on the central gears are distributed over several gear meshes. The geometrical conditions for an even spacing of the planets around the circumference of the central gears are discussed in section 43.

In drive trains with several planets, the power-flow between the central gears and the planet carrier is branched between the planets. Therefore, this type of power-flow shall be called "power branching."

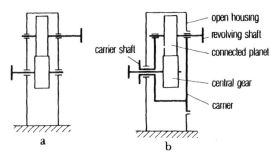

Fig. 6. Derivation of an open planetary gear train from a conventional gear train with parallel axes: *a*, conventional gear train with parallel axes; *b*, open planetary gear train.

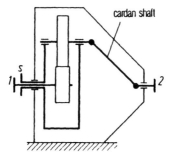

Fig. 7. Reverted planetary gear train derived from the open planetary gear train of fig. 6.

It is characteristic of power branching that the number of branches influences neither the speed characteristics nor the torques and powers of the central gears or the carrier shaft. Thus, theoretically, it does not influence the efficiency of planetary transmissions and need not be considered when the operating characteristics of revolving drive trains are analyzed. However, it is an important aspect of their design which influences the overall size of the drive train.

Power branching is not limited to planetary transmissions. As illustrated by figs. 8 and 9, it is also used in conventional transmissions of the type shown in figs. 3a and 4a. These transmissions are often incorrectly called

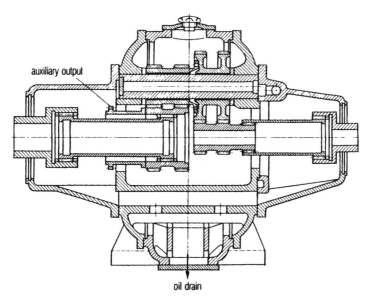

Fig. 8. Two-stage star transmission with positive speed-ratio according to Renk.

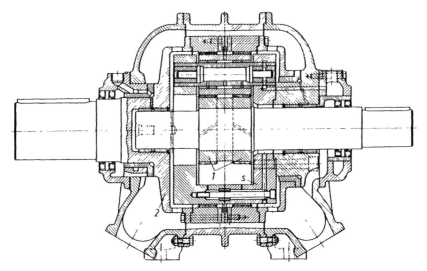

Fig. 9. Single-stage star transmission with negative speed-ratio manufactured by Voith KG, Heidenheim. Series RT, for rated torques $T_2 \approx 10^2$ to 10^6 Nm and $i = -1.3$ to -9.

planetary transmissions, although they do not possess the kinematic characteristics of revolving drive trains such as revolving planets. More appropriate are terms such as "branching transmissions" or, because of the appearance of the gear train when viewed in the axial direction, "star transmissions." However, both these terms are already in use.

Section 3. Degrees of Freedom of Transmissions

The degree of freedom of a physical system is defined as the number of independent variables which can and must be specified to describe clearly the state of the system. Although, in a drive mechanism, the forces and torques belong to these characteristic variables, the degree of freedom of a transmission is usually defined only by those variables which determine its mobility. Therefore, the number of degrees of freedom F of a transmission is usually understood to mean the number of independent speeds which can and must be specified to describe its state of motion unambiguously. Consequently, when dealing with drive trains, the degree of freedom F is also called the "degree of mobility F" [17, 20]. A drive train with only one degree of freedom or one degree of mobility is said to be "constrained."

Conventional transmissions whose shaft speeds have a permanently-fixed ratio (which depends on the design) have only *one* degree of freedom and, thus, only one speed can be freely chosen (figs. 3a, 4a, and 6a). However, revolving drive trains with three rotating shafts which evolved from conventional transmissions by allowing the housing to rotate freely, have one additional degree of freedom, or a total of two degrees of freedom (figs. 3b, 4b, 6b, and 44).

The degree of freedom of a transmission can be reduced in two ways:

1. By constraint, that is, by locking some of the shafts to the fixed housing.

2. By linkage, that is, by positively coupling some of the shafts.

In the first case, the speed n of a fixed shaft becomes zero. In the second case only one common speed can be freely chosen for two or more coupled shafts.

Thus, the constrained transmission with one degree of freedom (fig. 9) evolved from a revolving drive train with two degrees of freedom by constraint, that is, by anchoring the carrier s to the fixed housing. The three-shaft transmission depicted in fig. 10 was reduced to one degree of freedom

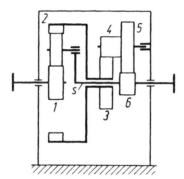

Fig. 10. The degree of freedom of the planetary gear train *1, 2, s* has been reduced to $F = 1$ by coupling *2* and *s* through the auxiliary gear train *3, 4, 5,* and *6*.

by linkage, by positively coupling two of its three shafts through an auxiliary gear train consisting of the gears *3, 4, 5,* and *6*. According to fig. 11, a constrained transmission with one degree of freedom can also originate from two epicyclic component transmissions, which together have four degrees of freedom as shown in fig. 11a. Two of their shafts, in this case shafts *2* and *2'*, and *s* and *1'*, are coupled and, in addition, a monoshaft, *s'* (fig. 11b), or a coupling shaft *22'* (fig. 11c) is fixed.

The degree of freedom F of a compound transmission, therefore, is given by the sum ΣF_i of the degrees of freedom of its component transmissions i minus the number of constraints h and linkages l, so that:

$$F = \Sigma F_i - h - l.$$

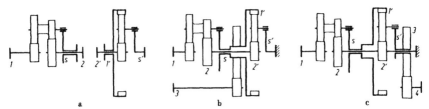

Fig. 11. Formation of a constrained drive train with one degree of freedom from two component gear trains each of which has two degrees of freedom: *a*, the two component gear trains with *F* = 2 each; *b*, compound gear train with *F* = 1, which arose by coupling shaft *2* with *2'* and *s* with *1'* and fixing the monoshaft *s'* in the housing; input at *1*, output at *3*; *c*, compound gear train with *F* = 1, which has the same couplings as gear train *b*, but coupling shaft *22'* is fixed, the input is at *1*, the output at *4*.

Consequently, the degree of freedom of the bicoupled transmissions shown in figs. 11b and 11c is found to be

$$F = (2 + 2) - 1 - 2 = 1.$$

We can thus formulate the generally valid R 5:

> *The degree of freedom F of a compound transmission is given*   R 5
> *by the sum of the degrees of freedom of its component trans-*
> *missions minus the number of its constraints and linkages.*

If more than two shafts of a compound transmission form a coupling, a further linkage is introduced with each additional shaft linked to it. Thus, a total of (*N*−1) linkages results when *N* shafts are positively coupled. For examples see fig 111.

By coupling simple revolving drive trains with fewer linkages or constraints, compound revolving trains with more than two degrees of freedom can be constructed, as shown in figs. 12 and 111q.

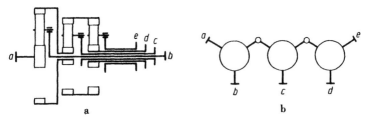

Fig. 12. Compound planetary transmission with 5 external shafts and 4 degrees of freedom: *a*, schematic, showing the construction; *b*, symbolic representation according to Wolf (see sec. 16).

These considerations lead to a simple distinction between simple revolving drive trains and conventional transmissions based on their respective degrees of mobility and concisely formulated in the following guide rule:

> *Conventional transmissions are always constrained transmis-*   R 6
> *sions with $F = 1$. However, simple revolving drive trains with*
> *three rotating shafts have two degrees of freedom, that is,*
> $F = 2$.

The general definition of the degree of freedom as introduced at the beginning of this section will be used in sections 24b and d, where the number of independent shaft torques also will be considered.

B. Simple Revolving Drive Trains, Terminology, Fundamental Principles, Types

Section 4. Terminology of Drive Trains, Shafts, and Gears in Planetary Transmissions

"Simple revolving drive trains" are simple planetary drive trains, or other revolving drive mechanisms which originated from simple conventional transmissions by allowing the latter's housings to rotate about central axes (sec. 2). Consequently, revolving drive trains may have three rotating shafts and frequently are referred to as "three-shaft transmissions" (or, if a central gear shaft is locked, "two-shaft transmissions").

"Simple planetary gear trains" are transmissions which consist of *one* planet carrier, one or more planets, and one or two central gears. Simple closed trains, or reverted planetary transmissions of the type shown in fig. 3b have one connected planet carrier, two connected central gears, and an arbitrary number of non-connected planets. Simple open-train planetary transmissions have one connected central gear, one connected planet carrier and one or more connected planets (figs. 6b, 104 and 105).

A revolving drive train whose carrier is stopped, or is assumed to be to facilitate its analysis, becomes kinematically equivalent to the conventional transmission with fixed axes of rotation from which it arose. It is called the "*basic* train" of the revolving drive. The concept of the basic drive train is of fundamental importance for the analysis of revolving drives.

A gear is called "connected" if it is rigidly coupled to an external shaft, and thus can transmit a torque into or out of the system.

The individual gears, and their respective shafts, are designated by the following symbols, which are used also as subscripts to identify other generally valid symbols:

1 either central gear (sun or ring); in open-train planetary transmissions, the connected central gear, or the connected planet; generally either shaft of the basic drive train,

2 the remaining central gear; in open-train planetary transmissions the connected planet, or the central gear; generally the other shaft of the basic drive train,

16

s the planet carrier (also called the spider or arm)
p a non-connected (idling) planet.

If a further differentiation is required between stepped or meshing planets, then

p_1 designates the planet which meshes with the central gear *1*,
p_2 designates the planet which meshes with the central gear *2*.

All equations, including subscripts, which will be derived for closed-train revolving drive mechanisms, are equally valid for open-train drives.

Basic drive train members are the same as those of the simple planetary transmissions, that is, the two rotating central gear shafts are arbitrarily labeled *1* and *2* while the housing or carrier-shaft is always labeled *s*, as shown in fig. 44.

Section 5. Basic Speed-Ratio and Basic Efficiency

As explained earlier (sec. 4), a three-shaft transmission reverts to its basic transmission when its carrier is locked (e.g., by connecting the carrier shaft to the transmission housing). The same is true when a three-shaft transmission operates in a state of motion where the carrier shaft is held immobile from the outside rather than by the housing. The simple conditions which exist in the fixed-carrier transmissions also exist in the three-shaft transmissions so that *this mode of operation completely characterizes each revolving drive train and, therefore, provides a basis for its further analysis.* It is defined by the speed-ratio and efficiency of the basic drive train which shall be called the "basic speed-ratio" and the "basic efficiency." Both must be distinguished from the speed-ratio and efficiency of the revolving drive train which shall be specified as "revolving carrier" speed-ratio and efficiency only when necessary to avoid confusion.

The basic speed-ratio is defined by the expression:

$$i_o = \left(\frac{n_1}{n_2}\right)_{n_s=0} . \tag{11}$$

Consequently,

$$\left(\frac{n_2}{n_1}\right)_{n_s=0} = \frac{1}{i_o} . \tag{12}$$

The basic efficiency is defined as

$$\eta_o = \eta_{12} = -\left(\frac{P_2}{P_1}\right)_{n_s = 0} \tag{13}$$

Since the efficiency is always smaller than 1, eq. (13) presumes that shaft *1* is the input shaft and shaft *2* the output shaft, a condition which is also expressed by the sequence of the subscripts of η.

In publications it is often presumed that the basic efficiencies of planetary transmissions are identical for either direction of the power-flow since conventional transmissions are symmetrical with respect to their losses. A theoretical investigation shows that this presumption indeed holds true with close approximation [27]. However, when the driver pushes the gear during the approach phase, the tooth-friction losses are somewhat larger than when the driver drags the gear during the recess phase. Therefore, loss-symmetry can theoretically exist only when the approach phase is equal to the recess phase, that is, when both sections of the line of action have equal lengths. The actual difference between the tooth-friction loss-factors of a spur gear set with addendum modifications of equal magnitude, but opposite signs, for a reversed power-flow direction, is shown in fig. 13. $\Delta \zeta$ is plotted as a function of the number of teeth z_P and the addendum modification factor of the pinion. If this addendum modification factor becomes $x_P = \pm 1$, then the pitch point lies approximately at the beginning or at the end of the path of contact and the asymmetry of the losses becomes a maximum. Even in this case, however, the difference between the tooth-friction

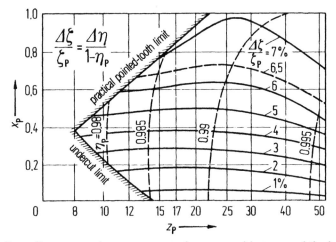

Fig. 13. Loss dissymmetry in a spur gear train expressed in terms of the loss factor $\zeta_P = 1 - \eta_P$; $x_P + x_G = 0$; $z_G/z_P = 8$; $\mu = 0.07$, $\Phi = 20°$; η_P = efficiency of the gear train when the pinion is the driver; η_G = efficiency of the gear train when the gear is the driver, $\Delta \eta = \eta_P - \eta_G = \zeta_G - \zeta_P = \Delta \zeta$ [27].

loss-factors $\Delta \zeta$ for a reversed power-flow is less than ten percent of the average tooth-friction loss-factor ζ. Therefore, the theoretical error resulting from the assumption of a perfect symmetry between the friction losses is less than $\pm 5\%$ of ζ. Thus, it is of the same order of magnitude as the errors inherent in efficiency measurements, and is smaller than the efficiency fluctuations caused by changes in the operating conditions (temperature, oil viscosity, lubrication, speed and load variations) and the manufacturing accuracy (tooth surface quality, hardness, material selection, size of the contact area). For all practical purposes it is, therefore, entirely justified to assume that the friction losses in basic transmissions are symmetrical, an assumption on which all further considerations of simple revolving drive trains can be based. Thus:

$$\eta_0 = (\eta_{12})_{\eta_s=0} \approx (\eta_{21})_{\eta_s=0} \qquad (14)$$

However, because of the very close approximation to complete equality of the middle and final terms of eq. (14), we may consider

$$\eta_0 = (\eta_{12})_{\eta_s=0} = (\eta_{21})_{\eta_s=0}$$

in all subsequent calculations such as that of eq. (41), for example.

Section 6. Determination of Basic Efficiency in Planetary Transmissions

Measurements on actual basic drive trains show that the power losses are composed of a load-dependent part which consists primarily of the tooth-friction losses P_{Lt}, and a load-independent part which represents the no-load or idling losses P_{Li}. These losses can be further separated into categories: constant losses, and the ventilation and splash losses which increase with increasing speed. The basic efficiency of a three-shaft drive train, therefore, is not constant; it even decreases to zero with decreasing output torque because of the inevitable no-load losses (fig. 14). More detailed information about the power-loss components has been provided by Niemann [33].

As compared to the conventional transmissions, planetary transmissions have two additional sources of losses, namely, the speed-dependent ventilation and splash losses caused by the rotation of the planet carrier, and the additional friction losses in the planet bearings caused by the centrifugal forces which act on the planets. The large centrifugal forces, which accompany the orbiting motion of the planets, also act on the rolling elements and retaining rings of the bearings and, hence, increase their internal friction. Thus, when the planets are supported in rolling element bearings it can be

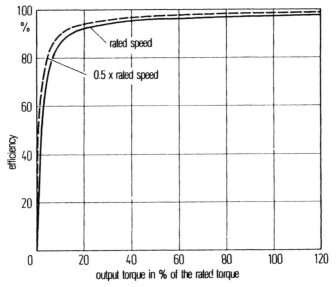

Fig. 14. The efficiency of a heavy-duty planetary transmission as a function of the output torque.

expected that the no-load losses of planetary transmissions differ from those of conventional transmissions. Nevertheless, the losses in rolling element bearings are small and can usually be neglected. However, the hydrodynamic friction losses in high speed revolving drive trains, in which journal bearings must be used, have a component which can no longer be neglected. Efficiency calculations for planetary transmissions, therefore, contain an even larger measure of uncertainty than those for conventional transmissions.

In spite of these numerous influences, limiting conventions which form a usable basis for a sufficiently accurate computation or measurement of the efficiency of planetary transmissions must be accepted before a basic efficiency can be defined. As a basis for the theoretical computation of the efficiencies (sec. 15), only those losses are considered which are associated with the rolling of the gear teeth on each other, that is, the tooth-friction losses ΣP_{Lt}, and the load-dependent components of the planet-bearing losses which are usually very small.

The basic efficiency η_0 or the basic loss factor ζ_0 refers to an operating condition of the basic transmission where this "rolling power loss" accounts for almost all of the losses.

Practically, this operating condition exists whenever the rated torques are transmitted at low speeds. Compared with the rolling power losses, the constant part of the idling losses then becomes so small that it can be neglected. The same is true for the speed-dependent parts, which increase with the speed raised to a power of 1.5 to 3, but still do not attain substantial values at low speeds.

Thus, to determine the basic efficiency η_0, which we want to introduce directly into the equations for the torque ratios and efficiencies of the planetary transmissions, we can start from the following recommendations:

1. The basic efficiency η_0 shall be defined as the efficiency which the basic transmission attains when the rated torque is transmitted at low pitch line velocities.

2. For measurements of the basic efficiency on actual transmissions a pitch line velocity of 10 m/sec is considered as "low."

3. For the theoretical calculation of the basic efficiency only the tooth-friction losses ΣP_{Lt} shall be considered, so that:

$$\eta_0 = 1 - \zeta_0 \quad \text{where} \quad \zeta_0 = -\frac{\Sigma P_{Lt}}{P_{in}}, \quad (15)$$

and P_{in} is the input power of the basic transmission.

Consequently, the efficiency of a planetary transmission, which has been calculated from the basic efficiency η_0 of the transmission, is valid also with practical accuracy at the rated operating torque and low speed. To estimate the efficiency at rated speed an additional loss factor $\zeta_{sp} \approx 0.005$ to 0.05 should be introduced to consider the ventilation and splash losses where the higher values apply to high speeds and splash lubrication with a partly immersed carrier.

Investigation of the Basic Efficiencies

The tooth-friction power loss P_{Lt} and consequently the tooth-friction loss factor ζ_t can be calculated for each individual gear stage from the equation

$$\zeta_t = -\frac{P_{Lt}}{P_t} = \mu_t f_L, \quad (16)$$

which has been given by Duda [27], provided the purely geometrical factors combined in the loss coefficient

$$f_L = \pi\left(\frac{1}{z_P} + \frac{1}{z_G}\right)(Z_a + Z_r)$$

and the tooth-friction coefficient μ_t are known. In eq. (16), P_t represents the power input to each individual stage, which is transmitted by rolling between the gear teeth.

Depending upon the magnitudes of the approach contact ratio m_a and the recess contact ratio m_r the approach portion Z_a and the recess portion Z_r of the line of action assume the following values:

$$\text{for } 0 \leqq m_{a,r} \leqq 1 \quad Z_{a,r} = \tfrac{1}{2} - m_{a,r} + m_{a,r}^2 ,$$

$$\text{for } \quad m_{a,r} > 1 \quad Z_{a,r} = m_{a,r} - \tfrac{1}{2} ,$$

$$\text{for } \quad m_{a,r} < 0 \quad Z_{a,r} = \tfrac{1}{2} - m_{a,r} .$$

The components m_a and m_r of the contact ratio can be calculated for any involute spur gear, helical gear, or annular gear stage, with or without addendum modification from

$$m_a = \frac{z_P}{2\pi} (\tan \Phi_{atP} - \tan \Phi_{to}) \tag{17}$$

and

$$m_r = \frac{z_G}{2\pi} (\tan \Phi_{atG} - \tan \Phi_{to}) \tag{18}$$

where

z_P is the number of pinion teeth
z_G is the number of gear teeth,
Φ_{atP} is the transverse pressure angle at the addendum circle of the pinion,
Φ_{atG} is the transverse pressure angle at the addendum circle of the gear, and
Φ_{to} is the transverse pressure angle at the operating pitch circle.

The total profile contact ratio then becomes:

$$m_p = m_a + m_r \tag{19}$$

$$\tan \Phi_t = \frac{\tan \Phi_n}{\cos \Psi}$$

$\text{inv } \Phi_t = $ ——— (corresponds to tangent, from Involutometry and Trigonometry Tables)

$$\text{inv } \Phi_{to} = \frac{2(x_P + x_G)\tan \Phi_n}{(z_P + z_G)} + \text{inv } \Phi_t$$

$\tan \Phi_{to} = $ ——— (corresponds to involute function, from Involutometry
and Trigonometry Tables)

$$\tan \Phi_{atP} = \sqrt{\left(\frac{d_{aP}}{d_{bP}}\right)^2 - 1}$$

$$\tan \Phi_{atG} = \sqrt{\left(\frac{d_{aG}}{d_{bG}}\right)^2 - 1}$$

Φ_t is the transverse pressure angle,
Φ_n is the normal pressure angle,   } at the nominal pitch point
ψ is the helix angle,
d_{aP} is the modified diameter at the addendum circle of the pinion,
d_{aG} is the modified diameter at the addendum circle of the gear,
d_{bP}, d_{bG} are the base circle diameters of pinion and gear, and
x_P, x_G are the addendum modification coefficients for the pinion P and the gear G.

For external gear stages these equations can be used directly. However, for internal gear stages the specific number of teeth of the annular gear must be prefixed with a negative sign (see sec. 9). The addendum modification factors x for both external and internal gear stages are positive whenever the generating tool is shifted in such a way that the tooth thickness at the root increases, that is, away from the center for an external gear and towards the center for an annular gear.

    The loss coefficient f_L can be determined directly from fig. 15 for gear stages which consist of standard involute gears without addendum modification, i.e., with addendum equal to the module (the reciprocal of the diametral pitch), and a pressure angle of 20°. For all other involute gears, f_L can be calculated from eqs. (16) to (19). For internal gear stages with involute gears and differences between the number of gear and pinion teeth down to $\Delta z = 1$ the coefficient f_L can be found directly from figs. 16 and 17. Fig. 17 has been plotted for an addendum modification factor $x_G = -x_P$ [26], which is the minimum required for trouble-free meshing, when the difference between the number of gear and pinion teeth $\Delta z < 8$. The values of f_L, for a range of gear and pinion sizes which are plotted in fig. 16, are based on this practical limiting curve. This graph shows that the geometrical loss coefficient f_L increases with decreasing difference in the number of gear and pinion teeth in the range from 8 to 1. The path of contact shifts so far along the line of action, as a consequence of opposite addendum modifications of gear and pinion teeth, that the pitch point finally falls outside the contact path. With increasing distance between contact point and pitch point, sliding velocities between the tooth flanks increase. Consequently, tooth-friction power losses increase. Therefore, the addendum modification

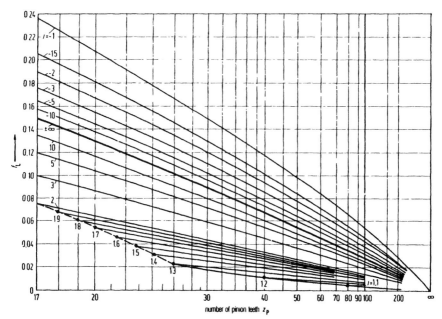

Fig. 15. Geometrical loss coefficient f_L for standard spur gear pairs without profile correction: z_P = number of teeth of the pinion; a negative speed-ratio i refers to external spur gear stages, a positive speed-ratio i to internal spur gear stages.

mandated by the practical limit curve of fig. 17 should not be substantially exceeded when annular gear stages with a minimum difference between the number of gear and pinion teeth are designed.

Since the addendum modification of the pinion is negative, undercutting of spur pinions can be avoided only when their minimum number of teeth z_{Pmin} increases with a decreasing difference Δz between the number of pinion and gear teeth. Therefore, fig. 17 also shows the minimum number of teeth z_{Pmin} necessary to avoid undercutting the pinion teeth (dashed horizontal lines), and to satisfy the condition that the addendum circle of annular gears must have a larger diameter than the base circle (dashed diagonal lines). Thus, the values of f_L given in fig. 17 are valid for pinions with a minimum number of teeth. For pinions with a larger number of teeth, f_L can be determined from fig. 16.

To calculate the power-loss factor ζ_t of a gear stage at rated load, the tooth-friction coefficient μ_t can be estimated on the basis of fig. 18 which shows test results obtained by Ohlendorf [14]. These tests were run with gear pairs which had a quality rating of 6 to 7 and carried a load of 250 Newtons per millimeter of face width. For approximate calculations it is recommended to use an average value $\mu_t = 0.05 \ldots 0.07$ [19, 29, 30, 41].

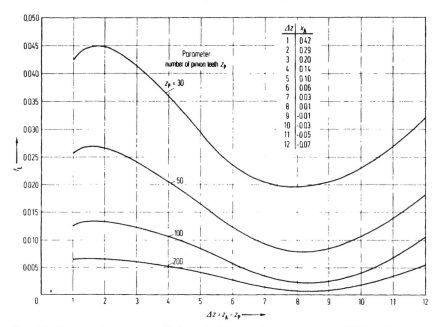

Fig. 16. Geometrical loss coefficient f_L for annular gear stages with involute spur gears with a minimum difference Δz between the number of gear and pinion teeth and profile correction factors $x_A = -x_P$ according to the practical limit curve given in fig. 17 or the inset table. The subscript A characterizes the annular gear, P the pinion.

A comparison shows that at high torques and low speeds the calculated power-loss coefficients of revolving and fixed-carrier transmissions agree sufficiently well with published test data to satisfy most practical requirements.

If, therefore, no better information is available, the basic efficiency η_0 of a planetary transmission which operates at rated torque and low speed, can be estimated as follows:

1. Standard involute gear trains without addendum modification

For any external, or annular gear stage, the tooth-friction loss factor can be calculated with eq. (16), where f_L can be determined from figs. 15, 16 or 17 and $\mu_t = 0.07$. The power-loss factors and efficiencies of the component gear stages *I, II, III,* ... of the basic transmission, which transmit the power in series, thus become:

$$\zeta_{tI} = f_{LI}\mu_{tI}, \qquad \zeta_{tII} = f_{LII}\mu_{tII}, \qquad \zeta_{tIII} = f_{LIII}\mu_{tIII}, \ldots$$

$$\eta_{tI} = 1 - \zeta_{tI}, \qquad \eta_{tII} = 1 - \zeta_{tII}, \qquad \eta_{tIII} = 1 - \zeta_{tIII}, \ldots.$$

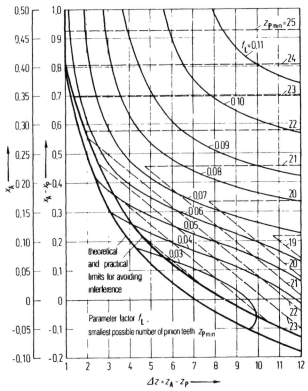

Fig. 17. Geometrical loss coefficient f_L for annular gear stages with involute gears and a minimum difference ΔZ between the number of gear and pinion teeth as a function of the profile correction factors $x_A = -x_P$, according to Duda [27] and Schäfer [26]. The dashed lines show the associated smallest possible numbers of teeth z_P of the pinion; subscript A $\hat{=}$ annular gear.

Therefore, with eqs. (9) and (10) the basic efficiency finally is obtained as

$$\eta_o = \eta_{tI}\eta_{tII}\eta_{tIII} \ldots \approx 1 - (\zeta_{tI} + \zeta_{tII} + \zeta_{tIII} + \ldots) . \qquad (20)$$

2. Involute gear trains with addendum modification

The friction loss factor ζ_t for each gear stage can be calculated from the transmission parameters, eq. (16), where $\mu_t = 0.07$ and the eqs. following (16) through (19) are used to calculate f_L. The basic efficiency η_o is then obtained from eq. (20) as shown in paragraph 1 above.

Since the efficiencies at rated load as calculated by this method only account for the tooth-friction losses, they must be modified for transmissions which operate at high speeds by including the loss factor ζ_{sp} which has been discussed earlier in this section.

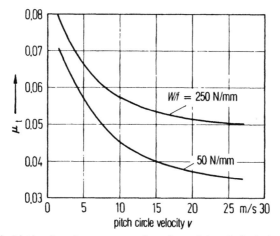

Fig. 18. Tooth-friction loss factor μ_t as a function of the pitch circle velocity v, for two different loads W per mm face width f, a lubricant viscosity of $1.3 \cdot 10^{-4}\,m^2/sec$ and a gear quality rating of 6 to 7; material *16 MnCr 5 F*; according to Ohlendorf [14].

Section 7. Positive- and Negative-Ratio Drives

The simple revolving drives form two major groups which have distinctly different operating characteristics—either a positive or a negative basic speed-ratio. The former, whose central gear shafts have the same sense of rotation when the carrier is fixed are called *"positive-ratio drives,"* the latter, *"negative-ratio drives."* All positive-ratio drives are kinematically similar. Furthermore, those with an equal basic speed-ratio i_o are "kinematically equivalent" with respect to the motions of their three shafts, *regardless of their internal construction.* The same is true for negative-ratio drives. Consequently, all possible configurations of the simple planetary drive trains are grouped in positive-ratio drives, depicted schematically by figs. 19 to 31, and negative-ratio drives as shown in figs. 32 to 43.

Figs. 19–31. Positive-ratio drive train types.

Fig. 19. $i_{o\,max} = 13.2$

Fig. 20. $i_{o\,min} = 0.076$

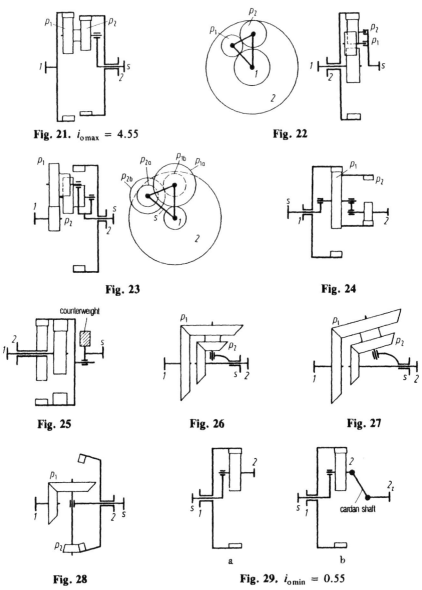

Fig. 21. $i_{o\,max} = 4.55$ Fig. 22

Fig. 23   Fig. 24

Fig. 25   Fig. 26   Fig. 27

Fig. 28   Fig. 29. $i_{o\,min} = 0.55$

Figs. 19–28, 29b. Reverted positive-ratio drive trains. **Figs. 29a, 30, 31.** Open positive-ratio drive trains. The speed ratios i_o given for some of the schematic representations are based on the assumptions that $z_{min} = 17$, with 3 planets, and $i_{max} = \pm 10$ per
gear pair where applicable, which represent geometrical limiting conditions. For gear
trains with stepped planets, equal root stresses and geometrical similarity of the
pinions have been assumed.

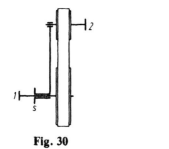

Fig. 30

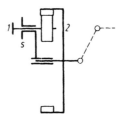

Fig. 31

Figs. 32–43. Negative-ratio drive train types.

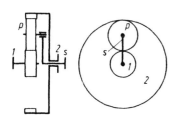

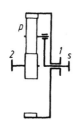

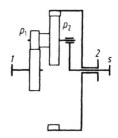

Fig. 32. $i_o = -1.2$ to -11.3

Fig. 33. $i_o = -0.09$ to -0.83

Fig. 34. $i_o = -0.22$ to -11.3. Drawn to scale for $i_o = -1$

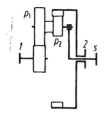

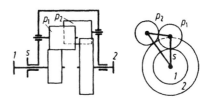

Fig. 35. $i_o = -11.3$ to -61

Fig. 36. $i_o = -0.09$ to -11.3

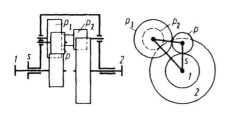

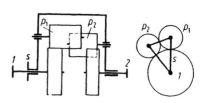

Fig. 37

Fig. 38

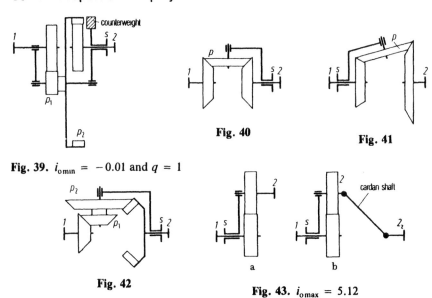

Fig. 39. $i_{o\,min} = -0.01$ and $q = 1$

Fig. 40

Fig. 41

Fig. 42

Fig. 43. $i_{o\,max} = 5.12$

Figs. 32–42. Reverted negative-ratio drive trains. **Fig. 43a.** Open negative-ratio drive train. The speed ratios i_o given for some of the schematic representations are based on the assumptions that $z_{min} = 17$, with 3 planets, and $i_{max} = \pm 10$ per gear pair where applicable, which represent geometrical limiting conditions. For gear trains with stepped planets, equal root stresses and geometrical similarity of the pinions have been assumed.

Section 8. Design Configurations of Simple Planetary Transmissions

The various design configurations of the simple planetary transmissions differ in their practical basic speed-ratio ranges, in their basic efficiencies, and in their design complexity. Of course, the limits of the basic speed-ratio ranges change with the chosen number of pinion teeth. Thus, they can sometimes substantially exceed the values given in figs. 19 to 42 if their design is based on assumptions other than those indicated. This is especially true for transmissions with stepped planets, when equal bending stresses at the roots of the teeth and geometrical similarity between the pinion gears are not demanded. If a specific basic speed-ratio can be realized with several kinematically-equivalent design configurations, the choice between them must be based both on design considerations and on their basic efficiencies. For practical applications, therefore, the simplest gear trains shown in figs. 19, 20, 32, 33, and 40 are used most frequently. Figs. 19 to 43 list the practical limits of the basic speed-ratios for the most common gear train configurations. Within these limits the basic speed-ratio ranges can, of course, be covered only in steps, since the gears must have integral numbers of teeth.

The *positive-ratio* gear trains shown in figs. 19 and 20 are identical except for the reversed designations of their central gear shafts *1* and *2*. Consequently, their basic speed-ratios are reciprocal with respect to each other. At $i_o = 1$, however, they are equal so that the two transmissions become kinematically equivalent and act like a direct coupling between their respective central gear shafts. For other gear train pairs only the corresponding reciprocal i_o-ranges are given, while the gear train schematics themselves are not repeated with merely reversed order of subscripts.

Gear trains as shown in fig. 21 have annular central gears and, because of the larger diameter of these annular gears, can transmit larger torques than gear trains of the type shown in fig. 19, although the planet sizes are identical in both designs. However, their possible speed-ratio range is smaller. Transmissions according to figs. 22 and 23 have only *one* annular gear, and meshing *planet pairs* have been arranged in order to obtain a positive basic speed-ratio i_o. If these planet pairs are stepped gears as shown in fig. 23, especially-high basic speed-ratios can be achieved.

Gear trains of the type shown in fig. 22 have a substantially smaller range of possible basic speed-ratios; in return, however, their axial dimensions become very small. This configuration is the *only* positive-ratio drive train whose gears lie in a single plane like the gears of the negative-ratio drive train shown in fig. 32. Because of their design complexity, positive-ratio drive trains with bevel gears (figs. 26 to 28) have little practical use.

Since a basic speed-ratio of $i_o = -1$ cannot be realized with the *negative-ratio* transmissions shown in figs. 32 and 33, they do not have a common speed-ratio range although they are identical except for the exchanged indices of their central gear shafts. These gear trains also have a smaller speed-ratio range than, for example, the simplest positive-ratio transmissions of fig. 19. Therefore, even rather complex configurations of negative-ratio transmissions have practical importance when they are able to cover the adjoining speed-ratio ranges.

Drive trains with bevel gears as shown in fig. 40, or spur gears of equal size and meshing planet pairs as shown in fig. 38, realize the basic speed-ratio $i_o = -1$ and are used as automotive differentials.

Transmissions with stepped planets as depicted in fig. 34 where $d_{p1} < d_{p2}$, can realize the speed-ratio range from -0.22 to -1.2 including the special basic speed-ratio $i_o = -1$, which cannot be covered by gear trains of the type shown in fig. 32. In the configuration of fig. 35, where $d_{p1} > d_{p2}$, they can extend the speed-ratio range of the basic design of fig. 32 beyond the range of approximately -11, to about -37 for $d_{p1} = d_{p2}$ and to about -60 for $d_{p2} = d_{min}$. To realize the speed-ratio range around $i \approx -1$, suitable transmissions generally have bevel gears as shown in figs. 41 and 42, or spur gear trains without annular gears, but with meshing planet pairs as shown in fig. 36. Without using annular gears, drive trains with meshing stepped planets as shown in fig. 37 can realize a substantially wider range of basic speed-ratios than gear trains of the type shown in fig. 36.

Figs. 29a and 43a show *open, positive-ratio* and *open, negative-ratio* gear trains respectively which find practical applications (e.g., in mixers and in spinning frames). The planets then become the input or output gears, while in all previously mentioned drive trains the planets were merely intermediate gears. Their absolute speeds, or angular velocities, must always be related to the *stationary* parts of the transmission, as if they were measured at the non-revolving connected stub shafts 2_z of figs. 29b and 43b. Both of these gear trains became reverted gear trains by arranging a universal joint shaft between the connected planet shafts *2* of figs. 29a and 43a and the central axis of the drive. They can well constitute economical solutions for drives whose carrier rotates or oscillates slowly.

Figs. 24, 25, 31, and 39 show that, in general, annular gears also can be used as planets. However, designs of these types have little technical significance because they are expensive to build, do not allow power branching, and offer no other compensating advantages.

As shown in fig. 30, an open positive-ratio drive train of belt, cable, or chain-drive type can become an open- or closed-train *revolving drive* when the arm is permitted to rotate. If gears are replaced by traction wheels, a wide variety of open- or closed-train *revolving drives* arise which sometimes have infinitely variable basic speed-ratios i_o, and are similar to the well-known traction drives. Finally, revolving hydrostatic drives may be mentioned (e.g., fig. 44). Depending on the tilt direction and angle of the cam

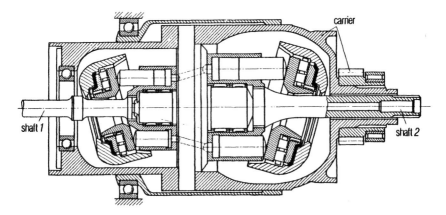

Fig. 44. Revolving hydrostatic transmissions developed by Allgaier.

(swash) plate, they can operate as positive- or negative-ratio drives with a variable basic speed-ratio i_0 as will be described later in section 33k.

It is emphasized that the rules and equations which shall be derived subsequently are also valid for all previously mentioned revolving drive trains. These include open drive trains where the connected planet shaft, instead of the missing connected central gear shaft, is labeled with *1* or *2* to identify the input or output respectively.

Simple planetary transmissions can be combined with other planetary transmissions, simple gear stages, or infinitely variable transmissions, in a variety of ways to suit special operating conditions or design requirements. Such compound planetary transmissions will be discussed in part III.

Section 9. Determination of Basic Speed-Ratio i_0

The basic speed-ratio i_0 of a multi-stage basic gear train can be calculated from the speed-ratios of its component stages. For each component stage which consists of the gears or friction wheels *a* and *b*, the speed-ratio is determined by the ratio of either the number of teeth z, or the pitch circle diameters D of *a* and *b*, so that

$$i_{ab} = \frac{n_a}{n_b} = -\frac{z_b}{z_a} = -\frac{D_b}{D_a} . \tag{21}$$

According to R 1, the speed-ratio of an external gear stage must become negative, and that of an annular gear stage, positive. Since the pitch circle diameters and the numbers of teeth are defined as

positive (+) for external gears and

negative (−) for annular gears,

R 1 requires that their ratios in eq. (21) be preceded by a negative sign.

Thus if the multi-stage basic transmission consists of the stages *I, II, III,* etc., its basic speed-ratio i_0 simply equals its overall speed-ratio so that

$$i_{1,N} = i_0 = i_I i_{II} i_{III} \ldots \qquad \text{or}$$

$$i_0 = \left(-\frac{z_{bI}}{z_{aI}}\right)\left(-\frac{z_{bII}}{z_{aII}}\right)\left(-\frac{z_{bIII}}{z_{aIII}}\right) \ldots \left(-\frac{z_{bN}}{z_{aN}}\right) .$$

The following basic speed-ratios are obtained for some of the gear trains shown in figs. 19–43. For positive-ratio gear trains of the type shown in fig. 19, which operate as two-stage basic transmissions with $n_{p1} = n_{p2}$.

$$i_o = \frac{n_1}{n_2} = \frac{n_1 n_{p2}}{n_{p1} n_2} = \left(-\frac{z_{p1}}{z_1}\right)\left(-\frac{z_2}{z_{p2}}\right) = \frac{z_2 z_{p1}}{z_1 z_{p2}} > 0 \; .$$

For negative-ratio gear trains of the type shown in fig. 32, which operate as basic transmissions with an idling planet p,

$$i_o = \frac{n_1}{n_2} = \frac{n_1 n_p}{n_p n_2} = \left(-\frac{z_p}{z_1}\right)\left(-\frac{z_2}{z_p}\right) = \frac{z_2}{z_1} < 0 \; ,$$

whereby the number of teeth substituted for $z_2 < 0$. For negative-ratio gear trains of the type shown in fig. 37, which operate as three-stage basic transmissions,

$$i_o = \frac{n_1}{n_2} = \frac{n_1 n_p n_{p2}}{n_p n_{p1} n_2} = \left(-\frac{z_p}{z_1}\right)\left(-\frac{z_{p1}}{z_p}\right)\left(-\frac{z_2}{z_{p2}}\right) = -\frac{z_2 z_{p1}}{z_1 z_{p2}} < 0$$

and, finally, for open or reverted positive-ratio gear trains as shown in figs. 29a or 29b,

$$i_o = \frac{n_1}{n_2} = -\frac{z_2}{z_1} > 0 \; ,$$

whereby the number of teeth substituted for $z_1 < 0$.

The final sign of i_o may be predetermined without reference to the formula, and prior to the selection of numbers of teeth. Simply scan the schematic diagram and count the number of external gear meshes; if odd, the sign is minus, otherwise plus. For chain drives, the sprockets may be treated like external gears, and the chain as an annular idler gear driven by one sprocket while simultaneously driving the other.

Belt and traction drives can be handled analogously. However, their speed-ratios must be calculated from the ratios of the pulley or friction wheel diameters rather than from the ratios of the number of gear teeth. The slip is usually neglected in the application of this method, but could easily be considered as long as it remains reasonably constant.

For planetary transmissions with bevel gears, the ratio between the number of teeth determines only the magnitude of the basic speed-ratio, its sign must be determined separately from the design configuration, since R 1 is defined only for parallel shafts. However, this is simple enough if it is observed that the basic speed-ratio of a revolving bevel gear stage with a simple planet is always negative, as is evident from figs. 40 and 41. The basic speed-ratio of a revolving bevel gear stage with a stepped planet is positive whenever both the pitch cones of the central gears and the pitch cones of the idlers face in the same direction (figs. 26 and 27) or if both face in opposite directions (fig. 28) where

$$i_o = + \left(\frac{z_{p1}}{z_1} \frac{z_2}{z_{p2}} \right) ;$$

it becomes negative if one pair of pitch cones faces in the same direction and the other does not, as shown in fig. 42, where

$$i_o = - \left(\frac{z_{p1}}{z_1} \frac{z_2}{z_{p2}} \right) .$$

II. SIMPLE REVOLVING DRIVE TRAINS

A. Working Principles

Section 10. General Remarks and Speed-Ratio Notation

The basic laws of the simple revolving drive trains define fundamental relationships between the speeds, the torques and the power of their shafts, as well as their efficiencies. They shall be derived for the general case of the three-shaft transmission and, consequently, are valid also for two-shaft epicyclic transmissions, and basic transmissions which are merely special cases —characterized by the fact that one of their shafts is locked, that is, either $n_1 = 0$, or $n_2 = 0$, or $n_s = 0$.

The laws of motion for revolving drive trains are independent of the transmitted torques. For this reason, all kinematic relationships—that is, the relationships between the speed-ratios of the shafts and gears—will be derived without consideration of the torques or powers and, therefore, remain valid for all operating conditions. Consequently, these relationships are independent also of the location of the input or output shafts.

The shaft torques satisfy the static equilibrium conditions of the system and therefore are independent of the shaft speeds as long as no accelerating torques occur. Consequently, they can be investigated without any consideration of the speeds. However, because of the friction losses the location of the input and output shafts must now be considered.

The analysis of compound revolving drive trains becomes clearer when the constant speed-ratios of those with only one degree of freedom are strictly distinguished from the variable speed-ratios of those drive trains which have more than one degree of freedom. Wolf, therefore, retained the symbol i only for constant, design-dependent speed-ratios which he specified by two subscripts in the order, numerator, denominator, so that, for example,

$$i_{21} = \frac{n_2}{n_1} = \frac{1}{i_{12}} \,.$$

For the speed-ratios of drive trains with more than one degree of freedom, which are independent of the design and the torque ratio of the considered shafts, he introduced the symbol k, so that, for example,

$$k_{s2} = \frac{n_s}{n_2} = \frac{1}{k_{2s}} \ .$$

This distinction is very useful in more detailed studies of revolving drive trains and shall be used henceforth.

Section 11. Speeds and Speed-Ratios

a) Speed Characteristics of Three-Shaft Revolving Drive Trains

The somewhat confusing relationship between the speeds in three-shaft drive trains was greatly clarified in 1841 when Willis showed that the complex general state of motion can be explained as the superposition of two very simple partial motions. Starting from this concept he derived eq. (22), which allows the determination of all of the speed-ratios of the revolving drive trains.

The first partial motion is the rotation of the central gears—turning in mesh with the planets—*relative* to the carrier. Consequently, it is equivalent to the motion of the basic gear train whose carrier is stationary and

$$\frac{n_1'}{n_2'} = i_o \ ,$$

where n_1' and n_2' are the speeds of the central gear shafts relative to the carrier. The first partial speed of the carrier, relative to itself, is $n_s' = 0$, while the total speed of the carrier n_s is not necessarily equal to zero.

The second partial motion is an equal rotation of the central gear shafts and the carrier, $n_1'' = n_2'' = n_s''$. No relative motions arise and the gear train rotates as a rigid coupling.

If the two partial motions are superimposed, the total speed of each shaft is obtained as the algebraic sum of its two partial speeds, that is,

$$n_s = n_s' + n_s'' = 0 + n_s'' \ ,$$
$$n_1 = n_1' + n_1'' = n_1' + n_s \ ,$$
$$n_2 = n_2' + n_2'' = n_2' + n_s \ .$$

Thus the basic speed-ratio i_o becomes:

so that,

$$i_o = \frac{n_1'}{n_2'} = \frac{n_1 - n_s}{n_2 - n_s} \ ,$$
$$n_1 - i_o n_2 - n_s(1 - i_o) = 0 \ . \tag{22}$$

The speed components $n_1' = n_1 - n_s$ and $n_2' = n_2 - n_s$ are also called the "rolling speeds" of the central gears 1 and 2, since they characterize their rolling velocities, that is, the velocities of their pitch circles relative to the pitch point. Referring to the second partial motion, the speed components n_1'', n_2'' and n_s are also called the "coupling speeds" of the shafts. Since, according to eq. (22), i_o determines relationships between all three shaft speeds of a revolving drive train and at the same time constitutes the speed-ratio of the basic train, it is appropriate to call it the "*basic* speed-ratio."

From eq. (22) the shaft speeds of a three-shaft transmission are obtained as follows:

$$n_1 = i_o n_2 + (1 - i_o) n_s ,\tag{23}$$

$$n_2 = \frac{n_1 - n_s(1 - i_o)}{i_o} ,\tag{24}$$

and

$$n_s = \frac{n_1 - i_o n_2}{1 - i_o} .\tag{25}$$

If each of these equations is then divided successively by each of the two n's of its right side, the speed-ratios assume the following form:

In explicit notation

In reduced notation according to Wolf

$$\frac{n_1}{n_2} = i_o + (1 - i_o)\frac{n_s}{n_2} , \qquad k_{12} = i_o + (1 - i_o)k_{s2} .\tag{26}$$

$$\frac{n_1}{n_s} = (1 - i_o) + i_o\frac{n_2}{n_s} , \qquad k_{1s} = (1 - i_o) + i_o k_{2s} .\tag{27}$$

$$\frac{n_2}{n_1} = \frac{1}{i_o} + \left(1 - \frac{1}{i_o}\right)\frac{n_s}{n_1} , \qquad k_{21} = \frac{1}{i_o} + \left(1 - \frac{1}{i_o}\right)k_{s1} .\tag{28}$$

$$\frac{n_2}{n_s} = \frac{\dfrac{n_1}{n_s} - (1 - i_o)}{i_o} , \qquad k_{2s} = \frac{k_{1s} - (1 - i_o)}{i_o} .\tag{29}$$

$$\frac{n_s}{n_1} = \frac{1 - i_o\dfrac{n_2}{n_1}}{1 - i_o} , \qquad k_{s1} = \frac{1 - i_o k_{21}}{1 - i_o} .\tag{30}$$

$$\frac{n_s}{n_2} = \frac{\dfrac{n_1}{n_2} - i_o}{1 - i_o} , \qquad k_{s2} = \frac{k_{12} - i_o}{1 - i_o} .\tag{31}$$

b) Speed-Ratios of Two-Shaft Transmissions

The speed-ratios of the three possible two-shaft transmission designs can be obtained from eqs. (23) to (25) by setting the appropriate speed on the right side of each equation equal to zero. Consequently, with $n_s = 0$ eqs. (26) and (28) yield:

$$\frac{n_1}{n_2} = i_{12} = i_o$$

$$\frac{n_2}{n_1} = i_{21} = \frac{1}{i_o}$$

Fixed-carrier speed-ratios of two-shaft transmissions (32)

With $n_2 = 0$ in (27) and (30):

$$\frac{n_1}{n_s} = i_{1s} = 1 - i_o \qquad (33)$$

$$\frac{n_s}{n_1} = i_{s1} = \frac{1}{1 - i_o} \qquad (34)$$

With $n_1 = 0$ in (29) and (31):

Rotating-carrier speed-ratios of two-shaft transmissions

$$\frac{n_2}{n_s} = i_{2s} = \frac{i_o - 1}{i_o} \qquad (35)$$

$$\frac{n_s}{n_2} = i_{s2} = \frac{i_o}{i_o - 1} \qquad (36)$$

According to eqs. (32) to (36), the speed-ratios of the two-shaft transmission depend solely on i_o and, thus, are constants which are determined by the design of each revolving drive train. For this reason, they shall be symbolized by i rather than k. Two-shaft revolving drives can be used as step-up or step-down drives like simple conventional transmissions. A graphical representation of eqs. (32) to (36) is given in fig. 45 where the speed-ratios of the two-shaft transmissions are plotted as functions of the basic speed-ratio i_o. Consequently, this diagram allows either the direct determination of the basic speed-ratio associated with a desired revolving-carrier speed-ratio and thus the determination of the transmission configuration, or vice versa.

If eqs. (33) to (36) are solved for i_o, the basic speed-ratio can be eliminated from any two of the equations so that each speed-ratio can be expressed in terms of any of the others. This procedure leads to worksheet 1 where each possible speed-ratio i is given in terms of each of the others. Worksheet 1 is provided as a practical aid for the layout of planetary transmissions.

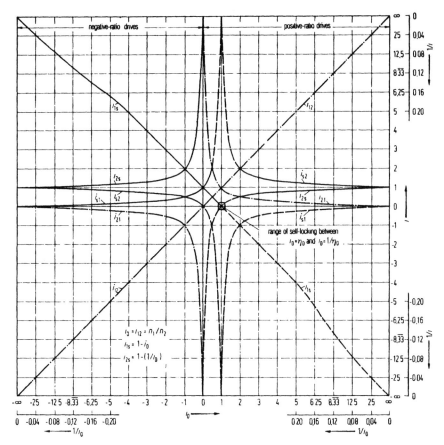

Fig. 45. Speed-ratios of simple revolving drives which operate in the two-shaft mode as a function of the basic speed-ratio i_o (see worksheet 1). The range in which self-locking can occur is shown larger than actual size. The ranges of the revolving carrier efficiencies are indicated by the type of line: ——— $\eta > \eta_o$, ---- $\eta < \eta_o$, and —·—·— $\eta = \eta_o$. With the analogous subscripts $1 \triangleq I$, $2 \triangleq II$, $s \triangleq S$, and $i_o \triangleq i_{1II}$, this diagram is valid also for constrained bicoupled transmissions.

For transmissions with three rotating connected shafts, each speed-ratio k can likewise be expressed in terms of all other speed-ratios k when eqs. (26) to (31) are evaluated in a similar manner. The resulting relationships are clearly tabulated in worksheet 2.

If any one of the speed-ratios has a value of $k = 1$, then all other speed-ratios become equal to 1. This can be easily verified from eqs. (26) to (31). Consequently, whenever two of the three connected shafts have the same speed, then all shafts rotate in unison, that is, the transmission operates at the "coupling point."

c) Relative Speeds of Planets

The speed $n_p' = (n_p - n_s)$ of the planets relative to the carrier which determines, for example, the design of the planet bearings can readily be obtained from the first partial speeds $n_1' = (n_1 - n_s)$ or $n_2' = (n_2 - n_s)$ of the central gears which have been introduced in section 11a. Since the first partial motion is equivalent to the motion of the basic transmission, the speed-ratios between the planets and the central gears can be calculated from eq. (21) like the speed-ratios of the basic transmission. For a gear train of the type shown in fig. 32 we thus obtain:

$$(n_p - n_s) = (n_1 - n_s)i_{p1} = (n_1 - n_s)\left(-\frac{z_1}{z_p}\right),$$

or

$$(n_p - n_s) = (n_2 - n_s)i_{p2} = (n_2 - n_s)\left(-\frac{z_2}{z_p}\right).$$

According to section 9 the numerical value substituted for z_2 must be preceded by a negative sign since gear *2* is an annular gear.

If the planets are stepped gears as shown in figs. 19, 27, 35, 42, etc., the speed-ratio can again be determined from the planet mesh with either gear *1* or gear *2*, so that

$$(n_{p1} - n_s) = (n_1 - n_s)i_{p1} = (n_1 - n_s)\left(-\frac{z_1}{z_{p1}}\right),$$

or

$$(n_{p2} - n_s) = (n_2 - n_s)i_{p2} = (n_2 - n_s)\left(-\frac{z_2}{z_{p2}}\right).$$

According to section 9 z_2 must be preceded by a negative sign if gear *2* is an annular gear as in fig. 35. These equations are valid also when bevel gears are used as planets as shown in fig. 40. However, no special sign convention is needed for bevel planets since the speeds of their nonparallel shafts must be known only for the layout of their bearings, which in general is not influenced by direction of rotation.

d) Generalization of the Willis Equation

The two independent partial motions defined by Willis, and described in section 11a, can be imparted to the external shafts of any coaxial three-shaft transmission with two degrees of freedom, even if their relationship to the gear train is unknown. If, as in fig. 46, we *arbitrarily* assign the symbols *a*, *b* and *c* to these shafts, we can lock shaft *a*, rotate shaft *b* by a known number of turns n_b, and count the resulting number of turns of shaft *c*, n_c. From this first partial motion we obtain a fixed-member speed-ratio,

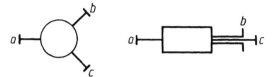

Fig. 46. Revolving drive train with an unknown position of the carrier shaft for the derivation of the generalized Willis equation.

$$i_{bc} = \frac{n_b'}{n_c'} \, ,$$

which, like the basic speed-ratio i_0, depends solely on the design of the unknown gear train. As a second partial motion we can then superimpose an equal speed n_a, that is, a coupling speed $n_a = n_b'' = n_c'' = n_a''$ on the external shafts, so that, analogous to the derivation of the Willis equation (22),

$$n_b = n_b' + n_b'' = n_b' + n_a \, ,$$

$$n_c = n_c' + n_c'' = n_c' + n_a \, ,$$

and, consequently,

$$i_{bc} = \frac{n_b'}{n_c'} = \frac{n_b - n_a}{n_c - n_a} \, ,$$

or   (37)

$$n_b - i_{bc} n_c - (1 - i_{bc}) n_a = 0 \, .$$

This generalized form of the Willis equation can then be solved for the three shaft speeds so that, analogous to eqs. (23) to (25),

$$n_b = i_{bc} n_c + (1 - i_{bc}) n_a \, , \qquad (38)$$

$$n_c = \frac{n_b - n_a (1 - i_{bc})}{i_{bc}} \, ,$$

$$n_a = \frac{n_b - i_{bc} n_c}{1 - i_{bc}} \, ,$$

Subsequently, a series of generalized equations can be derived which are completely analogous to eqs. (26) to (36). These equations can then be used to generate generalized worksheets 1 and 2. Thus, worksheet 1, as obtained from the generalized eqs. (32) to (36), assumes the form:

i	$f(i_{bc})$
i_{bc}	i_{bc}
i_{cb}	$1/i_{bc}$
i_{ba}	$1 - i_{bc}$
i_{ab}	$\dfrac{1}{1 - i_{bc}}$
i_{ca}	$1 - \dfrac{1}{i_{bc}}$
i_{ac}	$\dfrac{i_{bc}}{i_{bc} - 1}$

This single column represents the *complete* worksheet 1 since the general subscripts b, c, a can be consecutively replaced by the six permutations of the subscripts 1, 2, s, and thus yield the six columns of the familiar worksheet 1, although arranged in a somewhat different sequence.

If we had an opportunity to inspect the given drive train, and found that the carrier shaft had been assigned the subscript b, and the two central gear shafts the subscripts a and c, so that

$$a \triangleq 1, \quad b \triangleq s, \quad \text{and} \quad c \triangleq 2 ,$$

we could rewrite column 1 of the generalized worksheet 1 in terms of the familiar subscripts 1, 2 and s. We would then obtain the following table which we will recognize as column 6 of worksheet 1, read from the bottom to the top:

i	$f(i_{s2})$
i_{s2}	i_{s2}
i_{2s}	$1/i_{s2}$
i_{s1}	$1 - i_{s2}$
i_{1s}	$\dfrac{1}{1 - i_{s2}}$
i_{21}	$1 - \dfrac{1}{i_{s2}}$
i_{12}	$\dfrac{1}{i_{s2} - 1}$

Thus the completed generalized worksheet 1 could be converted term by term into the familiar worksheet 1, although arranged somewhat differently.

If we recall that the choice of the subscripts for the shafts of the unknown transmission was entirely arbitrary, we must realize that the association between the ordered sets {a, c, b} and {1, 2, s} is likewise arbitrary, and we could have converted the generalized worksheet 1 into a form of worksheet 1 by using *any* conversion key. As a matter of fact we would have obtained worksheet 1 in the correct order and the Willis equation in the form of eq. (22) if we had chosen the conversion key

$$b \triangleq 1, \quad c \triangleq 2, \quad a \triangleq s .$$

It should become clear that, by means of the reverse procedure, generalized forms of worksheets 1 and 2 can be obtained simply by direct substitution of general for specific subscripts in the published tables.

The speed-ratios are unaffected by the choice of the conversion key (which affects only the order of the entries of the worksheets). Thus, we can use immediately, and without restriction, any appropriate general form of the Willis equation, or of worksheets 1 and 2, when the position of the carrier shaft is unknown.

If the transmission losses can be neglected, that is, if we can assume that the basic efficiency $\eta_o = 1$, we can also apply a general subscript code to the torque equations so that i_o in eqs. (42), (44) and (45) could be replaced by a general expression such as i_{bc}. Cases in which the tooth-friction losses must be considered will be treated later in section 15g.

e) Kinematically-Equivalent Epicyclic Transmissions

Revolving drives are said to be kinematically equivalent when, as two-shaft transmissions, they coincide in all their *i*-ratios, and, as three-shaft transmissions, coincide in all their interrelationships between *k*-ratios. These conditions prevail when the drives have the same basic speed-ratio, since worksheets 1 and 2 show that all *i*-ratios and the relationships between *k*-ratios depend only on i_o.

We also learn from the generalized Willis equation, or worksheet 1, that two revolving drives are kinematically equivalent if a random basic or revolving speed-ratio of one coincides with a random basic or revolving speed-ratio of the other. Thus, for example, a negative-ratio transmission is kinematically equivalent to a positive-ratio transmission when its basic speed-ratio $i_{o'}$, is equal to the latter's revolving speed-ratio i_{1s}.

As a proof we shall distinguish the speed-ratios of the negative-ratio transmission from those of the positive-ratio transmissions by a prime ('). Thus, we find from worksheet 1 that with $i_o' = i_{1'2'} = i_{1s}$:

$$i_{1's'} = 1 - i_{o'} = 1 - i_{1s} = i_o = i_{12} ,$$

and

$$i_{2's'} = 1 - \frac{1}{i_{o'}} = 1 - i_{s1} = i_{s2} \ .$$

It follows, of course, that the corresponding reciprocal speed-ratios are also equal. The two equations confirm what was already evident from the identity: $i_{1'2'} = i_{1s}$, namely, that the shafts $1'$ and 1, $2'$ and s, and, consequently, s' and 2 of the two transmissions, correspond to each other as kinematically equivalent. Thus,

> *Two revolving drive trains are kinematically equivalent if the*   R 7
> *numerical values of any random pair of two-shaft speed-*
> *ratios, drawn from each train's complete set of six i-ratios,*
> *coincide.*

Section 12. Torques

The equilibrium conditions require that during a stationary operating condition the sum of all external torques which act on a revolving drive equals zero, that is,

$$T_1 + T_2 + T_s = 0 \ . \tag{39}$$

Obviously, this equation is satisfied only when the sum consists of positive and negative terms. Since, however, two of the three torques always have the same sign, the magnitude of their partial sum equals the magnitude of the third torque. Accordingly, that shaft which carries the sum of the torques is called the "summation shaft." The other two shafts are called the "difference shafts." Consequently, the absolute value of the torque carried by a difference shaft equals the difference between the absolute values of the torques carried by the summation shaft and the second difference shaft.

These relationships are summarized in two guide rules, which facilitate a quick analysis of even complicated revolving drive trains.

> *The torque of the summation shaft and the torques of the*   R 8
> *difference shafts have opposite signs.*

> *The torques of the two difference shafts have equal signs.*   R 9

When the carrier is locked and the basic speed-ratio i_o and the basic efficiency η_o are known, the torques of the two central gear shafts can be determined like the torques of a conventional transmission from eqs. (7) and (8), which are derived directly from the energy principle, and (14). Thus, if shaft 1 is the input shaft:

$$\left(\frac{T_2}{T_1}\right)_{P_{R1}>0} = -i_o \eta_{12} = -i_o \eta_o \ . \tag{40}$$

If shaft *2* is the input shaft:

$$\left(\frac{T_2}{T_1}\right)_{P_{R1}<0} = -i_o \eta_{21}^{-1} = -i_o \eta_o^{-1} \ . \tag{41}$$

The subscript P_{R1} refers to the "rolling-power" of shaft *1*, that is, the power which it transmits at the rolling speed $(n_1 - n_s)$ as defined in section 11a. When the carrier is locked the rolling-power is identical to the external power carried by shaft *1*, that is,

$$P_{R1} = T_1 n_1' = T_1(n_1 - n_s) = T_1 n_1 \ .$$

If the carrier is freed and an additional coupling speed n_s is superimposed on the three connected shafts, the rolling speeds and, consequently, the rolling-powers, remain unchanged as long as the external torques remain constant. Eqs. (40) and (41), therefore, are generally valid for all operating conditions.

Eqs. (40) and (41) are equal except for the sign of the exponent of η_o, which is positive for a positive rolling-power P_{R1} and negative for a negative rolling-power P_{R1}. Therefore, Brandenberger [5] suggested that eqs. (40) and (41) be combined into a single equation and that a general exponent K be assigned to the basic efficiency η_o. Instead, we shall subsequently use the exponent $r1$ to avoid a possible confusion with the speed-ratio k and to emphasize the fact that its sign is identical to the sign of the rolling-power of shaft *1*. The introduction of this exponent makes it possible to derive the torque and efficiency equations before the locations of the input and output shafts are specified. This is especially important for the analysis of compound and simple revolving drive trains when the location of the input and output shafts is not known beforehand, or if it changes during operation. Thus we obtain:

$$\frac{T_2}{T_1} = -i_o \eta_o^{r1} \ , \tag{42}$$

where

$$r1 = \frac{P_{R1}}{|P_{R1}|} = \frac{T_1(n_1 - n_s)}{|T_1(n_1 - n_s)|} = \pm 1 \ . \tag{43}$$

The preceding discussion can be summarized in the following guide rule:

The sign of the exponent $r1$ is always equal to the sign of the   R 10
rolling-power P_{R1} of shaft 1 and its magnitude is 1.

For practical applications the exponent $r1$ can be taken directly from worksheets 4 and 5 where it is tabulated as a function of the power-flow and the transmission type.

Since eq. (42) can be solved for either T_1 or T_2, it combines with eq. (39) either in the form of eq. (44) or eq. (45), that is, with

$$T_1 + T_s = i_o \eta_o^{r1} T_1 \, ,$$

$$\frac{T_s}{T_1} = i_o \eta_o^{r1} - 1 \qquad (44)$$

or with

$$T_2 + T_s = \frac{T_2}{i_o \eta_o^{r1}}$$

$$\frac{T_s}{T_2} = \frac{1}{i_o \eta_o^{r1}} - 1 \, . \qquad (45)$$

Eq. (42), (44), and (45) show that the torque ratios depend only on the basic speed-ratio i_o, the basic efficiency η_o and, because of $r1$, on the direction of the rolling power-flow but not on the shaft speeds. Thus we obtain the fundamental relationship,

$$T_1 : T_2 : T_s \approx f(i_o) = \text{constant} \qquad (46)$$

which is valid for all revolving drives as long as $\eta_o \approx 1$. If the losses can no longer be neglected so that $\eta_o \neq 1$, we obtain two slightly different, though constant, torque ratios which depend on whether the rolling-power P_{R1} flows from shaft I to shaft 2 or vice versa. The torque ratios then depend on the sign of $r1$. This can be expressed as a further guide rule:

> *As long as friction losses can be neglected, torque-ratios of*   R 11
> *revolving drive trains are determined solely by their basic*
> *speed-ratios. They are independent of shaft speeds.*

Probably more than any other rule, R 11 facilitates an intuitive understanding of the operating characteristics of simple and compound three-shaft transmissions. It says that these drive trains cannot transmit power if one of their shafts remains unconnected and thus carries zero torque. In this case, eq. (46) requires that the torques of the other two shafts are likewise zero and the transmission can merely idle. (The operating characteristics of epicyclic transmissions under self-locking conditions [sec. 15] have been purposely disregarded since this case constitutes the only, and, in practice, rarely encountered exception.) Eq. (46) also shows that during a stationary operating condition the torques of any three connected machines must bear the constant ratios, $T_1 : T_2 : T_3$. Therefore, they can only operate at those speeds where their operating characteristics satisfy the required torque ratios. If the torque characteristics of the three machines do not have a common operating point at which the required equilibrium torque ratios exist, then at least two of the machines will attain excessive speeds, or else one or all of them will stall. Conditions of this type will be treated in more detail in section 24. For now, the following guide rule can be formulated:

> *The external torques always must be matched to those of the*   R 12
> *revolving drive train. The reverse is impossible because the*
> *torques of the three connected shafts of the transmission bear*
> *an invariable ratio to each other.*

Section 13. Determination of the Summation Shaft from Basic Speed-Ratio i_O

From the torque relationships derived in section 12, the summation shaft can be found immediately when i_o and η_o are known. Thus, for example, eq. (42) shows that in negative-ratio drives, where

$$\eta_o^{rl} i_o < 0 \qquad \text{so that} \qquad \frac{T_2}{T_1} > 0 \, ,$$

the carrier shaft s is the summation shaft because the torque ratio T_2/T_1 becomes positive only when both torques have the same sign. R 9 states that in this case shafts *1* and *2* are difference shafts. Consequently, the carrier shaft s must be the summation shaft.

Analogously, we find for positive-ratio drives where

$$\eta_o^{rl} i_o > 1 \qquad \text{and therefore}$$

$$\frac{T_2}{T_1} < -1, \qquad \text{that is,} \qquad |T_2| > |T_1| \, .$$

that shaft *2* is the summation shaft and shafts *1* and *s* consequently are the difference shafts. If

$$0 < \eta_o^{rl} i_o < 1, \qquad \text{so that} \qquad -1 < \frac{T_2}{T_1} < 0 \qquad \text{and} \qquad |T_2| < |T_1| \, ,$$

shaft 1 is the summation shaft and, consequently, shafts 2 and s are the difference shafts. If $T_2/T_1 < 0$, either one of the two shafts can be the summation shaft; the other shaft and the carrier shaft are the difference shafts. Since the absolute value of the torque which is transmitted by the summation shaft must always be larger than the absolute value of the torque which is transmitted by either difference shaft, shaft 2 is the summation shaft whenever $|T_2/T_1| > 1$ and a difference shaft whenever $|T_2/T_2| < 1$.

This can be expressed in a further guide rule which states:

> *In negative-ratio drives, the carrier shaft always is the sum-*   R 13
> *mation shaft; but in positive-ratio drives the central gear shaft*
> *carrying the larger absolute torque is the summation shaft.*

In most publications which analyze the operating characteristics of revolving drives, matters are simplified by neglecting the friction losses so that $\eta_0 = 1$. For most practical revolving drives this is entirely permissible. However, some positive-ratio drives with basic speed-ratios of approximately 1, or more precisely,

$$\eta_0 < i < \frac{1}{\eta_0} ,$$

can become self-locking so that the summation shaft either cannot be identified or changes its position. This case will be discussed in more detail in section 15 where the problem of self-locking is analyzed. Disregarding this special case, however, we can write $\eta_0 \approx 1$ and derive the simple relationships between the basic speed-ratio i_0 and the positions of the summation shaft which are tabulated in worksheet 3. Worksheet 3 also lists the ranges of the speed-ratios i_{1s} and i_{2s} which exist with the three possible positions of the summation shaft.

Section 14. Powers and Power-Flow

a) Signs

A revolving drive train, whose three shafts are free to rotate, can operate in two general modes, that is, with either one input shaft and two output shafts or two input shafts and one output shaft. Whether a particular shaft is an input shaft or an output shaft depends on the sign of its power as explained in section 1 and summarized in R 2 and R 3.

An input shaft always introduces a positive power into the drive train and, consequently, the torque T and the speed n carry the same signs:

$$P_{\text{in}} = T_{in} n_{\text{in}} > 0 ;$$

while an output shaft always introduces a negative power into the drive train so that T and n carry opposite signs and

$$P_{out} = T_{out} n_{out} < 0 \ .$$

(Whenever the power equations $[P = T \cdot 2\pi \cdot n]$ is written without the constant factor 2π to simplify the notation, the power P has the dimension $[Nm/\text{sec} \cdot 1/2\pi]$.) Any shaft of a revolving drive train, whether it is the carrier shaft, a central-gear shaft, the summation shaft, or a difference shaft, can be the sole input shaft or the sole output shaft. The sole input shaft as well as the sole output shaft carries the total power and, therefore, shall be called the *total-power shaft*. Analogously, the other two shafts, each of which consequently carries only a fraction of the total power, shall be called the *partial-power shafts*.

The energy principle can then be restated in the form of the following guide rule:

> *If the total-power shaft is an input shaft, then both partial-*   R 14
> *power shafts are output shafts and vice versa.*

If a revolving drive has two input shafts, they may turn in opposite directions. If one of these is arbitrarily defined as positive, the other must, of course, be negative. Since the accepted sign conventions define input powers as positive, the drive torque of the former shaft must be positive and that of the latter negative. Consequently, we cannot generally define the direction of rotation of an input shaft as positive.

b) Power Division

In a revolving drive with only one input shaft, the transmitted power is divided between the two output shafts. The ratio of the two partial powers is then uniquely determined by the torques and speeds of the output shafts and thus is dependent on the transmission type and its momentary state of motion. The mode which characterizes this particular operating condition shall subsequently be called "power division." However, power branching, as defined in section 2, does not affect the operating characteristics of epicyclic transmissions. This fundamental difference must be emphasized at this point, especially since most publications do not clearly distinguish between the concepts of power division and power branching, but rather use them interchangeably.

Any epicyclic transmission can be operated in the power division mode and may then be called a (power) division drive. In the literature it is sometimes called a "differential," although this term should be reserved for simple epicyclic transmissions with two degrees of mobility, which also includes transmissions operating in the power summation mode that will be discussed in the following section 14c.

c) Power Summation

If a revolving drive has two input shafts, their powers are added in the drive train and then transmitted to the output shaft. Transmissions which operate in this mode are called summation drives. This term also characterizes an application or an operating condition for which basically any revolving drive is well suited.

d) Rolling-Power and Coupling-Power

The two partial motions of the revolving drive trains which have been defined by Willis and are described in section 11a transmit their associated partial powers by two different principles as will be subsequently shown. Thus, in the first partial motion, the power is transmitted from the input shaft to the output shaft solely through action between the gear teeth as in a basic transmission and, therefore, this partial power shall be called the "rolling" power P_R of the connected shafts. In a basic transmission, the rolling-power is equal to the total power transmitted by the connected shafts.

If an equal speed is now superimposed on all three connected shafts of the revolving drive, then the partial speeds $(n_1 - n_s)$ and $(n_2 - n_s)$ of the central gear shafts relative to the carrier remain unaffected and an observer located on the carrier could detect no change in the speed of the gears. He would further observe that the relative speeds between the gears, and thus the velocities with which the central gears and the planets roll with each other, remain unchanged. This leads to the other important observation that the sliding velocities between the tooth profiles, which cause friction losses, likewise remain constant.

If, during the described superposition of an equal speed on all three connected shafts, the shaft torques remain unchanged, then the tangential forces which act at the pitch circles of the gears also remain constant. Thus we can conclude that the rolling-powers, that is, the products between the tangential forces and the pitch circle velocities, are not affected by the second partial motion; neither are the tooth-friction losses.

This consideration is equally valid for reverted and open revolving gear trains.

Thus, the rolling-power represents a partial power which is transmitted by the central gear shafts 1 and 2 of the revolving gear train and can be obtained for each of these shafts by calculating the product of their speeds relative to the carrier and their torques. That is,

$$P_{R1} = T_1(n_1 - n_s)$$

and

$$P_{R2} = T_2(n_2 - n_s) \ . \tag{47}$$

Within the gear train the rolling-power "flows" from the central gear shaft which transmits the positive rolling-power to the central gear shaft transmitting the negative rolling-power. These shafts correspond to the input and output shafts respectively, of the basic gear train. In this process, the tooth-friction power loss P_{Lt} is converted into heat by friction. Like the housing of a conventional gear train, the carrier does not participate in the transmission of the rolling-power.

When the revolving gear train operates solely in the second partial motion, that is, without a relative motion between the carrier and the central gears, it behaves like a rigid coupling. The three connected shafts rotate with the same speed n_s and power is transmitted between the members of the gear train without losses. Therefore, this power shall be called the "coupling-power" P_s.

Because the coupling powers of the three shafts are transmitted at the same speeds, they must always bear the same constant ratio, with respect to each other, as the associated torques, that is:

$$P_{s1} : P_{s2} : P_{ss} = T_1 n_s : T_2 n_s : T_s n_s = T_1 : T_2 : T_s \ . \tag{48}$$

Thus the coupling-powers of the two difference shafts have the same sign, and the sum of their absolute values equals the absolute value of the coupling-power of the summation shaft. According to the sign conventions for the powers, this means that within the gear train a positive coupling-power always "flows" from the summation shaft to the two difference shafts, or vice versa. In contrast to the rolling power-flow, the coupling power-flow is without losses, and its direction need not be considered when the efficiency of the gear train is analyzed.

The superposition of the two relative powers must be executed separately for each of the connected shafts, and with eqs. (47) and (48) we obtain:

$$P_1 = P_{s1} + P_{R1} = T_1 n_s + T_1(n_1 - n_s) = T_1 n_1 \tag{49}$$

$$P_2 = P_{s2} + P_{R2} = T_2 n_s + T_2(n_2 - n_s) = T_2 n_2 \tag{49a}$$

$$P_s = P_{ss} + 0 = T_s n_s \tag{49b}$$

Thus, the power transmitted by each of the connected shafts of a revolving gear train constitutes the sum of its coupling-power and its rolling-power.

If, therefore, the principle of conservation of energy is applied to these partial motions, it follows that the sum of the rolling powers, including their associated friction losses, and the sum of the coupling-powers must equal zero, that is:

the coupling-powers $\qquad P_{s1} + P_{s2} + P_s = 0 \qquad$ and

the rolling-powers $\qquad P_{R1} + P_{R2} + P_{Lt} = 0 \qquad$ so that

the shaft powers $\qquad P_1 + P_2 + P_s + P_{Lt} = 0 \ . \qquad\qquad$ (50)

For convenience, the total and partial powers of the three connected shafts of a revolving drive are compiled in table 1 as functions of their torques and speeds.

TABLE 1. SUMMARY OF EQUATIONS FOR EXTERNAL AND PARTIAL SHAFT POWERS
OF A REVOLVING DRIVE TRAIN

SHAFT	EXTERNAL POWER	COUPLING-POWER	ROLLING-POWER
1	$P_1 = T_1 n_1$	$P_{s1} = T_1 n_s$	$P_{R1} = T_1(n_1 - n_s)$
2	$P_2 = T_2 n_2$	$P_{s2} = T_2 n_s$	$P_{R2} = T_2(n_2 - n_s)$
s	$P_s = T_s n_s$	$P_s = T_s n_s$	$P_{Rs} = 0$
	external power	= coupling-power	+ rolling-power

As can be seen from eqs. (49) and (49a), the rolling and coupling-powers of each of the connected shafts *1* and *2* can be positive or negative and flow in the same or opposite directions depending on the magnitude and the sign of their speeds. This important property of each revolving drive shall be investigated in the following mental experiment on the negative-ratio drive shown in fig. 47.

In this experiment we want to determine the two partial powers of shaft *1* at a constant positive input torque T_1, and a constant positive input speed n_1, but at different carrier speeds. Figs. 47a to 47d illustrate the steps of the

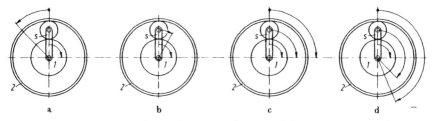

a   b   c   d

Fig. 47. Rolling-power and coupling-power in a revolving gear train at different states of motion: *a*, $n_s = 0$; *b*. $k_{s1} = i_{s1}$ so that $n_2 = 0$; *c*, operation at the coupling point where $n_1 = n_2 = n_s$; *d*, $n_s > n_1$ causes the rolling-power at the input shaft *1* to become negative.

experiment, and the lengths of the arrows symbolize the tangential velocities of the connected shafts.

a) We assume that the carrier is initially locked so that the total input power is transmitted as rolling power by the gears, that is, $P_1 = P_{R1}$. Since the basic speed-ratio i_0 of the gear train is negative, eq. (32) implies that the speed-ratio i_{12} is likewise negative and, thus, shafts 1 and 2 rotate in opposite directions, or, if we define n_1 as positive, the output speed n_2 is negative. The coupling-power $P_{s1} = T_1 n_s = 0$ because the carrier is at rest.

b) The carrier is now rotated in the same direction as the input shaft 1. This decreases the latter's partial speed $(n_1 - n_s)$ relative to the carrier, and thus its rolling-power. The reduction of the rolling-power, however, is compensated by the arising coupling-power so that

$$P_1 = P_{R1} + P_{s1} \quad \text{where} \quad \begin{cases} P_{R1} = T_1(n_1 - n_s) > 0 \\ P_{s1} = T_1 n_s > 0 \end{cases}$$

Since the torques of shaft 1 and the carrier have opposite signs, but both shafts move in the same direction, the carrier transmits a negative power and is an output shaft. If eq. (24) is rewritten with the appropriate negative i_0 and n_2, it assumes the form:

$$-|n_2| = -\frac{n_1}{|i_0|} + \left(\frac{1 + |i_0|}{|i_0|}\right)n_s , \quad \text{and}$$

it can readily be seen that the absolute speed n_2 of the output shaft 2 decreases as n_s increases. It reaches a value of zero when

$$n_1 = (1 - i_0)n_s , \quad \text{or}$$

$$\frac{n_1}{n_s} = k_{1s} = (1 + |i_0|) = i_{1s}$$

and then reverses its direction of rotation. When shaft 2 reverses the sign of its speed, the sign of its power also changes which means that shaft 2 now becomes an input shaft. Consequently, the carrier shaft is the sole output shaft. It is obvious that the relative speed $(n_2 - n_s)$ decreases as n_s increases. It can be verified on the basis of section 11c that the same holds true for $(n_p - n_s)$.

c) If the carrier speed n_s is still further increased until it becomes equal to the input speed n_1, the relative speed $(n_1 - n_s)$, as well as the other relative speeds, becomes zero so that $n_2 = n_s = n_1$. Thus, the partial powers become:

$$P_{R1} = T_1(n_1 - n_s) = 0 , \qquad P_{s1} = T_1 n_s = P_1 ;$$

that is, the transmission operates at the "coupling point" and the total external power is transmitted as coupling-power.

d) If the carrier speed is finally increased beyond the value of the input speed n_1, the rolling speed $(n_1 - n_s)$ and, consequently, the rolling power P_{R1} of shaft I must also change their signs. Thus all relative speeds and their associated rolling-powers have assumed signs opposite to those they had in steps a) and b). The fact that the rolling-power of shaft I is negative implies that its coupling power exceeds its external input power P_1, that is:

$$P_1 = P_{R1} + P_{s1} \begin{cases} P_{R1} = T_1(n_1 - n_s) < 0, \\ P_{s1} = T_1 n_s > P_1 > 0. \end{cases}$$

Thus, the coupling-power P_{s1} of shaft I represents an input power, while the rolling-power P_{R1} is an output power. Therefore, we may imagine that the two partial power-flows in the transmission are opposed to each other. This may be easier to understand if we view all relative speeds from the standpoint of an observer located on the carrier. Since the carrier now rotates faster than gear I, we observe that the latter rotates backwards, that is, against the still positive external input torque. Based on this observation we must define shaft I as an output shaft and we must conclude that its rolling-power is negative.

e) Illustration of Power Transmission within a Simple Revolving Drive Mechanism

At this point we may interject a discussion which should further facilitate an understanding of the function of the revolving drives.

The external power of a single input shaft can be divided between two output shafts, or the external power of two input shafts can be transferred to a single output shaft. Such power-flows can occur only with coupling-power, since it alone can be transmitted between all three shafts. However, according to eqs. (48), (42), (44), and (45), these coupling-powers must bear an invariable ratio relative to each other, which is determined by the basic speed-ratio i_0 and the basic efficiency η_0, that is, by the design of the transmission. As pointed out in eq. (48) and the text immediately following it, a positive coupling-power always flows from the summation shaft to the two difference shafts, or vice versa. If the external torques of a given transmission are presumed to be constant, then the coupling-powers are proportional to the absolute value of the carrier speed and independent of the speeds of the gears. The rolling-power, however, which can flow only between the shafts I and 2, is proportional to the partial speed $(n_1 - n_s)$ or $(n_2 - n_s)$ of each gear relative to the carrier. At any constant carrier speed its magnitude and direction can be arbitrarily changed by changing the speed of shaft *1* or shaft *2*. Thus, by superposition of the constant flow of coupling-power and the variable flow of rolling-power, any arbitrary external power-flow can be generated solely by an appropriate choice of the speed-ratio between any two connected shafts.

If we review these power-flows for all practical transmission designs, we observe two fundamentally different possibilities for the superposition of the power components. One of these possibilities applies to all positive, the other to all negative-ratio drives.

In negative-ratio drives the carrier shaft *s* always constitutes the summation shaft. If it is also an input shaft, its coupling-power flows to the central gear shafts *1* and *2*, if an output shaft, it sums the coupling-powers of shafts *1* and *2*. In each of these modes the rolling-power which can only be transferred between the central gear shafts, can either flow from shaft *1* to shaft *2*, or vice versa. Consequently, there exist four distinct internal power-flow modes which are illustrated by the schematics a to d in fig. 48. If the carrier shaft is also the sole input shaft, or one of the two input shafts, the transmission operates with power division as shown in fig. 48, I. As an alternative the carrier shaft can be the sole output shaft or one of the two input shafts. The transmission then operates with power summation as shown in fig. 48, II. In either case, the partial power which flows between any connected shafts cannot exceed the external power at any terminus. Consequently, the rolling-power which is associated with the internal friction losses must always be smaller than the transmitted external power. Based on the transmission of an equal external power, therefore, the revolv-

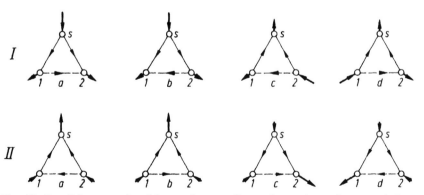

Fig. 48. Possible external and internal power-flows of negative-ratio transmissions: *I a–d,* for power division; *II a–d,* for power summation.

      external power { ⟶   total-power shaft
                      ⟶   partial-power shaft
      internal power { ⟶—   coupling-power
                      – –⟩– –   rolling-power

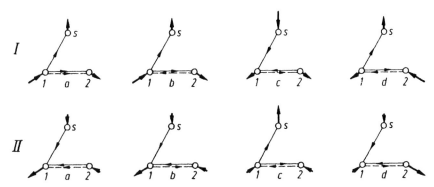

Fig. 49. Possible external and internal power-flows for a positive-ratio transmission whose shaft *1* is the summation shaft: *I a–d,* for power division; *II a–d,* for power summation.

external power { —→ total-power shaft
{ —→ partial-power shaft
internal power { —▶— coupling-power
{ – –▶– – rolling-power

ing carrier efficiency of a negative-ratio transmission is always higher than its basic efficiency η_o.

In positive-ratio drives, one of the two central gear shafts always constitutes the summation shaft. Thus a coupling-power plus a rolling-power can flow between the summation shaft and the second central gear shaft. Since, however, rolling-power cannot be transmitted between a central gear shaft and the carrier shaft, no direct power-flow exists between the second central gear shaft and the carrier shaft. The resulting possible power-flow modes for positive-ratio gear trains with power division and power summation are shown in fig. 49. These schematics are drawn for the case that shaft *1* is the summation shaft. If shaft *2* becomes the summation shaft, the correct power-flows can be obtained by simply exchanging the symbols *1* and *2*.

Fig. 49 shows that in the power-flow modes Ia and IIa, where the gear train operates with power division and power summation respectively, all partial power-flows including the flow of rolling-power, are smaller than the total external power.

In both of these cases, the rolling and the coupling-power between shafts *1* and *2* flow in the same direction and thus have the same signs.

In the power-flow modes which are described by the schematics b, c, and d of figs. 49 I and 49 II the flow of rolling-power and the flow of coupling-power between shafts *1* and *2* oppose each other. Thus, the rolling-power is smaller than the total external power only when the summation shaft becomes either the sole input shaft or the sole output shaft, as shown in the schematics Ib and IIb. However, if the second central gear shaft, which acts as a difference shaft, is the sole input shaft, or the sole output shaft as shown in figs. 49 Id and IId, then the rolling-power must be larger than the total external power. Therefore, in these modes, the positive-ratio gear trains operate with larger losses, that is, with a lower efficiency than in their basic mode. In the power-flow modes Ic and IIc of fig. 49, where the carrier shaft is the sole input shaft or the sole output shaft, the rolling-power may be larger or smaller than the external power, depending on whether it is larger or smaller than the power at the carrier shaft.

f) Inner Futile Power in Positive-Ratio Drives

If, as shown in figs. 49a to 49d, the flow of rolling- and coupling-power between the two central gear shafts is opposed, then only the difference between absolute values constitutes an effective shaft power. This means, that in these modes of operation, the smaller of the two partial powers always is opposed to the effective power-flow. Consequently, it does not contribute to the effective power and, therefore, may be called a futile power. This futile power must be compensated by an equal amount of the opposing larger partial power which is also lost as an effective power and thus represents another futile power. However, the two opposing futile power-flows, that is, the flow of futile coupling- and rolling-power, remain active in full magnitude. The friction losses which are associated with the flow of futile rolling-power lower the overall efficiency of the gear train and contribute to the generation of heat.

g) Power Characteristics of Positive- and Negative-Ratio Drives

The previous considerations reveal the following principles which characterize the power transmission in revolving gear trains and which we shall express in the form of two further guide rules:

> *In negative-ratio gear trains, the rolling-power always is*   R 15
> *smaller than the external power.*

Consequently, their efficiencies are always higher than their basic efficiencies. No futile power arises in negative-ratio drives.

> *In positive-ratio gear trains, the rolling-power can be larger,*   R 16
> *smaller, or equal to the external power, depending on the*
> *existing power-flow mode.*

Depending on the momentary state of motion, their efficiencies, therefore, can be smaller than, equal to, or larger than their basic efficiencies. In six of the eight possible power-flow modes which are illustrated in fig. 49, a futile power-flow exists.

The occurrence of a futile power in some modes of operation has sometimes unjustifiably discredited the positive-ratio gear trains since the possible drop of the efficiency below the basic efficiency erroneously has been considered as a necessary consequence of the futile power. However, whether the overall efficiency becomes larger or smaller than the basic efficiency depends exclusively on whether the absolute value of the rolling power is smaller or larger than the external power. This should be obvious if it is recalled that the basic efficiency is the efficiency with which a pure rolling-power is transmitted. In any case, it is immaterial whether the rolling-power is a futile power or not. A review of fig. 49 shows that in a majority of the possible power-flow modes the efficiency of a positive-ratio gear train is rather higher than its basic efficiency, depending on which of the external shafts is the total-power shaft.

If, as in figs. 49a and 49b, one of the two central gear shafts is the summation shaft and at the same time the total-power shaft, then the rolling-power is always smaller than the total power.

If, as in fig. 49d, the other gear shaft, which is a difference shaft, is the total-power shaft, then the rolling-power is always larger than the total power.

If, as in fig. 49c, the carrier shaft is the total-power shaft, then the rolling-power may be larger or smaller than the total power which acts on the carrier shaft. Which of these possibilities occurs depends only on the shaft speeds.

The mode of the power-flow of a revolving gear train, whose basic speed-ratio i_0 and basic efficiency η_0 are given, depends only on the speeds of its shafts. Since, by definition, even a partial power within the gear train flows from a shaft with a positive coupling- or rolling-power to a shaft with a negative coupling- or rolling-power, the internal power-flow at any known speed can be determined easily from the equations which are summarized in table 1.

If, in two-shaft transmissions, one of the two speeds n_1 or n_2 becomes zero, then the number of possible power-flow modes is substantially reduced, since those modes in which the locked central gear shaft serves as the total power shaft are no longer feasible. Since a pure coupling-power can be transmitted only when all three shafts have the same speed, two-shaft positive-ratio gear trains can only operate in those modes which are characterized by the occurrence of futile power-flows. However, depending on their speed-ratios, the rolling-powers can be larger or smaller than the external powers.

According to the equations which are summarized in table 1, even the locked central gear shaft of any two-shaft transmission must transmit the full rolling-power and an equal and opposite coupling-power.

h) Role of Internal Power-Flows in Analysis of Revolving Gear Trains

The purpose of these discussions was to facilitate a clear understanding of the operation of revolving gear trains. However, in a practical analysis it is not necessary to consider the flow of internal power which does not even enter into purely kinematical investigations. It is true that in an analysis of the efficiencies the direction of the flow of rolling-power, as well as the basic efficiency, must be considered. However, this is done automatically when the sign of the rolling-power of shaft *I* is determined and then the sign of the exponent $r1$ is calculated from eq. (43).

Section 15. Efficiency and Self-Locking Capability

a) General Remarks Concerning Efficiency

We assume that the losses in a planetary gear train are like those in a conventional gear train (secs. 5 and 6). We exclude losses due to centrifugal forces and include only tooth-friction losses and load-dependent friction losses of the planet bearings. The power-loss in such a gear train is directly proportional to its rolling-power. If we denote the positive (input) rolling-power with $P_{R,in}$ and the negative (output) rolling-power with $P_{R,out}$, then the power-loss of a planetary gear train according to eqs. (3) and (5) becomes:

$$P_L = -P_{R,out} - P_{R,in} = \eta_0 P_{R,in} - P_{R,in} = -P_{R,in}(1 - \eta_0) \ .$$

If this power-loss is related to the total input power of the three-shaft transmission, then it follows that

$$-\frac{P_L}{P_{in}} = \frac{P_{R,in}}{P_{in}}(1 - \eta_0) = 1 - \eta \ .$$

Thus

$$\eta = 1 - \frac{P_{R,in}}{P_{in}}(1 - \eta_0) \ , \tag{51}$$

and, according to eq. (5),

$$\zeta = \frac{P_{R,in}}{P_{in}} \cdot \zeta_0 \tag{52}$$

where

$$\zeta_0 = 1 - \eta_0 \ .$$

Thus, the efficiency of a planetary transmission can be calculated when η_0 and the input power and and positive rolling-power are known. To simplify we shall subsequently develop and tabulate equations which can be immediately applied, and depend only on the basic speed ratio i_o and one of the speed-ratios k of the gear train.

b) Efficiencies of Three-Shaft Transmissions

According to eq. (5), the efficiency of a transmission in general is defined as

$$\eta = -\frac{\text{output power}}{\text{input power}} = -\frac{P_{out}}{P_{in}} \; .$$

Since, in revolving drives with three rotating shafts, the input or output power is always composed of two partial powers, the efficiency equation assumes the form:

$$\eta = -\frac{P_{out}}{P_{in}} = -\frac{\Sigma T_{out}\omega_{out}}{\Sigma T_{in}\omega_{in}} = -\frac{\Sigma T_{out}n_{out}}{\Sigma T_{in}n_{in}} \; . \tag{53}$$

Since each shaft can be the sole input or output shaft, each three-shaft transmission can operate with six different external power-flows which must be described by six different efficiency equations. Therefore, we must know the operating conditions of a revolving drive before we can write down the proper efficiency equation.

The six possible external power-flows and their efficiencies are summarized in fig. 50 where each η is characterized by a sequence of subscripts indicating the associated external power-flow. Equations, with shafts denoted by the letters a, b, or c, are valid for operating conditions where the same shaft is either the sole input or the sole output shaft and, therefore, are reciprocal to each other. Thus, it is sufficient to derive the efficiency equations for the three cases of power division. The equations for the three cases of power summation are then obtained as their reciprocals.

If we assume that the basic speed-ratio i_o and the basic efficiency η_0 of a revolving drive are known, we can expand its efficiency equation for the power-flow of fig. 50, 1a by substituting eqs. (42) and (44) for T_2 and T_s, so that:

$$\eta_{1<\,^2_s} = -\frac{P_2 + P_s}{P_1} = -\frac{T_2 n_2 + T_s n_s}{T_1 n_1}$$

$$= -\frac{-T_1 i_o \eta_0^{r1} n_2 + T_1 (i_o \eta_0^{r1} - 1) n_s}{T_1 n_1} \; .$$

After T_1 is eliminated, this equation can be written in the notation which has been introduced in section 10 and then becomes:

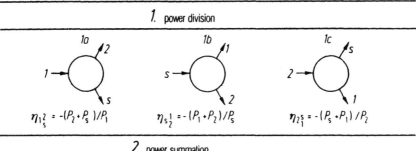

Fig. 50. Six possible efficiency equations for three-shaft revolving drive trains. Eqs. (1a), (2a); (1b), (2b): and (1c), (2c) are reciprocal to each other.

$$\eta_{1<\frac{2}{s}} = i_o \eta_o^{rl} k_{21} - i_o \eta_o^{rl} k_{s1} + k_{s1} \ .$$

According to eq. (28), the speed-ratio k_{21} can be expressed in terms of k_{s1}. Therefore, since:

$$k_{21} = \frac{1}{i_o} + \left(1 - \frac{1}{i_o}\right)k_{s1} \ ,$$

the efficiency equation simplifies to:

$$\eta_{1<\frac{2}{s}} = \eta_o^{rl} + (i_o \eta_o^{rl} - \eta_o^{rl})k_{s1} - i_o \eta_o^{rl} k_{s1} + k_{s1} \ ,$$

$$\eta_{1<\frac{2}{s}} = \eta_o^{rl} + k_{s1}(1 - \eta_o^{rl}) \ . \tag{54}$$

As observed earlier, the efficiency for the power-flow depicted in fig. 50, 2a can be obtained merely by taking the reciprocal of eq. (54), so that:

$$\eta_{s>1}^2 = \frac{1}{\eta_o^{rl} + k_{s1}(1 - \eta_o^{rl})} \ . \tag{55}$$

Analogously, the power-flow described by fig. 50, 1c can be obtained by expanding $\eta_{2<\frac{1}{s}}$ with eqs. (44) and (42). Thus

$$\eta_{2<\frac{1}{s}} = -\frac{T_1 n_1 + T_s n_s}{T_2 n_2} = \frac{k_{12} + i_o \eta_o^{rl} k_{s2} - k_{s2}}{i_o \eta_o^{rl}} \ .$$

Since according to eq. (26) $k_{12} = i_o + (1 - i_o)k_{s2}$, this efficiency becomes:

$$\eta_{2<\frac{1}{s}} = \frac{i_o + k_{s2} - i_o k_{s2} + i_o \eta_o^{rl} k_{s2} - k_{s2}}{i_o \eta_o^{rl}},$$

$$\eta_{2<\frac{1}{s}} = \frac{1}{\eta_o^{rl}} + \left(1 - \frac{1}{\eta_o^{rl}}\right) k_{s2}. \tag{56}$$

For a power-flow as illustrated by fig. 50, 2c, the efficiency is again obtained by taking the reciprocal of eq. (56) which leads to:

$$\eta_{s>2}^{\frac{1}{}} = \frac{1}{\frac{1}{\eta_o^{rl}} + \left(1 - \frac{1}{\eta_o^{rl}}\right) k_{s2}}. \tag{57}$$

The efficiency for the power-flow of fig. 50, 1b is found as previously described by expanding the given equation with eqs. (42) and (44). Thus,

$$\eta_{s<\frac{1}{2}} = -\frac{T_1 n_1 + T_2 n_2}{T_s n_s} = \frac{i_o \eta_o^{rl} k_{2s} - k_{1s}}{i_o \eta_o^{rl} - 1},$$

and since according to eq. (29)

$$k_{2s} = \frac{k_{1s} - 1 + i_o}{i_o}$$

we find that:

$$\eta_{s<\frac{1}{2}} = \frac{\eta_o^{rl}(k_{1s} - 1 + i_o) - k_{1s}}{i_o \eta_o^{rl} - 1},$$

$$\eta_{s<\frac{1}{2}} = \frac{\eta_o^{rl}(i_o - 1) + k_{1s}(\eta_o^{rl} - 1)}{i_o \eta_o^{rl} - 1}. \tag{58}$$

Finally, the reciprocal of eq. (58), that is,

$$\eta_{2>s}^{\frac{1}{}} = \frac{i_o \eta_o^{rl} - 1}{\eta_o^{rl}(i_o - 1) + k_{1s}(\eta_o^{rl} - 1)}, \tag{59}$$

gives the efficiency for the power-flow which is indicated by fig. 50, 2b.

Since with eqs. (26) through (31) (or worksheet 2) each of the speed-ratios k can be expressed in terms of each of the other speed-ratios, each of the six efficiency equations can be written in six different forms which, of course, must all yield identical results. If, for instance eq. (30) is used to express k_{s1} in terms of k_{21}, then eq. (54) assumes the form:

$$\eta_{1<\overset{2}{s}} = \eta_o^{r1} + \frac{i_o k_{21} - 1}{i_o - 1} (1 - \eta_o^{r1}) \ .$$

Analogously, its reciprocal

$$\eta_{\overset{2}{s}>1}^{} = \frac{i_o - 1}{\eta_o^{r1}(i_o - 1) + (i_o k_{21} - 1)(1 - \eta_o^{r1})}$$

is one of the six alternate forms of eq. (55). By continuing this process we can finally obtain 36 or 18 reciprocal pairs of efficiency equations which describe the six possible power-flows.

Worksheet 4 summarizes these six efficiency equations in terms of the speed-ratio k_{12} between the input shaft *1* and the output shaft *2* for the three basic speed-ratio ranges $i_o < 0$, $0 < i_o < 1$ and $i_o > 1$. The appropriate exponent $r1$ is already considered in these equations. However, for other purposes such as torque calculations, it has been explicitly included in worksheet 4. For those power-flows whose speed-ratio range includes the coupling point $k_{12} = 1$, two efficiency equations are given. There is one for each of the ranges above and below the coupling point $k_{12} = 1$, since according to R 17 the exponent $r1$ changes its sign whenever the speed-ratio passes through the coupling point.

If, instead of k_{12} another speed-ratio of the transmission, for example, k_{1s}, is known or given, then k_{12} must first be expressed in terms of k_{1s} (by using worksheet 2) before it can be inserted into the appropriate efficiency equation (to avoid confusion, and since worksheet 2 greatly facilitates this task, only the twenty-four equations in terms of k_{12}, rather than all of the 6 × 24 = 144 efficiency equations in terms of all of the six possible speed-ratios, have been listed in worksheet 4).

For practical use, worksheet 4 is organized in such a way that the appropriate group of rows can be chosen by first matching the known basic speed-ratio of the transmission with the basic speed-ratio range in column 1. By matching the known speed-ratio k_{12} with column 2, the correct row and the associated total-power shaft can then be identified. Finally, the power-flow and thus the appropriate efficiency equation can be determined if any one of the three shafts is known to be an input or output shaft.

If, for instance, the efficiency of an arbitrary three-shaft transmission with a basic speed-ratio $i_o = 0.4$ and a speed-ratio $k_{s2} = 1.6$ is to be determined, then the second group of rows, where $0 < i_o < 1$ is applicable. From worksheet 2, box 6, we find that

$$k_{12} = k_{s2}(1 - i_o) + i_o = 1.6 \ (1 - 0.4) + 0.4 = 1.36 \ .$$

In worksheet 4, this value falls into the range of $k_{12} > 1$, where shaft *1* is the total-power shaft. Thus the power-flow is either $1<\overset{2}{s}$ or $\overset{2}{s}>1$. If now the design of the transmission indicates that shaft 2 must be an input shaft, then the power flow is $\frac{2}{s} > 1$, and the efficiency becomes

$$\eta_s^2 > 1 = \frac{k_{12}(1 - i_o)}{\eta_o i_o(1 - k_{12}) + k_{12} - i_o} = \frac{1.36 \cdot 0.6}{\eta_o \cdot 0.4(-0.36) + 1.36 - 0.4}$$

$$= \frac{0.816}{0.96 - 0.144\eta_o}$$

If we are interested further) (e.g., in the torques of this transmission), we find in worksheet 4 that the correct exponent is $r1 = +1$, which allows us to evaluate eqs. (42), (44), and (45).

In the rare case when none of the three shafts can be safely identified as an input or output shaft, the signs of the speed and torque and thus the sign of the power must be determined for at least one of the shafts from the state of operation. The appropriate efficiency equation can then be identified with the help of R 2.

These equations show that the efficiency of a three-shaft transmission depends as much on a speed-ratio k, that is, on its momentary state of motion, as on i_o and η_o. This distinguishes the revolving drives with three rotating shafts from the simple speed-reducing drives, whose efficiencies, under the assumption of section 5, can be considered as constant and, thus, as independent of the speeds. The efficiency of a three-shaft transmission, therefore, must be calculated individually for each of its states of motion.

In all these equations the efficiency becomes 1 when the speed-ratio k assumes a value of 1, that is, at the coupling point where all three speeds are equal and the rolling power and its associated losses become equal to zero in eq. (51). If, during a change of the transmission speed, the shaft speeds pass through the coupling point, then all the speed-ratios k momentarily assume a value of 1, which means that all speed differences, including the rolling speed ($n_1 - n_s$) and the exponent $r1$, eq. (43), change their signs. This fact is expressed in the following guide rule:

If, during a speed change, the shaft speeds pass through the   R 17
coupling point, then the rolling power-flow changes its direction and the exponent $r1$ changes its sign.

Unlike $r1$ and, thus, the efficiency, the external power-flow established by the position of the total-power shaft, and the signs of the three shaft powers do not change at the coupling point.

c) Determination of External Power-Flow

The external power-flow of a simple three-shaft transmission which determines the appropriate efficiency equation can usually be identified from the positions of the prime mover and the driven machinery. If the power-flow mode is not obvious (e.g., when the transmission is connected to two motors, one of which could operate as a generator), the power-flow can be determined from any speed-ratio k, if i_o and at least one input or output shaft are known.

According to R 2, the speeds and torques have the same signs for input shafts, and opposite signs for output shafts. Therefore, the signs of all torques of a transmission are determined when the sign of one of the torques and the position of the summation shaft are given. The power-flow can then be influenced only by a suitable choice of the shaft speeds. Consequently, we can specify the operating range of each power-flow by the upper and lower limits of the speed-ratios k.

As an example, we shall determine the range of the power-flow $1 <^2_s$ for the three transmission types whose basic speed-ratios are $i_o < 0$, $0 < i_o < 1$ and $i_o > 1$. The input power $P_1 > 0$ so that we know:

$$T_1 > 0, \quad \text{and} \quad n_1 > 0 .$$

The resulting signs of the torques and speed-ratios for all three transmission types are summarized in table 2.

TABLE 2. SIGNS OF SPEEDS, TORQUES, AND SPEED-RATIOS
OF THREE-SHAFT TRANSMISSIONS WITH POWER-FLOW $1 <^2_s$

TRANSMISSION TYPE	NEGATIVE-RATIO DRIVE	POSITIVE-RATIO DRIVE	POSITIVE-RATIO DRIVE	SOURCE
i_o:	<0	$0 \ldots 1$	>1	
SUMMATION SHAFT	s	1	2	WORKSHEET 3
T_1	>0	>0	>0	Definition
n_1	>0	>0	>0	
T_2	>0	<0	<0	R 8, R 9
n_2	<0	>0	>0	R 2
T_s	<0	<0	>0	R 8, R 9
n_s	>0	>0	<0	R 2
$k_{s1} = n_s/n_1$	>0	>0	<0	
$k_{s2} = n_s/n_2$	<0	>0	<0	
$k_{12} = n_1/n_2$	<0	>0	>0	
Column	1	2	3	4

Since the sign of T_1 is given, we can obtain the signs of the output torques T_2 and T_s, when the location of the summation shaft is known, by using R 8 and R 9. According to R 2, the signs of the output speeds n_2 and n_s are then simply opposite to the signs of the torques so that the signs of the speed-ratios k_{s1}, k_{s2}, and k_{12} also can be determined.

According to worksheet 2, the two speed-ratios k_{s1} and k_{12} are related by the equation,

$$k_{12} = \frac{i_o}{1 - k_{s1}(1 - i_o)} \ . \tag{60}$$

When i_o is negative, k_{12} is likewise negative as can be verified from table 2, column 1. Therefore, the denominator on the right-hand side of eq. (60) must be positive, that is

$$1 - k_{s1}(1 - i_o) > 0 \ , \qquad \text{and, consequently,}$$

$$k_{s1} < \frac{1}{1 - i_o} \ , \qquad \text{or, according to eq. (34),}$$

$$k_{s1} < i_{s1} \ .$$

Since, according to column 1 of table 2,

$$k_{s1} > 0 \qquad \text{at the same time,}$$

the speed-ratio k_{s1} for the given power-flow, can only lie between the limits

$$0 < k_{s1} < i_{s1} \ . \tag{61}$$

If these limits are substituted back into eq. (60) we find the range of k_{12}, which is

$$-\infty < k_{12} < i_o \ , \qquad \text{that is,} \qquad k_{12} < i_o \ . \tag{62}$$

Since $k_{s2} = (k_{12} - i_o)/(1 - i_o)$ [worksheet 2, box 31, or eq. (31)] we can obtain the range of the third speed-ratio from the range of the speed-ratio k_{12} as given by eq. (62).

Thus:

$$-\infty < k_{s2} < 0 \ , \qquad \text{that is,} \qquad k_{s2} < 0 \ .$$

If these considerations are repeated for the positive-ratio transmission whose basic speed-ratio $0 < i_o < 1$ and whose speed-ratios k are given in column 2 of table 2, then eq. (60) must have a positive denominator, that is:

$$0 < k_{s1}(1 - i_o) < 1 \ .$$

With $\quad i_{s1} = \dfrac{1}{1 - i_o} \ , \quad$ eq. (34), this becomes $\quad 0 < k_{s1} < i_{s1} \ .$

When these limits are substituted into eq. (60) we obtain first

$$\infty > \left(k_{12} = \frac{i_o}{1 - k_{s1}(1 - i_o)} \right) > i_o \, , \qquad \text{that is,} \qquad k_{12} > i_o$$

and then with $\quad k_{s2} = \dfrac{k_{12} - i_o}{1 - i_o} \, , \qquad$ eq. (31), $\qquad k_{s2} > 0 \, .$

Finally, for the positive-ratio transmission of column 3 in table 2 the basic speed-ratio $i_o > 1$ and the denominator of eq. (60) is positive, while k_{s1} is negative, that is $0 > k_{s1}$.

Thus

$$1 - k_{s1}(1 - i_o) = 1 + k_{s1}(i_o - 1) > 0 \, ,$$

$$k_{s1} > \left(-\frac{1}{i_o - 1} = \frac{1}{1 - i_o} = i_{s1} \right) ,$$

so that:

$$0 > k_{s1} > i_{s1} \qquad \text{and}$$

$$\infty > \left[k_{12} = \frac{i_o}{1 - k_{s1}(1 - i_o)} \right] > i_o \, , \qquad \text{that is,} \qquad k_{12} > i_o \, ,$$

and

$$0 > \left(k_{s2} = \frac{k_{12} - i_o}{1 - i_o} \right) > -\infty \, , \qquad \text{that is,} \qquad k_{s2} < 0 \, . \qquad (63)$$

The identical speed-ratio limits are found when these considerations are repeated for the reciprocal power-flow $\frac{2}{s} > 1$, whose torques or speeds have opposite signs but are otherwise equal to those of the just-investigated power-flow $1 < \frac{2}{s}$. This correspondence is due to the fact that although the signs of all the torques or speeds are reversed, the signs of the speed-ratios in columns 1 through 3 of table 2 remain the same.

Thus, if we know the basic speed-ratio i_o and an arbitrary speed-ratio k of a three-shaft transmission, then, following the above considerations, we can identify its total-power shaft. If we further know an input or output shaft of the transmission, then its complete external power-flow is determined by R 14.

The above example shows that for a given external power-flow the limits of each operating range are characterized by the fact that one of the partial-power shafts stops so that the revolving drive becomes a two-shaft transmission. At this point, therefore, the speed-ratio between the two rotating shafts is either the basic speed ratio i_o or one of the other two-shaft speed-ratios i. The momentarily stopped shaft changes its direction of rotation and thus changes either from an input shaft to an output shaft, or vice versa, when the transmission subsequently goes to another mode of operation with a different power-flow.

Fig. 51 summarizes the three-shaft transmission power-flow modes, their associated borderline two-shaft speed-ratios, and the shafts which are stopped at these limits. It is valid for both positive- and negative-ratio drives so that the values and the signs of the tabulated speed-ratios i can be calculated from eqs. (32) to (36) only when the transmission type has been specified by its basic speed-ratio i_0. In the two three-shaft power-flow modes adjacent to an i-ratio, signs of corresponding k-ratios are the same as the sign of the i-ratio. In the third operating range k assumes the opposite sign. Thus, after crossing a boundary line the sign of k remains the same only if the limit ratio at the line is an i-ratio.

Fig. 52 shows the ranges of the speed-ratios k for each power-flow. It is subdivided for the three transmission types, so that the signs of k can be included. This figure also contains the speed-ratio limits of the previous examples as calculated from eqs. (61) to (63). However, fig. 52 could have been derived immediately, and without these calculations from fig. 51, if

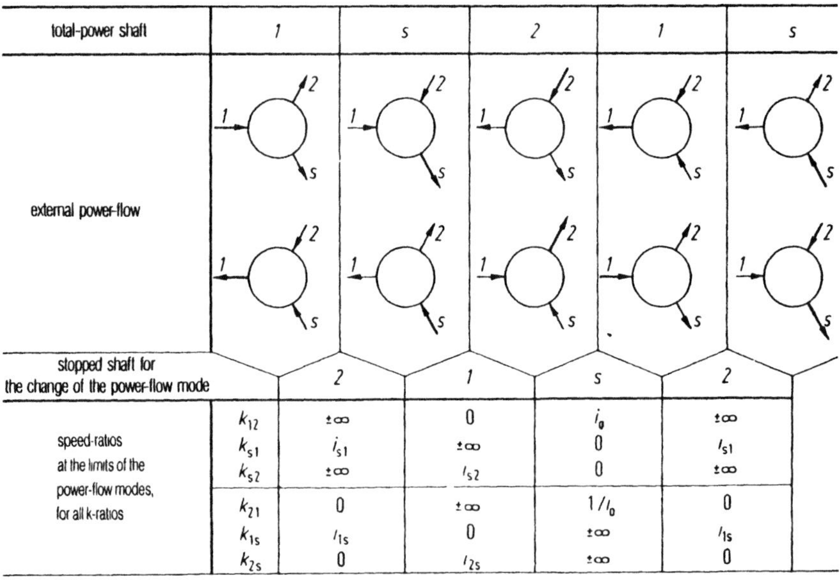

total-power shaft	1	s	2	1	s
external power-flow					
stopped shaft for the change of the power-flow mode	2	1	s	2	

speed-ratios at the limits of the power-flow modes, for all k-ratios				
k_{12}	$\pm\infty$	0	i_0	$\pm\infty$
k_{s1}	i_{s1}	$\pm\infty$	0	i_{s1}
k_{s2}	$\pm\infty$	i_{s2}	0	$\pm\infty$
k_{21}	0	$\pm\infty$	$1/i_0$	0
k_{1s}	i_{1s}	0	$\pm\infty$	i_{1s}
k_{2s}	0	i_{2s}	$\pm\infty$	0

Fig. 51. Speed-ratios at the limits between the k-ratio operating ranges with different external power-flows. According to eqs. (32)–(36) the signs of the speed-ratios i depend on the basic speed-ratios i_0 of the considered transmission types. The speed-ratios k in the two operating ranges which are adjacent to a limit ratio $= i$ have the same sign as this i. In the third operating range k assumes the opposite sign.

total-power shaft		s	2	1
power summation		(diagram)	(diagram)	(diagram)
power division		(diagram)	(diagram)	(diagram)
negative-ratio drives with: $i_0 < 0$	k_{12}	> 0	$i_0 < k_{12} < 0$	$< i_0$
	k_{s1}	$> i_{s1}$	< 0	$0 < k_{s1} < i_{s1}$
	k_{s2}	$> i_{s2}$	$0 < k_{s2} < i_{s2}$	< 0
	k_{21}	> 0	$< 1/i_0$	$1/i_0 < k_{21} < 0$
	k_{1s}	$0 < k_{1s} < i_{1s}$	< 0	$> i_{1s}$
	k_{2s}	$0 < k_{2s} < i_{2s}$	$> i_{2s}$	< 0
positive-ratio drives with: $0 < i_0 < 1$	k_{12}	< 0	$0 < k_{12} < i_0$	$> i_0$
	k_{s1}	$> i_{s1}$	< 0	$0 < k_{s1} < i_{s1}$
	k_{s2}	$< i_{s2}$	$i_{s2} < k_{s2} < 0$	> 0
	k_{21}	< 0	$> 1/i_0$	$0 < k_{21} < 1/i_0$
	k_{1s}	$0 < k_{1s} < i_{1s}$	< 0	$> i_{1s}$
	k_{2s}	$i_{2s} < k_{2s} < 0$	$< i_{2s}$	> 0
positive-ratio drives with: $i_0 > 1$	k_{12}	< 0	$0 < k_{12} < i_0$	$> i_0$
	k_{s1}	$< i_{s1}$	> 0	$i_{s1} < k_{s1} < 0$
	k_{s2}	$> i_{s2}$	$0 < k_{s2} < i_{s2}$	< 0
	k_{21}	< 0	$> 1/i_0$	$0 < k_{21} < 1/i_0$
	k_{1s}	$i_{1s} < k_{1s} < 0$	> 0	$< i_{1s}$
	k_{2s}	$0 < k_{2s} < i_{2s}$	$> i_{2s}$	< 0

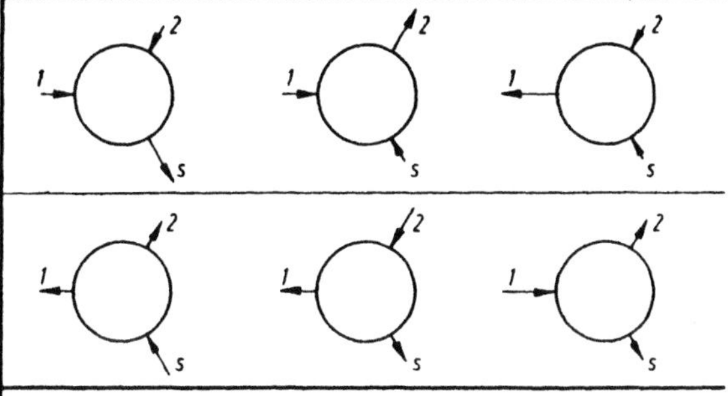

Fig. 52. Coordination between the k speed-ratios and the power-flows in simple revolving drives. With the analogous indices $1 \triangleq I$, $2 \triangleq II$, $s \triangleq S$, and $i_0 \triangleq i_{III}$, this figure also applies to bicoupled transmissions with two degrees of mobility.

the signs of the listed two-shaft speed-ratios i had been determined from eqs. (32) to (36) for all three transmission types.

Fig. 52 is especially useful for the analysis and the layout of compound transmissions. Its organization has been retained in worksheet 4 which can be used to calculate the efficiencies of simple revolving drives. Furthermore, the following two guide rules which greatly facilitate the examination of compound transmissions, can be derived from fig. 52:

> *The position of the total-power shaft and thus the external*   R 18
> *power-flow of a revolving drive train changes if the direction*
> *of rotation of one or two of its three shafts reverses.*

and:

> *At the limit between two different power-flow modes, a three-*   R 19
> *shaft transmission operates as a two-shaft transmission.*

The location of the total-power shaft as a function of the speed-ratios k and the basic speed-ratio i_0 can also be found directly from the diagrams of figs. 63 to 65.

d) Efficiencies of Two-Shaft Transmissions

The efficiencies of two-shaft transmissions can be obtained as special cases from the efficiencies of the three-shaft transmissions when one of the three shaft speeds is set to zero.

For $n_s = 0$ (or stopped shaft s), $k_{s1} = 0$ in eq. (54), yielding the basic efficiency as defined in section 5. Thus:

$$\left. \begin{array}{ll} \eta_{12} = \eta_0^{r1} & \text{where} \quad r1 = +1 \\[2mm] \eta_{21} = \dfrac{1}{\eta_0^{r1}} & \text{where} \quad r1 = -1 \end{array} \right\} \quad \eta_{12} = \eta_{21} = \eta_0 \ .$$

For $n_1 = 0$ (or stopped shaft 1), $k_{1s} = 0$ in eq. (58), yielding the revolving carrier efficiencies:

$$\eta_{s2} = \frac{i_0 - 1}{i_0 - \dfrac{1}{\eta_0^{r1}}} \ ,$$

$$\eta_{2s} = \frac{i - \dfrac{1}{\eta_0^{r1}}}{i_0 - 1} \ .$$

For $n_2 = 0$ (or stopped shaft 2), $k_{1s} = i_{1s} = 1 - i_0$ in eq. (58), so that

$$\eta_{s1} = \frac{i_0 - 1}{i_0 \eta_0^{r1} - 1} \ ,$$

$$\eta_{1s} = \frac{i_o\eta_o^{r1} - 1}{i_o - 1} \ .$$

As defined earlier, the first subscript of η identifies the input shaft, while the second subscript identifies the output shaft. The exponents $r1$, which can be determined from eq. (43), always assures that $\eta < 1$. Its sign depends only on the direction of the power-flow and the location of the summation shaft and, thus, on i_o. Thus, for the three possible transmission types whose summation shafts, of course, have different locations, the four general revolving carrier efficiencies yield a total of twelve efficiency equations for which the appropriate exponents $r1$ are now determined. These twelve equations are tabulated in worksheet 5. The exponents $r1$ are already included in those equations but are, nevertheless, also listed. These equations show that the efficiencies of the two-shaft transmissions no longer depend on the speeds. However, they still depend on the respective locations of the input and the output shafts. Revolving drives, therefore, are not loss-symmetrical (see sec. 5) even when they operate as two-shaft transmissions. The efficiency equations of worksheet 5 do not include the splash and ventilation losses, the planet-bearing friction losses due to centrifugal force, etc. (see sec. 6) which are always present, however, in actual transmissions.

As in the three-shaft transmissions, the efficiencies of the two-shaft transmissions with negative basic speed-ratios are always higher than their basic efficiencies.

The efficiency of the two-shaft positive-ratio drives, however, can be higher, equal, or lower than the basic efficiency. While it is obviously equal when the carrier shaft is stopped, it depends only on i_o and the direction of the external power-flow as to whether it becomes higher or lower when another shaft is stopped. In fig. 45, therefore, the speed-ratio ranges of the two-shaft transmissions over which $\eta > \eta_o$ or $\eta < \eta_o$ can be identified by using different line types. Thus, this diagram contains all the information needed for the preliminary selection of a revolving two-shaft gear train with a given speed-ratio.

e) Self-Locking

The term "self-locking" characterizes an internal state of a mechanism where the external driving forces or torques cause internal friction forces of such magnitude that the mechanism cannot be moved. It is especially characteristic for self-locking that these friction forces grow in the same proportion as the driving foces, so that even theoretically infinite input torques cannot operate the mechanism. Consequently, its output forces or torques are equal to zero.

Thus, a mechanism becomes self-locking when its power loss due to internal friction exceeds the input power and, consequently, its power demand cannot be satisifed by the smaller input power. Therefore, eq. (4) assumes the form

$$-\frac{P_L}{P_{in}} > 1 , \quad \text{or} \quad \frac{P_L}{P_{in}} < -1 ,$$

so that eq. (5) yields a negative efficiency

$$\eta = -\frac{P_{out}}{P_{in}} = \frac{P_{in} + P_L}{P_{in}} = 1 + \frac{P_L}{P_{in}} < 0 .$$

If the friction power-loss is just equal to the input power, the efficiency $\eta = 0$. Even if the mechanism does not lock under these conditions, its output power $P_{out} = 0$ and its output torque $T_{out} = 0$ so that it operates at its "self-locking limit."

Thus, a revolving drive can become self-locking whenever its efficiency becomes zero, or negative, and an analysis of the self-locking conditions reduces to an analysis of the efficiencies.

A review of the equations (54) to (59), however, shows that the efficiency can become zero only when the carrier shaft is the sole output shaft as assumed in eq. (59), and when

$$i_o \eta_o^{rl} = +1 ,$$

that is, in *positive*-ratio drives with a basic speed-ratio of approximately 1. Under these conditions two cases are possible:

1) If i_o is slightly larger than 1, then $k_{ls} < 0$ as can be verified with fig. 52. Thus, if we further define the torque and speed of the input shaft l as positive, that is, if $T_1 > 0$ and $n_1 > 0$, then $n_s < 0$. Consequently,

$$T_1(n_1 - n_s) > 0 \quad \text{and according to eq. (43)} \quad r1 = +1$$

Since $k_{ls} < 0$, the denominator of eq. (59) is always positive, but the numerator becomes negative, when

$$i_o \eta_o < 1 , \quad \text{that is,} \quad 1 < i_o < \frac{1}{\eta_o} .$$

In this case, $\eta_{2>s}^1$ becomes negative and the transmission locks.

2) If i_o is slightly *smaller* than l, than $0 < k_{ls} < i_{ls} < 1$. With $i_{ls} < 1$, $n_s > n_1$ if we define again $T_1 > 0$ and $n_1 > 0$. Thus,

$$T_1(n_1 - n_s) < 0 \quad \text{which implies that} \quad r1 = -1 .$$

Consequently, the efficiency $\eta_{2'>s}^{1} = 0$ when

$$i_o \eta_o^{-1} = \frac{i_o}{\eta_o} = 1 , \qquad \text{that is, when} \qquad i_o = \eta_o .$$

Eq. (33) then yields

$$i_{1s} = 1 - \eta_o .$$

If this value of the upper limit is substituted for k_{1s}, in eq. (59), we find that the denominator of eq. (59) remains negative in the entire speed range $0 < k_{1s} < 1 - \eta_o$ which is indicated in fig. 52. Thus, the efficiency $\eta_{2'>s}^{1}$ becomes negative whenever the numerator becomes positive, that is, when

$$i_o \eta_o^{r1} = \frac{i_o}{\eta_o} > 1 , \qquad \text{or} \qquad \eta_o < i_o < 1 .$$

This analysis shows that positive-ratio gear trains can become self-locking if the carrier shaft is the sole output shaft and at the same time

$$\eta_o < i_o < \frac{1}{\eta_o} .$$

This result does not depend on the magnitude of k_{1s} and, thus, is valid also for two-shaft transmissions whose shaft 1 or shaft 2 is locked. However, eqs. (55) and (57) show that a negative efficiency cannot occur when shaft 1 or shaft 2 is the sole output shaft. For two-shaft transmissions, the range of self-locking is indicated in fig. 45.

Birkle [25] has shown that three-shaft transmissions with two output shafts never become self-locking since, as long as a finite input torque acts on the transmission, the equilibrium conditions $\Sigma T = 0$ can be satisfied only when at least one of the output torques is not equal to zero. If s is the sole *input* shaft, then the efficiency remains positive even if the basic speed-ratio is approximately equal to 1, and the drive, thus, is capable of self-locking. This can be verified by an analysis with eq. (58). In conclusion, we can formulate the following guide rule:

> *Only those positive-ratio, two-shaft or three-shaft transmis-* R 20
> *sions with a basic speed-ratio* $\eta_o < i_o < 1/\eta_o$ *are capable of*
> *self-locking. However, these transmissions do not actually*
> *lock unless the carrier shaft is the sole* output *shaft.*

A locked transmission drive can be set in motion if the carrier shaft is driven in the direction of its blocked output speed, making the sign of its "output power" positive. The required drive torque can be calculated from eqs. (42) and (44) or from worksheet 6 with $r1$ from worksheet 4. Thus for a self-locking transmission with a basic speed-ratio $1/\eta_o > i_o > 1$, $T_1 > 0$ and $r1 = +1$ (case 1), we obtain:

$$T_s = T_1(i_o\eta_o - 1) < 0 \quad \text{and} \quad T_2 = -T_1 i_o\eta_o < 0$$

so that the required drive torque T_s can be determined when the numerical values of i_o, η_o and T_1 are known. As compared to a non-self-locking positive-ratio transmission with a basic speed-ratio $i_o > 1/\eta_o$, where according to worksheet 3, shaft 2 is the summation shaft, this torque T_s has the opposite sign and, consequently, shaft 1 becomes the summation shaft, and shafts 2 and s the difference shafts. If, however, such a potentially self-locking transmission is driven through the carrier shaft and, thus, according to R 20 cannot lock, then shaft 2 remains the summation shaft.

In the above mentioned case 1, the input powers P_1 and P_2 must be positive. Thus, the speeds are $n_1 > 0$ and $n_2 < 0$, that is $k_{12} < 0$. From this we find, with worksheet 2, that $k_{s1} < 0$. Thus $n_s < 0$ and indeed $P_s > 0$. We see that a self-locking three-shaft transmission, which is moved by driving its blocked carrier shaft, has three input shafts and, according to eq. (50), the sum of the three input powers equals the friction power-loss. Analogously, this is true for a self-locking two-shaft transmission.

f) Partial-Locking

If a transmission capable of self-locking is driven at only one of the two central gear shafts, so that the carrier and the other central gear shaft form the output, then a complete locking of all three shafts is not possible. However, since the transmission is otherwise capable of self-locking it assumes a special state of operation which we shall describe as partial-locking.

To analyze partial-locking we may assume that the transmission is jammed by self-locking which means that shaft s is the sole output shaft. Then, $n_1 = n_2 = n_s = 0$, $T_1 = -T_2$, and $T_s = 0$. Now a coupling speed (e.g., $n_s > 0$) can still be "superimposed" on all three shafts so that

$$P_1 = T_1 n_1 = T_1 n_s > 0 ,$$

$$P_2 = T_2 n_2 = -T_1 n_s > -P_1 < 0 ,$$

$$P_s = T_s n_s = 0 n_s = 0 .$$

Therefore, such a three-shaft transmission whose carrier shaft is not connected can still operate with a power-flow $1 <_s^2$, or $2 <_s^1$ and with $P_s = 0$, and transmit power as a coupling while it is internally jammed by self-locking. Since, under these conditions $k_{s1} = 1$, eq. (54) shows that the efficiency becomes $\eta = 1$ which means that shaft 2 transmits the total power without losses.

If from now on the carrier shaft is assumed to be connected to a power-consuming machine, three different states of operation may arise:

1. If the connected machine requires a higher input torque than T_s provided by the carrier shaft, then it stalls and $n_s = 0$. The transmission operates in its basic mode and its efficiency $\eta = \eta_0$ while the carrier torque T_s can be calculated from eq. (44) or worksheet 6.

2. If the connected machine requires a smaller torque than T_s provided by the carrier shaft, then the transmission remains internally locked and rotates as a coupling with its three shafts having the same speeds $n_1 = n_2 = n_s > 0$. Then $k_{12} = 1$, and according to worksheet 4, $\eta = 1$.

Proof: As pointed out in the previous section e), a positive-ratio transmission is capable of self-locking when:

a) i_0 is slightly larger than 1, or, referring to R 20, $1 < i_0 < 1/\eta_0$. Assume that initially $n_s < n_1$, so that $0 < k_{s1} < 1$, and eq. (43) yields an exponent $r1 = +1$. Then $T_s = T_1(i_0\eta_0 - 1)$ from worksheet 6, and since $i_0 < 1/\eta_0$, $T_s < 0$. Consequently, $P_s < 0$, that is, shaft s is an output shaft and the connected machine is accelerated to $n_s = n_1$. This speed cannot be exceeded because for $n_s > n_1$, according to eq. (43), $r1 = -1$ and $T_s = T_1(i_0/\eta_0 - 1) > 0$. This would make $P_s > 0$, but this is impossible because no input power is available from the power-consuming machine.

b) i_0 is slightly smaller than 1, that is, $\eta_0 < i_0 < 1$. Assume again that initially $0 < k_{s1} < 1$, then $r1 = +1$ and $T_s = T_1(i_0\eta_0 - 1)$. With $i_0 < 1$, $T_s < 0$, and $P_s < 0$ so that shaft s is an output shaft which accelerates its connected machine to $n_s = n_1$. It cannot exceed n_1 because then $k_{s1} > 1$ and eq. (43) would yield $r1 = -1$, for which worksheet 6 would show $T_s > 0$ and hence $P_s > 0$. Again this is impossible because no input power is available from the connected power-consuming machine.

3. If the output shaft s assumes any speed between zero and n_1 when the connected machine, for instance, a centrifugal pump, absorbs all of the available torque $T_s = T_1(i_0\eta_0^{r1} - 1)$, then it rotates in the input direction n_1. The efficiency for this state of operation lies between the efficiencies of the two extreme states of operation previously described. That is, $\eta_0 < \eta < 1$, as can be verified with worksheet 4. From this we can conclude: if in a transmission capable of self-locking, shaft 1 or 2 is an input shaft and the other together with the carrier shaft s are output shafts, and the connected machine can absorb all the torque provided by the carrier shaft, then the carrier shaft can reach only a speed between zero and the input speed and in the same sense of rotation as the given input speed n_1 or n_2.

Worksheet 4 can be used to calculate efficiencies for transmissions capable of self-locking, in spite of the fact that in some cases the tabulated power-flow may be incorrect for the ranges of i_o and k_{12} and the values of the exponent $r1$ which apply to these cases.

In a partially locked gear train, therefore, the carrier shaft can transmit an output power only at speeds which lie between zero and a speed inclusive of the coupling speed. As soon as the carrier shaft must rotate faster than, or opposite to the direction of the input shaft, it must be driven. Such an operating condition, which is very instructive, can be obtained as follows:

With the aid of an infinitely variable transmission, which is assumed to be free of losses, a part of the output power of shaft 2 could be used to drive the carrier in the constrained direction, that is, in the direction in which it would turn in the absence of friction. We thus obtain the state of motion which, in the loss-free gear train, corresponds to the power-flow $1 <^2_s$ (fig. 52). It is characteristic for partially-locked gear trains that the driving carrier shaft, which is normally an output member, must be driven and, thus, transmits a "negative output power" as soon as it is required to rotate faster than the input shaft, or in the opposite direction.

In this hypothetical arrangement, only the remaining output power of shaft 2 is an effective power. The overall efficiency $\eta_{1 < \frac{2}{s}}$ of the gear train can be calculated from eq. (54) when the speed-ratio k_{s1} is known. Conversely, if we let the efficiency equal zero, then we can find the maximum speed-ratio k_{s1} for which the total output power of shaft 2 just equals the negative output power of the carrier shaft s. Such a mode of operation is conceivable only for self-locking transmissions, and only when the carrier shaft is one of the two output shafts. For a transmission not capable of self-locking, the assumption $\eta_{1 < \frac{2}{s}} = 0$ leads to a speed-ratio k_{s1} which lies outside the limits which characterize this power-flow in fig. 52.

The described operation in the range of partial-locking, where the carrier must be driven in the constrained direction, seems to lead to contradictions. Based on the existing speed-ratios, fig. 52 indicates that the carrier shaft must be considered as an output shaft, as in a loss-free gear train. Therefore, we could be tempted to consider the carrier shaft as an input shaft, but this would be incorrect since fig. 52 demands that for a power-flow $^1_s > 2$, the carrier shaft must be driven against its existing "output" direction, or that both central gear shafts must change their sense of rotation. Indeed, the power of the carrier shaft, which previously has been described as "negative output power," is not an effective power since it is not transmitted through the drive mechanism to the other two shafts. Its torque does not effect a change in the direction of the load on the tooth flanks, but merely restores the sliding motion between previously jammed gears.

The same considerations apply when shafts *1* and *2* are exchanged so that the power-flow becomes $2 <^3_1$ whose efficiency can be calculated from eq. (56). Therefore, we shall summarize what is important for the determination of the efficiencies in the following guide rule:

> *The total-power shaft and, thus, the external power-flow of a* R 21
> *revolving drive train can be determined from its basic speed-*
> *ratio i_o and an arbitrary speed-ratio k. Its self-locking capa-*
> *bility need not be considered.*

If we obtain the previously considered, partially-locked state of operation by transmitting a part of the output power through an infinitely variable auxiliary transmission (which, however, is no longer loss-free) to the constrained carrier shaft, then we obtain a compound transmission. According to what has been said before, this compound transmission becomes self-locking when the speed-ratio of the auxiliary transmission reaches a value k_{s2}. At this point, the power in its output shaft just equals the "negative output power" of the carrier shaft. Transmissions of this type will be discussed later in part III.

g) Generalization of Efficiency Equations

By extending the considerations of section 11d to the powers, we find that the efficiency equations likewise can be freed from their previously chosen dependence on the carrier and the central gear shafts. Thus, we can lock the shaft a of the transmission introduced in fig. 46, whose carrier shaft position may be unknown, and then measure the loss factor $\zeta_{bc} = 1 - \eta_{bc}$ so that its power-loss $P_L = -P_b \zeta_{bc}$ for this operating condition becomes known. If we assume that ζ_{bc} is independent of the speeds and torques, as we assumed earlier in section 6 for ζ_o, then the power-loss P_L does not change if we impart an additional equal speed on all three shafts, that is, if we superimpose a loss-free coupling power on our first state of motion where $n_a = 0$. We can then find the efficiencies for all other states of motion by following the procedures which we have used in the earlier part of this section 15. The only difference is that we can no longer assume loss-symmetry when the position of the carrier shaft is unknown, so that

$$\eta_{bc} \neq \eta_{cb} \; .$$

Finally, we can rewrite all previously derived efficiency equations with the aid of an arbitrary index key so that, for example, $i_o \triangleq i_{bc}, \eta_o = \eta_{12} \triangleq \eta_{bc}$ when $a \triangleq s, b \triangleq 1$ and $c \triangleq 2$. However when $r1 = -1$, then $\eta_o^{-1} = 1/\eta_{21}$ $\triangleq 1/\eta_{cb}$. For the chosen subscript system, eq. (43) takes the form:

$$r1 = \frac{T_b(n_b - n_a)}{|T_b(n_b - n_a)|} \; .$$

When worksheets 4 and 5 (see sec. 33) are used the dissymmetry of the friction losses has already been considered by the unequivocal indexing of the basic efficiencies $\eta_{I\,II} \triangleq \eta_{12}$ and $\eta_{II\,I} \triangleq \eta_{21}$. However, in order to determine η_{cb}, we must exchange the input and the output shaft and then perform a second efficiency measurement.

We shall use this generalization later in part III to analyze the simple bicoupled transmissions. However, for the analysis of simple revolving drive trains we shall normally assume a known position of the carrier shaft, since this makes it particularly easy to determine the basic speed-ratio and the basic efficiency.

B. Graphical Analysis

Section 16. Symbolic Representations of Epicyclic Transmissions According to Wolf

In the previous chapter it has been shown that the function of a revolving drive can be analyzed without consideration of each detail of its design. Rather, a kinematic analysis can be performed when only the basic speed-ratio i_0 is known, which, according to worksheet 3, also determines the position of the summation shaft.

Wolf [13], therefore, proposed a symbolic representation of the revolving drives which contains only the previously mentioned kinematical characteristics, as shown in fig. 53. A circle represents the transmission and three radial lines the three shafts. Because of their special importance, the summation shaft is symbolized by a double line, and the carrier shaft by an extension of the single or double line into the interior of the circle. According to worksheet 3, the location of the summation shaft also distinguishes negative-ratio drives from positive-ratio drives with $i_0 < 1$ or $i_0 > 1$. Cross-

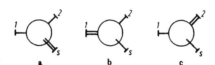

Fig. 53. Symbols for revolving drive trains according to Wolf [13]: a, negative-ratio transmission $i_0 < 0$; b, positive-ratio transmission $0 < i_0 < 1$; c, positive-ratio transmission $i_0 > 1$.

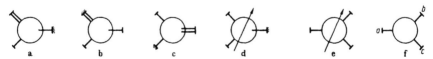

Fig. 54. Symbolic representation of the operating conditions of revolving drive trains: a, basic transmission with a positive speed-ratio; b–c, two-shaft transmissions; d, basic transmission with an infinitely variable speed-ratio; e, three-shaft transmission with an infinitely variable basic speed-ratio i_0, e.g., as shown in fig. 44; f, three-shaft transmission with unknown or freely selectable position of carrier and summation shaft.

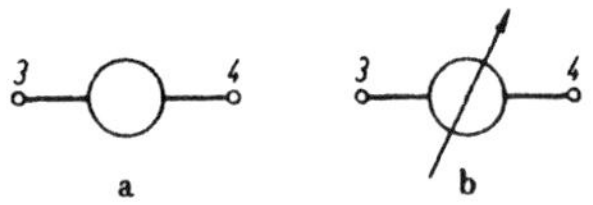

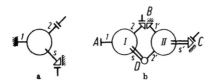

Fig. 55. Symbolic representation of a conventional transmission: *a*, with a constant speed-ratio i_{34}; *b*, with an infinitely variable speed-ratio i_{34}.

hatching at a shaft flange, as shown in figs. 54a, b, and c, indicates that this shaft is fixed to the housing and, therefore, cannot rotate. If the carrier shaft is locked as shown in fig. 54a, then the Wolf symbol depicts a basic transmission. A locked central gear shaft as shown in figs. 54b and c represents a two-shaft revolving drive. When either the speed-ratio of a basic transmission or the basic speed-ratio of a three-shaft transmission is infinitely variable, an arrow is drawn across the transmission symbol as shown in figs. 54d and 54e respectively. A revolving drive train whose carrier and summation shafts have an unknown location, or can be freely chosen, is depicted with three equal shafts as shown in fig. 54f. In contrast to the practice for well-defined transmissions, these shafts are denoted by lower case letters (e.g., *a, b, c*) rather than by *1, 2,* and *s*.

A simple reduction drive with fixed axes may be symbolized by a circle with only two shafts as shown in fig. 55, although basically it should be represented by the symbols shown in figs. 54a or 54d. However, in practical applications of these transmission symbols it seems preferable that conventional transmissions with fixed axes, which are sometimes separately connected to the input or output shafts of revolving drive trains, be immediately identified as such.

Simple reduction drives which are components of a compound revolving drive train should always by symbolized by fig. 54a, where the carrier shaft

Fig. 56. Symbolic representation of the possible coupling conditions for three-shaft transmissions: *a*, transmission whose shaft *1* can be changed from freely rotating to locked, whose shaft *2* can be freely rotating or connected, and whose shaft *s* can be freely rotating, connected, or locked; *b*, change-gear, which consists of the two gear trains *I* and *II* and shows the following coupling conditions: *A* is rigidly connected (input or output); at *D, s* and *2'* are rigidly coupled but not externally connected (free coupling shaft); *C* can be freely rotating (idle condition), connected (input or output), or locked; at *B*, if *2* and *1'* are coupled, the coupling shaft can be free, connected, or locked; if *2* and *1'* are not coupled, shaft *2* may be either freely rotating or connected and shaft *1'* may be either freely rotating or locked.

corresponds to the housing (see sec. 2).

The coupling conditions of the shafts can be symbolized as shown in fig. 56a. Especially for compound transmissions, the analysis of power-flow, shaft torques, and speeds is greatly simplified when the Wolf-symbols are inspected, rather than the actual layouts of the gear trains.

As an illustration, fig. 56b shows a change gear whose shafts can be connected in several different ways.

Section 17. The Kutzbach Speed Diagram
for Planetary Transmissions

a) General Remarks

Kutzbach, who gave very modern impulses to mechanical engineering by combining kinematic and design concepts, also developed a graphical method for the determination of the speeds in planetary transmissions [3, 7]. This method is especially suitable to illustrate the motions in compound planetary transmissions. In connection with the representation of the compound transmissions, and the internal coupling points of compound planetary transmissions as suggested by Helfer [31], the Kutzbach diagram has gained renewed importance for the synthesis of compound planetary transmissions as will be explained in more detail in section 40. However, an analysis of the motions of simple or compound planetary transmissions can be accomplished faster by the previously described analytical methods and the graphs which will be introduced in the next sections.

b) Speed Diagram for a Simple Negative-Ratio Transmission

The Kutzbach speed diagram of a revolving drive train is based on a schematic representation of the transmission. The pitch circles are drawn to scale and the number of the known or assumed speeds equals the number of degrees of freedom. As an example, figs. 57a to 57d show a negative-ratio drive and the design of its speed diagram. In the schematic representation of fig. 57b the pitch circles are drawn to scale and then projected onto the ordinate of diagram fig. 57c, while the tangential velocities u of the gears and the carrier at section AB are plotted along the abscissa. It is expedient but not necessary to consider the carrier as stationary, so that $n_s = 0$ and also $u_s = 0$.

If the central gear l is assumed to have an arbitrary tangential velocity u_1, then a radial velocity line $l'\,0$ can be drawn from the origin 0 of the coordinate system through the point l which determines the angle δ_1. Since the planet p meshes with gear l at the radius r_1, its tangential velocity at r_1 is also u_1. However, the center of the planet p at the radius r_s has the same tangential velocity as the planet carrier, which by assumption is $n_s = 0$. Thus, a velocity line for the planet can be drawn through the points $1'$ and s' and extended until it intersects the radius line r_2 in point $2'$. At the radius r_2, the planet p meshes with the ring gear 2 and therefore its velocity at this point is equal to the tangential velocity u_2 of the ring gear 2. Starting from the point of intersection $2'$, the radial velocity line of gear 2 can now be drawn to the origin 0 which determines the angle δ_2. Finally, the radial velocity line and the angle δ_p of the planet are obtained by drawing a parallel to the velocity line $1'2'$ through the origin.

From the speed diagram, the angular velocities ω and the speeds n of the gears can now be found as follows:

$$\omega_1 = \frac{u_1}{r_1} = \tan \delta_1 = \frac{n_1}{R},$$

$$\omega_2 = \frac{u_2}{r_2} = \tan \delta_2 = \frac{n_2}{R},$$

$$\omega_s = \frac{u_s}{r_s} = \tan \delta_s = \frac{n_s}{R},$$

$$\omega_p = \frac{u_p}{r_p} = \frac{u_1 - u_s}{r_1 - r_s} = \tan \delta_p = \frac{n_p}{R}$$

where R is a scale factor for the speed vectors. R can be arbitrarily chosen since it cancels out when the speed-ratios are taken, so that:

$$n_1 : n_2 : n_s : n_p = \tan \delta_1 : \tan \delta_2 : \tan \delta_s : \tan \delta_p .$$

Because of the assumption $u_s = n_s = 0$, it follows that $\tan \delta_s = 0$.

The radial velocity lines project the speeds of all gears in the same scale onto a line g which is parallel to the abscissa so that the ratio of the speeds $n_1 : n_2 : n_p$ equals the ratio of the distances between the points 1, 2, and p, and the point 0 on line g. Since by assumption $n_s = 0$, the point s of the carrier coincides with the point 0. The speeds on the right and left hand side of the point 0 have opposite signs. As can be readily understood, the distances between two arbitrary speed points thus represent their relative speeds, so that all distances measured from point 2 represent speeds relative to gear 2, or if gear 2 is locked while all relative speeds remain the same, the operating condition of the two-shaft transmission.

The determination of the speeds can be substantially simplified when a nomogram is constructed above line g as shown in fig 57d.

This can be accomplished by first drawing a baseline parallel to g on

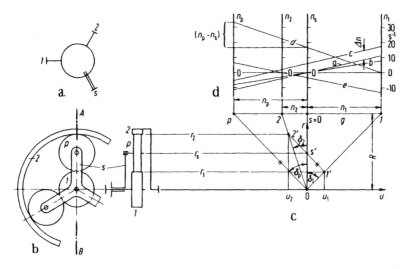

Fig. 57. *Top,* Kutzbach diagram for a simple negative-ratio transmission: *a,* Wolf-symbol, *s* is the summation shaft because $i_o < 0$; *b,* schematic representation drawn to scale; *c,* Kutzbach velocity diagram; *d,* Kutzbach speed diagram and nomogram where the operating characteristics *a* and *b* represent the basic mode, *c* the three-shaft mode, *d* and *e* the two-shaft mode of the transmission.

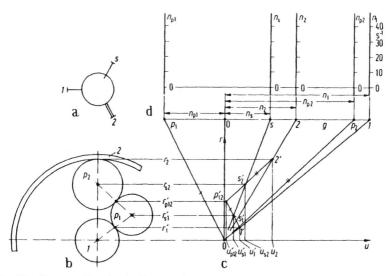

Fig. 58. *Bottom,* Kutzbach diagram for a positive-ratio transmission with meshing planets: *a,* Wolf-symbol, *2* is the summation shaft because $i_o > 1$; *b,* schematic representation of the transmission drawn to scale; *c,* Kutzbach diagram; *d,* nomogram.

which the perpendicular speed scales n_1, n_2, n_s, and n_p are erected in points *1, 2, s,* and *p*. According to the rules of geometric similarity, any straight line through the point $n_s = 0$ intersects the speed scales in the ratio of their respective speeds and thus determines a particular operating condition of the gear train. If the speed scales are divided arbitrarily, but equally, then the speeds of all gears for this operating condition can be read directly from the points of intersection on the speed scales. Alternately, a straight line can be drawn through any two given speed points and the speeds of all other gears can then be read directly from their respective scales.

The lines *a* and *b* through $n_s = 0$ characterize two states of operation where the planetary transmission operates as a basic transmission with two different input speeds. If all the speeds indicated by line *a* are increased by an identical amount Δn, then the resulting parallel line *c* represents an operating condition where all three shafts of the transmission rotate. Thus, line *c* graphically illustrates the result of Willis' mental experiment, where a coupling speed $\Delta n = n_s$ is superimposed on the rolling speeds of the basic transmission. The lines *d* through the point $n_1 = 0$, and *e* through the point $n_2 = 0$, represent different operating conditions of two-shaft transmissions.

The speeds which are directly read from the n_p-scale are absolute speeds like those of all other speed scales so that the speeds of the planets relative to the carrier, which determine the size of the planet bearings, must be obtained from the distance $(n_p - n_s)$. Thus, the steepest speed line of the nomogram, which in the given example is line *d*, indicates the highest bearing speed. This is expressed in the following guide rule:

> *The operating condition with the steepest speed-line is associated with the highest speed of the planet bearings.*   R 22

c) Speed Diagram for a Transmission with Meshing Planet Pairs

Fig. 58 shows the construction of a Kutzbach diagram for a positive-ratio drive of the type shown in fig. 22. In this transmission the axes and pitch points of the gears do not lie in a single plane and, therefore, we must start from a front view of the transmission which shows the pitch circles drawn to scale. As in the previous example we proceed from the horizontal projection lines through the planet axes and the pitch points. The projection lines r', from those axes and pitch points which do not lie in the symmetry axis of the gear train, project foreshortened radii onto the ordinate axis of the speed diagram. The points of intersection $1'$, $s1'$, etc., between these projection lines and the radial speed lines, accordingly project onto the abscissa only the horizontal components u' of the respective tangential velocities. However, since r' and u' are foreshortened by the same amount, their ratio $u'/r' = u/r = \tan \delta$ remains unaffected.

To construct the speed diagram of fig. 58c we will use an alternate method and begin with an arbitrary choice of the speeds n_1 and n_s on line g and then proceed in the following sequence of steps: the horizontal center-line u, and the five radius lines r_1' to r_2, are projected from the front view of the transmission to the right hand side onto the ordinate axis of the speed diagram, fig. 58. This ordinate axis can be erected at an arbitrary point $u = 0$. The line g is then drawn parallel to and with a suitable distance from the abscissa u of the diagram. The arbitrarily chosen speeds n_1 and n_s determine the points 1 and s on line g so that the radial lines 01 and $0s$ can now be drawn. Their points of intersection $1'$ with r_1' and s_1' with r_{s1}' determine the horizontal components of the tangential velocities u_1' of the pitch circle of gear 1 and u_{s2}' of the center of the planet p_1. The extension of the speed line $1's_1'$ to its point of intersection p_2' with the radius line r_{p12}', yields the horizontal component u_{p12}' of the pitch circle velocity u_{p12}' of the planet p_1. If the line $p_{12}'s_2'$ is extended beyond s_2', it intersects the radius line r_2 in point $2'$ whose distance from the ordinate represents a tangential velocity u_2 of the pitch circle of the ring gear. A radial line $02'$ from the origin of the diagram through point $2'$ intersects g in point 2 which determines the speed of the ring gear. Radial lines through the origin, and parallel to $1'p_{12}'$ and $p_{12}'2'$, finally intersect line g in the points p_1 and p_2 whose distance from 0 represents the absolute speeds of the planets p_1 and p_2. Thus, all speeds can now be read from line g, or alternately, the speed scales of the nomogram can be erected on g.

From the nomograms 57d and 58d we derive a further guide rule:

The speed scale of the summation shaft always lies between   R 23
the speed scales of the two difference shafts.

If only the speed scales of the three shafts *1, 2,* and *s* are considered and a speed line is drawn through the point *0* of the intermediate scale *s* in fig. 57d, or *2* in fig. 58d, then it becomes evident that the other two shafts, that is, the input and the output shafts, have opposite directions of rotation. According to R 2, therefore, their torques must have the same sign, and according to R 9 they must be difference shafts. Consequently, the shaft represented by the intermediate speed scale always is the summation shaft. This leads to the following basic guide rule:

The speed of the summation shaft always lies between the   R 24
speeds of the difference shafts.

For a two-shaft transmission this means that:

A two-shaft transmission has a negative speed-ratio when the   R 25 *summation shaft is locked, and a positive speed-ratio when one of the difference shafts is at rest.*

The location of the speed scales, however, does not indicate which one of the three shafts is the carrier shaft. For instance, by an appropriate choice of the dimensions of the negative-ratio drive shown in fig. 57, and the kinematically-equivalent positive-ratio drive of fig. 58, their speed nomograms become identical, except for speed scales of the planets. However, their shafts 2 and s are interchanged.

d) Speed Diagram for a Simple Positive-Ratio Drive with Stepped Planets

The speed diagram for a planetary transmission with stepped planets as shown in fig. 59 can be constructed by taking the following steps:

1. Draw to scale, a schematic representation of the gear train as shown in fig. 59b. Extend the horizontal centerline u and mark an arbitrary point 0 in which the ordinate axis r is then erected. At a suitable distance, draw a line g parallel to the abscissa u.

2. Assume that the carrier is locked. This determines the origin $s = 0$ of the corresponding speed scale g. Starting from this origin, mark off an arbitrary distance n_1. The endpoint 1 of this distance is the origin of the speed scale n_1.

3. Find the points of intersection $1'$ between the radial line 01 and the horizontal line r_1, s' between the line $0s$ along the ordinate axis and the horizontal line r_s, and $2'$ between the line $s'1'$ and the horizontal line r_2. A radial line from the origin through point $2'$ intersects g at the origin of the speed scale 2. Finally, a radial line through the origin, parallel to $s'1'$, determines the foot of the speed scale at p on g.

The speeds n_2 and n_p are now determined by the distances between their speed scales and the origin. If necessary, the nomogram of fig. 59d can be drawn with arbitrary but equal scales.

e) Transmission Synthesis with Aid of Kutzbach Diagram

When the design procedure is reversed, Kutzbach's method can be used to determine graphically the diameters of the individual gears. As an example, the rated speeds n_1, n_2, and n_s of a positive-ratio transmission of the type shown in fig. 19, may be given. First, a suitable origin is chosen on a line g. Then, a scale factor is assigned to the given speeds which can be represented by distance from the origin, on g with the end points 1, 2, and s shown in fig. 60a. Next, a perpendicular line r is erected on g in the origin and a line u is drawn parallel to g at an arbitrary distance. Line u becomes the horizontal symmetry axis of the gear train. Its point of intersection with line r is the

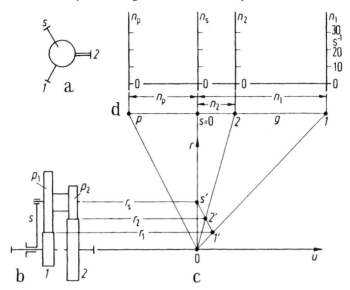

Fig. 59. Kutzbach diagram for a positive-ratio gear train with stepped planets: *a*, Wolf symbol, *2* is the summation shaft because $i_o > 1$; *b*, scale drawing; *c*, Kutzbach diagram; *d*, speed nomogram.

origin of the speed diagram through which the radial speed lines *01*, *02* and *0s* can now be drawn. The points of intersection *1'*, *2'*, and *s'*, or *1"*, *2"*, and *s"* between three lines parallel to *u* (two of them arbitrary) and the three radial speed lines, determine gear sizes which realize the given speed-ratios. Two additional design conditions may now be imposed to obtain an optimum design. First the smallest possible gear *1* can be determined from a stress calculation. Its pitch circle radius r_1 then determines the point of intersection *1'*. Next, a series of straight lines with different slopes can be drawn through *1'* which intersect *0s* in *s'*, *s"*, etc., and thus determine the centers of a series of planets from which the smallest structurally possible planet p_2 can be selected. This leads to the smallest gear train which satisfies the given conditions.

As an example, fig. 60 shows two different, but relative to the shafts *1, 2,* and *s,* kinematically equivalent gear train solutions whose planet speeds are not the same. This should be obvious from the differing slopes of the speed lines *1's'* and *1"s"*. The smaller gear train may be preferable for high-speed applications since the centrifugal loads on its planet bearings are lower at equal carrier speeds due to the smaller center distance r_s and the smaller planet mass. However, as indicated by the slope of the respective speed lines, the smaller gear train operates with higher planet speeds which can be confirmed when the speed diagram is completed by the construction of the two planet scales.

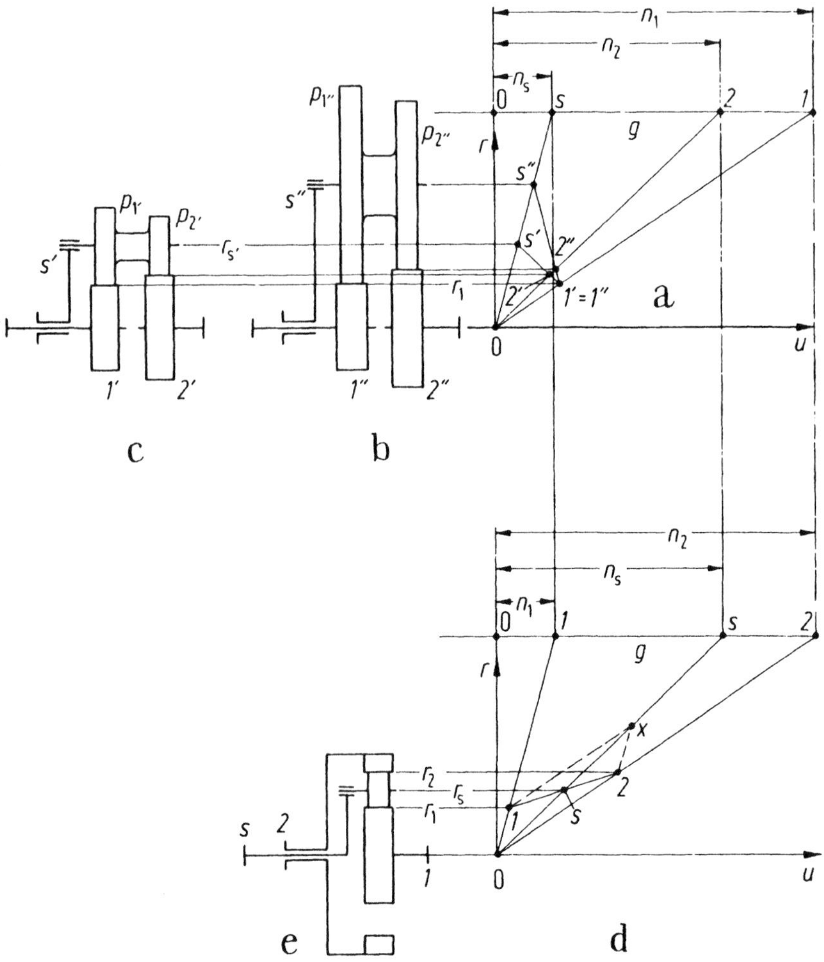

Fig. 60. Graphical transmission synthesis with the aid of a Kutzbach diagram: a, Kutzbach diagram for the given speeds of a positive-ratio transmission; $b-c$, two possible versions of a derived positive-ratio gear train; d, Kutzbach diagram for a negative-ratio gear train with the same speeds; e, scale drawing of a derived negative-ratio gear train.

A negative-ratio transmission for a given speed range can be designed when the procedure which has been described in the previous example (fig. 57) is being followed analogously but in reverse order, as shown in figs. 60d and 60e. Since, in this type of negative-ratio transmission with simple planets, the planet axis always divides the distance between the radii r_1 and r_2 of the central gears into equal parts, the line $1s2$ of the Kutzbach diagram must have a slope such that it is equally divided at its point s. If thus, for example, an arbitrary radius r_1 is chosen for the sun gear, then the distance $1s2$ can be found as the diagonal of the parallelogram 0-1-X-2-0 which, in fig. 60d, has been completed with dashed outlines. Therefore, for a given speed, this additional constraint leads to an infinite number of geometrically similar negative-ratio gear trains with simple planets.

However, a planetary gear train can be designed faster and more accurately if, as described in section 11, the basic speed-ratio i_0 is calculated from the given speeds and then the basic train is laid out like any conventional reduction drive. Fig. 45 and worksheets 1 or 2 can be used to simplify this procedure.

f) Kutzbach Diagram for Compound Planetary Transmissions

Compound transmissions which are composed of several simple planetary transmissions also can be analyzed graphically with the aid of a Kutzbach diagram. The resulting speed nomograms, however, remain valid only as long as the couplings between the individual component transmissions remain the same. An example will be discussed in section 40a, fig. 156. If, therefore, a planetary change gear is analyzed graphically, a new Kutzbach diagram must be drawn whenever a speed-ratio change involves a change of the coupling conditions between the transmission stages.

Section 18. Speed-Ratio Graphs

The use of speed-ratio graphs substantially facilitates the analysis as well as the synthesis of epicyclic transmissions. Fig. 61 is such a graph for *two-shaft transmissions* which shows the relationships between the speed-ratio i and the basic speed-ratio i_0 as obtained from eqs. (32) to (36).

Semi-inverted coordinates which change to a reciprocal scale when the speed-ratios exceed the values of ± 1 have been used for fig. 61 as well as for figs. 63 to 65, so that the total speed-ratio ranges from 0 to $\pm \infty$ can be covered accurately. In the speed-ratio ranges between 0 and ± 1, the coordinate values themselves can be interpolated linearly, while beyond these ranges the reciprocals of the coordinate values must be linearly interpolated. It should take only a short time to become familiar with this very practical method of plotting speed-ratio graphs which, in the form of fig. 61, were first used by Seeliger [18].

The relationship between the speed-ratios of *three-shaft transmissions*, which are described by the eqs. (26) to (31), are plotted in figs. 63 to 65. Each is constructed for a different combination of the three possible speed-ratios k_{12}, k_{1s}, and k_{2s}, or their reciprocal values. Thus, in practical applications, that graph must be chosen which contains the two speed-ratios which are of immediate interest. The graphs also indicate the speed-ratio ranges in

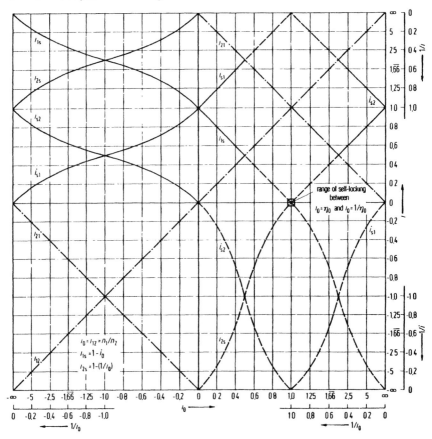

Fig. 61. The speed-ratios of simple revolving drive trains which operate in the two-shaft mode as a function of the basic speed-ratio i_o: ———— $\eta > \eta_o$; – – – – $\eta < \eta_o$; – · – · – · – $\eta = \eta_o$. Between 0 and ±1, the coordinate values can be linearly interpolated, while between ±1 and ±∞ their reciprocals can be linearly interpolated. With the analogous subscripts $1 \triangleq I$, $2 \triangleq II$, $s \triangleq S$, $i_o \triangleq i_{I\,II}$, the diagram is valid also for constrained bicoupled transmissions.

which shaft *1, 2,* or *s* operates as the total-power shaft.

The graphs are based on eqs. (26), (27), and (28), which describe straight lines whose constants depend only on the parameter i_o. For eq. (28), as an example, fig. 62 shows that the curves which represent a constant basic speed-ratio i_o, in terms of its coordinate values k_{21} and k_{s1}, are straight lines which always pass through the coupling point K where $k_{21} = k_{s1} = 1$. At a limit where $k_{s1} = 0$, that is, for a stopped carrier and $n_s = 0$, the straight line passes through the limit point $1/i_o = i_{21}$ on the ordinate axis. At a limit where $k_{21} = 0$, that is, when $n_2 = 0$ the straight line passes through the limit

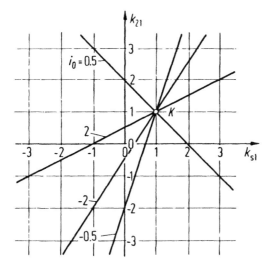

Fig. 62. Interdependence of the speed-ratios k_{21} and k_{s1} for a constant basic speed-ratio i_o as parameter, according to eq. (28).

point $1/(1 - i_o) = i_{s1}$ on the abscissa. Thus, the graph can be constructed rapidly, when for each desired ratio i_o, a straight line is drawn through the coupling point and a corresponding limit point on one of the coordinate axes.

Section 19. Power-Ratio Graphs

For an estimate of the total efficiency of an epicyclic transmission, it is frequently of interest to know the ratio between the rolling-power and the external input power, that is, to know whether the rolling-power will be larger than the external power and thus, according to eq. (51), the total efficiency will be lower than the basic efficiency η_o or vice versa.

Figs. 66 and 67 show the power-ratios for two-shaft transmissions as given in [25]. In these transmissions, the ratio between the rolling-power and the external power depends only on the basic speed-ratio, that is, on the design of the transmission. The basic efficiency η_o, which exerts a substantial influence only when self-locking is approached, is assumed to be $\eta_o = 1$ so that the rolling-powers of the central gear shafts become equal, that is $|P_{R1}| = |P_{R2}| = |P_R|$.

With table 1 and eq. (34), the curve P_R/P_{in} for $n_2 = 0$ in fig. 67 is obtained from the equation

$$\frac{P_R}{P_1} = \frac{T_1(n_1 - n_s)}{T_1 n_1} = 1 - \frac{n_s}{n_1} = \frac{i_o}{i_o - 1} = i_{s2} \ . \tag{64}$$

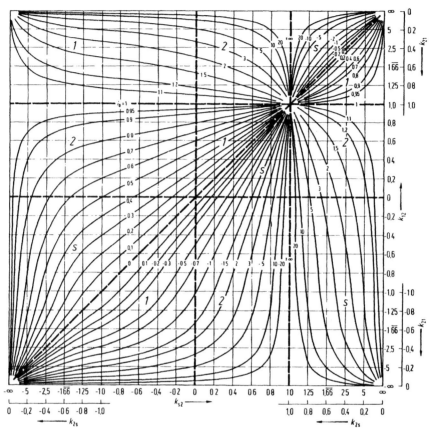

Fig. 63. Interdependence of the speed-ratios k_{12} and k_{s2} for constant basic speed-ratios i_o as parameter. Between 0 and ± 1, the coordinate values can be linearly interpolated, while between ± 1 and $\pm \infty$ their reciprocals can be linearly interpolated; 1, 2, and s denote the total-power shaft within the ranges which are outlined by heavily dashed lines. With the analogous subscripts $i_{1\,II} \triangleq i_o$; $k_{1\,II} \triangleq k_{12}$; $k_{S\,II} \triangleq k_{s2}$; $I \triangleq 1$; $II \triangleq 2$; $S \triangleq s$, the diagram is valid also for bicoupled transmissions with two degrees of mobility.

With table 1 and eq. (36), P_R/P_{in} for $n_1 = 0$ in fig. 66 becomes:

$$\frac{P_R}{P_2} = \frac{T_2(n_2 - n_s)}{T_2 n_2} = 1 - \frac{n_s}{n_2} = \frac{1}{1 - i_o} = i_{s1} \ . \tag{65}$$

Figs. 66 and 67 show the power-ratios $|P_R/P_{in}|$, and the associated two speed-ratios i from fig. 45, as functions of the basic speed-ratio i_o. If, during the layout of a two-shaft gear train, a particular speed-ratio i is chosen,

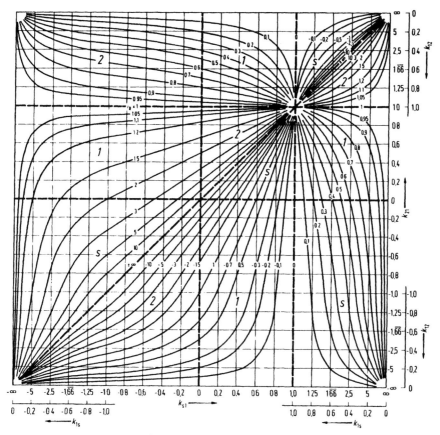

Fig. 64. Interdependence of the speed-ratios k_{21} and k_{s1} for constant basic speed-ratios i_o as parameter. Between 0 and ± 1, the coordinate values can be linearly interpolated while between ± 1 and $\pm \infty$ their reciprocals can be linearly interpolated; *1, 2*, and *s* denote the total-power shaft within the ranges which are outlined by heavily dashed lines. With the analogous subscripts $i_{1\,\text{II}} \hat{=} i_o$; $k_{1\,\text{II}} \hat{=} k_{12}$; $k_{\text{SI}} \hat{=} k_{s1}$; I $\hat{=}$ 1; II $\hat{=}$ 2; $S \hat{=} s$, the diagram is valid also for simple bicoupled transmissions.

then its power-ratio can be immediately determined from figs. 66 and 67. Consequently, its efficiency can be estimated with the aid of eqs. (51) and (52).

In three-shaft transmissions, the ratio between the rolling-power and the external power depends on the speed-ratios k and on the basic speed-ratio i_o, since the internal power-flow between the three rotating shafts changes whenever the state of motion changes. Since the speed-ratios between the three shafts depend only on each other when the basic speed-ratio i_o is given, it is sufficient to refer the power-ratio P_R/P_{tot} to any one of these

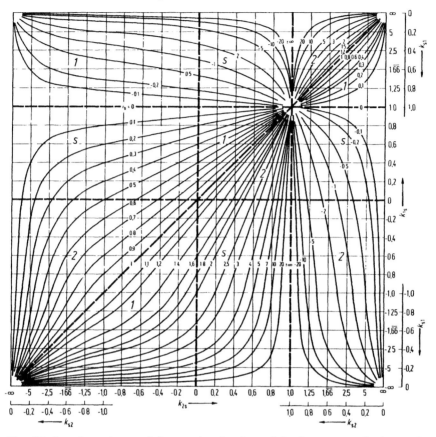

Fig. 65. Interdependence of the speed-ratios k_{1s} and k_{2s} for constant basic speed-ratios i_o as parameter. Between 0 and ± 1, the coordinate values can be linearly interpolated, while between ± 1 and $\pm \infty$ their reciprocals can be linearly interpolated; 1, 2, and s denote the total-power shaft within the ranges which are outlined by heavily dashed lines. With the analogous subscripts $i_{1\,II} \triangleq i_o$; $k_{1S} \triangleq k_{1s}$; $k_{II\,S} \triangleq k_{2s}$; $I \triangleq 1$; $II \triangleq 2$; $S \triangleq s$, the diagram is valid also for simple bicoupled transmissions.

speed-ratios. The conversion to the other speed-ratios can then be easily accomplished with the aid of worksheet 2 or figs. 63 to 65. If the friction losses are neglected so that $\eta_o = 1$, we can express the power-ratio in either one of the following forms:

$$\frac{P_R}{P_{tot}} = \frac{P_R}{P_1} \quad \text{or} \quad \frac{P_R}{P_{tot}} = \frac{P_R}{P_2} \quad \text{or} \quad \frac{P_R}{P_{tot}} = \frac{P_R}{P_s},$$

depending on whether shaft 1, 2, or s operates as the total-power shaft. Which

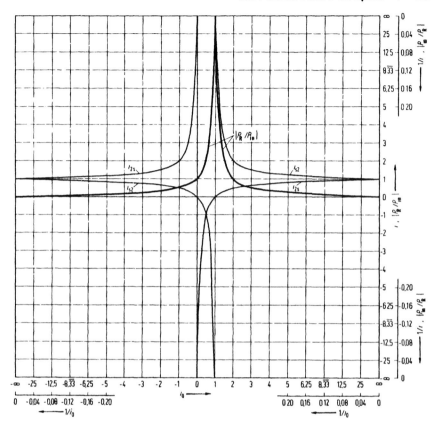

Fig. 66. Ratio between the absolute values of the rolling-power P_R and the input power P_{in} as well as the associated speed-ratios of the simple revolving gear train with a locked shaft I. The friction-losses are neglected so that $|P_R| = |P_{RI}| = |P_{R2}|$. With the analogous subscripts I $\hat{=}$ 1, II $\hat{=}$ 2, S $\hat{=}$ s, and P_R = series power, the diagram is valid also for bicoupled transmissions.

of these applies to a particular case can be decided with the help of fig. 52 when i_0 and, for example, k_{12} are known. Because the friction losses have been neglected, the absolute values of the rolling-powers are $|P_{R1}| = |P_{R2}| = |P_R|$. If, for example, shaft I is the total-power shaft, then we obtain from table 1,

$$\frac{P_R}{P_1} = \frac{T_1(n_1 - n_s)}{T_1 n_1} = 1 - k_{s1}$$

which can be expressed in terms of k_{21} with the aid of eq. (30). Since $k_{21} = 1/k_{12}$, we finally obtain

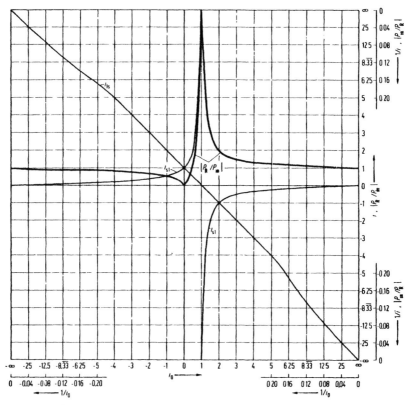

Fig. 67. Ratio between the absolute values of the rolling-power P_R and the input power P_{in} as well as the associated speed-ratios of the simple revolving gear train with a locked shaft 2. The friction-losses are neglected, so that $|P_R| = |P_{R1}| = |P_{R2}|$. With the analogous subscripts I $\triangleq$ 1, II $\triangleq$ 2, S $\triangleq$ s, and P_R = series power, the diagram is valid also for bicoupled transmissions.

$$\frac{P_R}{P_1} = \frac{i_0(1 - k_{12})}{k_{12}(1 - i_0)} \; .$$

According to fig. 52, this equation is valid for negative-ratio drives as long as $k_{12} < i_0$, and for positive-ratio drives when $k_{12} > i_0$.

If shaft 2 is the total-power shaft, we find analogously from table 1 and with eq. (31) that

$$\frac{P_R}{P_2} = \frac{T_2(n_2 - n_s)}{T_2 n_2} = 1 - k_{s2} = \frac{1 - k_{12}}{1 - i_0} \; .$$

For negative-ratio transmissions, these equations are valid in the range $i_0 <$

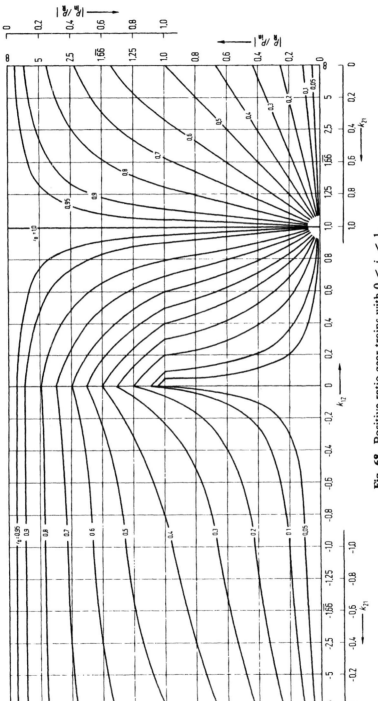

Fig. 68. Positive-ratio gear trains with $0 < i_0 < 1$.

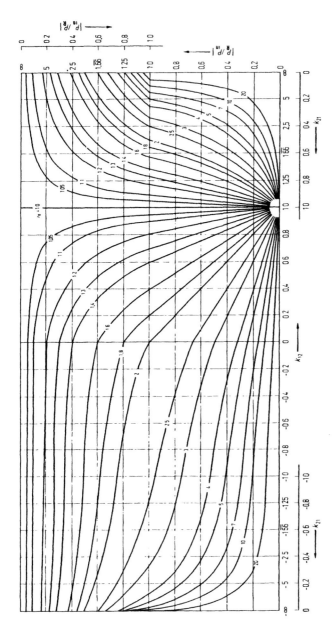

Fig. 69. Positive-ratio gear trains with $1 < i_o < \infty$.

Ratio between the rolling-power $|P_R|$ and the input power P_{in} for positive-ratio gear trains which operate with three rotating shafts for constant basic speed-ratios i_o as a function of the speed-ratio k_{12}. $|P_R| = |P_{R1}| = |P_{R2}|$. The other two speed-ratios k_{1s} and k_{2s} which are associated with the speed-ratio k_{12} can be found from figs. 63 and 64 or can be calculated with worksheet 2. With the analogous subscripts I $\hat{=}$ 1; II $\hat{=}$ 2; $P_R \hat{=}$ series power, the diagrams are valid also for bicoupled transmissions.

$k_{12} < 0$ and for positive-ratio transmissions in the range $0 < k_{12} < i_o$.

If, finally, shaft s is the total-power shaft then we obtain for $|P_R| = |P_{R2}|$ from table 1, and with eqs. (45), (36), and (29) that

$$\frac{P_R}{P_s} = \frac{T_2(n_2 - n_s)}{T_s n_s} = i_{s2}(k_{2s} - 1) = \frac{i_o}{1 - i_o} \cdot \frac{1 - k_{12}}{k_{12} - i_o} ,$$

which is valid for negative-ratio drives when $k_{12} > 0$ and for positive-ratio drives when $k_{12} < 0$.

These equations are plotted for positive-ratio drives in figs. 68 and 69, and in fig. 70 for negative-ratio drives. All these graphs again have semi-inverted scales so that all possible operating ranges can be represented.

C. Design and Operating Characteristics of Simple Planetary Transmissions

Section 20. Practically Achievable Basic Speed-Ratios

Theoretically, all the simple planetary transmission designs which are shown in figs. 19 to 43, could be designed with any basic speed-ratio. Practically, however, the range of the basic speed-ratios of each type is limited by the following aspects:

a) For economical reasons, an external gear stage should not exceed a speed-ratio of $i = -4$ to -8. For higher speed-ratios, multi-stage transmissions are smaller, lighter, and cheaper. This applies equally to planetary transmissions whose largest single-stage speed-ratio, therefore, should not normally exceed these values which thus also limit i_o.

b) If three or more planets are arranged as densely as possible around the circumference of the carrier, then the smallest possible sun gear, the largest ring gear and, thus, the largest possible ratio between their diameters results from the condition that the addendum circles of the planets may not touch each other.

c) In specific cases, other restrictions, which shall not be discussed at this point, may be imposed by design conditions such as weight, volume, or loads on the planet bearings due to the centrifugal forces.

The limits of the basic speed-ratios, as given in figs. 19 to 43, are valid for the most frequently encountered cases where three planets are arranged around the sun gear with a minimum distance between their addendum circles (which is equal to the reciprocal of the diametral pitch) and a sun gear with $z = 17$ teeth. For planetary transmissions with two sun gears it has been assumed that the smaller of the two sun gears has 17 teeth. In gear trains with stepped planets as shown, for example, in figs. 19 and 35, the pinion gears of each stage are geometrically similar and carry equal loads. Thus, the ratio between their (pitch circle) diameters equals the cube root of their torque ratio.

For the most frequently encountered types of planetary transmissions (that is, for negative-ratio transmissions as shown in fig. 33, positive-ratio transmissions as shown in fig. 19, and open revolving drives as shown in figs. 29 and 43), the largest possible basic speed-ratios i_o are given in figs. 71

and 72 as functions of the number of planets which are arranged around the circumference of the sun gear.

In a planetary transmission with bevel gears, as shown in fig. 41, an increasing basic speed-ratio i_o causes the planet's axis to become more nearly horizontal. Thus, the planetary gear train of fig. 33 can be considered as the limiting case of the bevel gear train of fig. 41 and, therefore, has the same maximum basic speed-ratio.

Section 21. Efficiency of the Two-Shaft Transmission

a) Influence of Design on Efficiency

The following three measures are suited to achieve high efficiencies in planetary transmissions:

1. Try to achieve low tooth-friction losses by applying such well-known gear design methods as the use of high quality tooth profiles and the selection of suitable materials and lubrication methods. Especially, the sliding motion between the tooth profiles must be minimized by using annular gear stages and gear stages with as many teeth as possible [27].

2. As indicated by eq. (10), the number of gear meshes which the power-flow must pass in series should be as small as possible in order to limit the sources of friction power-losses and thus obtain a high basic efficiency (parallel gear meshes which occur with power branching, theoretically, do not influence the efficiency).

3. Keep the rolling power P_R and thus the tooth-friction losses low as compared to the input power P_{in}, since according to eq. (51) the efficiency increases with a decreasing power ratio P_R / P_{in}.

Fig. 15 shows that for uncorrected gear trains, the tooth-friction losses, which in power transmissions constitute the bulk of the losses, decrease substantially with an increasing number of teeth. It also shows that the tooth-friction losses in internal gear meshes where $i > 0$ are much smaller than in external gear meshes where $i < 0$. Although the curves consider only the influence of geometrical factors, they facilitate qualitative evaluations and the comparison between gear trains of different designs. See also section 6.

The first of the measures which are recommended above leads to a preference for gears with a large number of teeth and designs which utilize annular gear trains.

The second measure rules out all designs with meshing planets (as shown, e.g., figs. 22, 23, 36, 37, etc.) since they have one more gear mesh than the simplest reverted gear trains. Not considering open planetary gear trains such as those shown in figs. 29 and 43, the rest of the gear trains always have two gear meshes through which the power flows in series. Therefore, their efficiencies are always lower than those of the single-stage transmis-

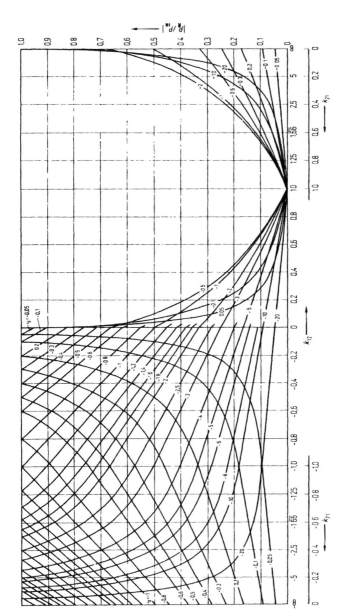

Fig. 70. Ratio between the rolling-power $|P_R|$ and the input power P_{in} for negative-ratio gear trains which operate with three rotating shafts, for constant basic speed-ratios i_0 as a function of the speed-ratio k_{12}. The friction-losses are neglected so that $|P_R| = |P_{R1}| = |P_{R2}|$. The other two speed-ratios k_{1s} and k_{2s} which are associated with the speed-ratio k_{12} can be found from figs. 63 and 64. With the analogous indices $I \triangleq i$, $II \triangleq 2$, and $P_R =$ series-power, the diagram is valid also for bicoupled transmissions.

sions with external gears. This disadvantage can be compensated by the third of the recommended measures only when the rolling-power fraction can be kept to a minimum, that is, when a small value of $|P_R/P_{in}|$ can be realized. However, as figs. 66 and 67 show, the ratio $|P_R/P_{in}|$ depends on i_o and as soon as the speed-ratio is specified, it may no longer be freely chosen.

b) Comparison between Planetary and Single-Stage Conventional Transmissions

For the sake of a basic investigation, and with reference to fig. 15, we may assume that the loss factor $\zeta_o = -P_L/P_{in}$ of an external gear stage is simply two times that of an annular gear stage. Then we can make the following comparison. According to eq. (5) the efficiency of a single external gear stage St is:

$$\eta_{St} = 1 + \frac{P_L}{P_{in}} \; .$$

The efficiency of a planetary gear stage Pl, however, where the power must successfully flow through the two gear meshes a and b, is:

$$\eta_{Pl} = 1 - \frac{P_{R,in}}{P_{in}}(1 - \eta_o) = 1 - \frac{P_{R,in}}{P_{in}} \cdot \left(-\frac{P_{La} + P_{Lb}}{P_{in}} \right) .$$

This can be verified by substituting eq. (5) into eq. (51). As defined in section 1, the power losses P_{La} and P_{Lb} are output powers and, therefore, negative so that the numerator $-(P_{La} + P_{Lb})$ becomes positive, and $\eta_{Pl} < 1$ as expected.

With the previous assumption we thus obtain the following valuations of the efficiencies: for a positive-ratio gear train with two annular stages, as shown in figs. 21 or 24,

$$\left(\frac{P_{La} + P_{Lb}}{P_{in}} \right)_{Pl} \approx \frac{\frac{1}{2}P_L + \frac{1}{2}P_L}{P_{in}} = \left(\frac{P_L}{P_{in}} \right)_{St}$$

where P_L is the power loss of an external gear stage as introduced above. Thus, for this type of planetary gear train, the efficiency is approximately equal to the efficiency of the external gear stage, that is,

$$\eta_{Pl} \approx \eta_{St} \quad \text{when} \quad \frac{P_{R,in}}{P_{in}} = 1 \; .$$

Following the same considerations, we find that for a positive-ratio gear train with two external stages as shown in fig. 19,

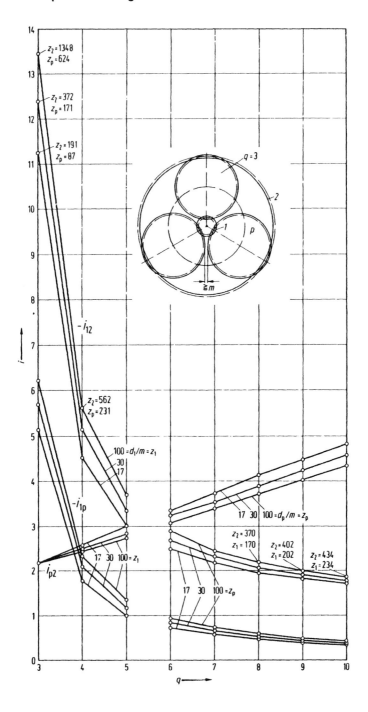

$$\eta_{Pl} \approx \eta_{St} \quad \text{when} \quad \frac{P_{R,in}}{P_{in}} = \frac{1}{2}.$$

For a negative-ratio gear train with one external and one internal stage as shown in figs. 32, 34 and 35 we obtain

$$\eta_{Pl} \approx \eta_{St} \quad \text{when} \quad \frac{P_{R,in}}{P_{in}} = \frac{1}{1.5}.$$

For all other speed-ratio ranges in which $P_{R,in}/P_{in}$ is smaller than these values, planetary gear trains under otherwise equal conditions should achieve a higher efficiency ($\eta_{Pl} > \eta_{St}$) than single stage external gear trains. For the three previously considered planetary gear trains, these speed-ratio ranges can be found from figs. 66 and 67. However, for convenience, they are summarized in table 3.

This comparison can be refined if, instead of using simplified assumptions, the ratios of the tooth-friction losses of the compared transmissions are accurately determined from fig. 15.

Table 3 shows that the efficiencies of the simple planetary gear trains are higher than the efficiencies of the conventional single-stage gear trains only when their direct or reciprocal speed ratios i lie in a range of approximately 1 to 3. The efficiencies of negative revolving speed-ratios are always lower than the efficiencies of the conventional single-stage gear trains.

Thus, it can be expected that in the practically utilized speed-ratio ranges of i (or $1/i$) $\approx -1 \ldots -7$ for single-stage drive trains with external gears, and i (or $1/i$) $\approx +1.1 \ldots +7$ for single-stage drive trains with annular gears, conventional transmissions achieve higher efficiencies than even the most advantageous revolving drive designs with the same speed-ratios. Planetary gear trains can exceed the efficiencies of conventional gear stages with external gears which have numerically equal but negative speed-ratios only in the mentioned speed-ratio range of approximately $+1$ to $+3$. Consequently, multistage conventional gear trains will also attain higher efficiencies than planetary gear trains with the same number of series-connected stages.

Fig. 71. The largest speed-ratios i_{12} and i_{1p} and the smallest speed-ratio i_{p2} for planetary gear trains as shown in figs. 29, 32, 33, and 43 as a function of the number of planets which are arranged around the circumference of the carrier. These speed-ratios are obtained from geometrical considerations and are valid if $3 \leqq q \leqq 5$ and gear 1 has a smallest number of teeth $z = 17$ or 30 or 100, or if $6 \leqq q \leqq 10$ and the planet p has the smallest number of teeth $z = 17$ or 30 or 100. The number of teeth for the other gears are shown only when they become excessively large. The minimum distance between the addendum circles of the planets is assumed to be equal to the module m ($= 1/P_d$).

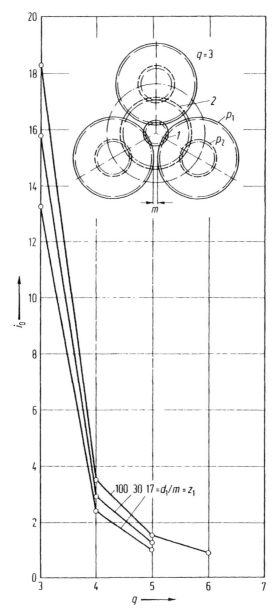

Fig. 72. The largest basic speed-ratio i_o of a positive-ratio gear train as shown in fig. 19 as a function of the number of planet sets. The diagram is valid for gear trains with $z_1 = 17, 30, 100$ and for geometrically similar pinions I and p_2 which have equal root stresses. The distance between the addendum circles of the planets p_1 is equal to the module m $(= 1/P)$.

TABLE 3. SPEED-RATIO RANGES WITHIN WHICH SIMPLE PLANETARY
GEAR TRAINS ACHIEVE EQUAL OR HIGHER EFFICIENCIES COMPARED TO
SIMPLE CONVENTIONAL GEAR TRAINS

TYPE	FIG.	$\dfrac{P_{R,in}}{P_{in}}$	SPEED-RATIO	SPEED-RATIO RANGE FROM	TO
Positive-ratio gear train	21, 24	≤ 1	i_{s2}	+1	+2
			i_{s1}	+1	+2
Positive-ratio gear train	19	$\leq 1/2$	i_{s2}	+1	+1.5
			i_{s1}	+1	+1.5
Negative-ratio gear train	32, 34, 35	$\leq 1/1.5$	i_{2s}	+1	+3
			i_{1s}	+1	+3

A negative-ratio planetary gear train can attain a higher efficiency only in the speed-ratio range of $i \approx +7 \ldots +11$, which it can still cover with a single stage. However, conventional gear trains normally, but not necessarily, are built with two stages. These considerations show that planetary gear trains do not necessarily achieve higher efficiencies than simple conventional drive trains with external gears even if their overall efficiencies are higher than their basic efficiencies η_0. Only if their speed-ratios lie in the narrow range between 0.5 and 2, that is, in the neighborhood of the coupling point, can they reach particularly high efficiencies of more than 99 per cent and thus exceed the efficiency of single-stage conventional gear trains with external gears.

The speed-ratio range, in which planetary gear trains are superior, shrinks even further when their efficiencies are compared with those of conventional drive trains containing annular gear stages.

The presented theoretical comparison, which is based solely on the tooth-friction losses, cannot unconditionally be applied to real transmissions, since, in practice, the idling losses can be neglected least when the theoretical efficiency approaches a value of 1 (see sec. 6).

c) Comparison between Planetary Transmissions

Eq. (51) states quite clearly that in order to select the planetary gear train which, for a given speed-ratio, achieves the highest efficiency, we must find the gear train with the highest basic efficiency η_0 and the lowest rolling power ratio $P_{R,in}/P_{in}$. The possibility of influencing the basic efficiency by design measures has been discussed earlier in section 21a and figs. 66 and 67 provide information about the rolling power-ratio $P_{R,in}/P_{in}$, which can be replaced by P_R/P_{in} when the friction losses are neglected.

In each of these diagrams, one of the three-shaft speed-ratios, together with its reciprocal value and the corresponding power-ratio P_R/P_{in}, have been plotted as functions of the basic speed-ratio. If we compare the power-ratios which correspond, for example, to each of four equal positive speed-ratios, it becomes evident that a negative-ratio transmission always has a smaller P_R/P_{in} than a positive-ratio transmission covering the same speed-ratio range. For example, the speed-ratio $i = +3$ can be realized with a negative-ratio gear train as $i_{1s} = 3$ for an $i_o = -2$ or as $i_{2s} = 3$ for an $i_o = -0.5$. In both cases $P_R/P_{in} = 2/3$. If we select a positive-ratio transmission, we can obtain $i_{s1} = 3$ with $i_o = 2/3$ or alternately $i_{s2} = 3$ with $i_o = 1.5$. Subsequently, we find that in both cases $P_R/P_{in} = 2$. If we use a positive-ratio gear train with a locked carrier and a basic speed-ratio $i_o = 3$, then, according to the definition given in section 14d, we have a power-ratio $P_R/P_{in} = 1$.

In those instances in which positive-ratio gear trains operate with a negative revolving speed-ratio, the power-ratio always becomes larger than 1, that is, $P_R/P_{in} > 1$, while negative-ratio gear trains which operate with a locked carrier have a power-ratio of only 1, that is, $P_R/P_{in} = 1$.

According to R 7, all of the planetary gear trains which have been compared in this example, are kinematically equivalent, since in each case one of the possible speed-ratios has the common value $i = +3$. As a result of this comparison we can formulate the following guide rule:

The rolling-power ratio $|P_R/P_{in}|$ is always higher in a positive-   R 26
ratio gear train than in a kinematically-equivalent negative-
ratio gear train.

This guide rule is generally valid also for planetary gear trains with three rotating shafts as can be checked by following the considerations which have been outlined in section 15g.

Section 22. Two-Shaft Transmissions with Extreme Speed-Ratios

By interchanging the input and output shaft, the same transmission can be used to realize extremely large or extremely small speed-ratios. It can be seen from figs. 45 and 61 that such extreme speed-ratios can be obtained in two ways:

1. The basic speed-ratio (negative or positive) is designed to be extremely large. As line i_{1s} in figs. 61 and 45, or box 13 in worksheet 1, indicate, the largest possible speed-ratio i_{1s} differs only by a value of 1 from the positive or negative basic speed-ratio i_o. Consequently, the possible extreme values of the two-shaft speed-ratios are approximately equal to the basic speed-ratios whose practically feasible limits for the various design configurations of figs. 19 to 43 are about ± 50, or $\pm 1/50$. Fig. 67 shows, that for these

speed-ratios the rolling power-ratio $P_R/P_{in} \approx 1.0$ and consequently the efficiencies assume the good value $\eta = \eta_o$. Because of their simplicity and good efficiency, gear trains of the types shown in figs. 19 and 35 are well suited for high speed-ratios within the given range. An executed example of a gear train according to fig. 35 is shown in fig. 73.

2. The basic speed-ratio of positive-ratio transmissions is designed to be less than but nearly equal to 1. It is clear from eqs. (34) and (36) that as the speed-ratios $\pm i_{s1}$ and $\pm i_{s2}$ become the higher, the closer i_o approaches a value of 1. At $i_o = 1$, they become infinitely large and can no longer be practically realized.

The speed-ratio i_o of approximately 1 can be realized in spacesaving configurations with almost any positive-ratio transmission design since they do not require large differences in the gear diameters. Open positive-ratio transmissions also can be used. Most frequently however, extremely high speed-ratios are realized with gear trains as shown in figs. 19 and 21 and figs. 74 and 75.

Their possible speed-ratios can be obtained with worksheet 1 and section 9. Since

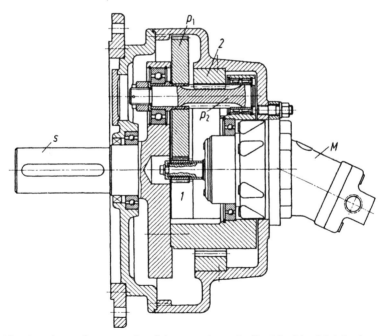

Fig. 73. Negative-ratio gear train of the type shown in fig. 35 with a high basic speed ratio $i_o = -49.3$. The transmission is designed to step-down the output speed of an axial piston hydraulic motor M (series K, Heinrich Desch KG): $i_{1s} = (1 - i_o) = +50.3$, $T_{s\,max} = 4{,}500$ Nm.

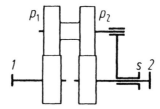

Fig. 74. Positive-ratio gear train with external gears for extreme speed-ratios: i_o is approximately +1.

$$i_{s1} = \frac{1}{1 - i_o} \quad \text{and} \quad i_o = \frac{z_{p1} z_2}{z_1 z_{p2}},$$

$$i_{s1} = \frac{1}{1 - \dfrac{z_{p1} z_2}{z_1 z_{p2}}}.$$

In the simplest design solution, both planets have an equal number of teeth and, consequently, can be combined into a single wide-face gear. If at the same time z_2/z_1 is close to 1, so that $z_2 = z_1 \pm 1$, the speed-ratio becomes:

$$i_{s1} = \frac{1}{1 - \dfrac{z_2}{z_1}} = \frac{1}{1 - \dfrac{z_1 \pm 1}{z_1}} = \mp z_1,$$

and is thus limited to the range with the most expedient number of teeth. To obtain equal center distances, the tooth profiles of the central gears must be corrected. The same ratio but not the same simplicity results when the transmission has equal central gears and planets whose number of teeth are $z_{p1} = z_{p2} \pm 1$.

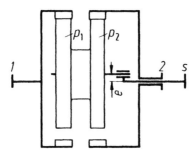

Fig. 75. Positive-ratio gear train with annular gears for extreme speed-ratios: i_o is approximately +1. The eccentricity e equals the orbit radius r_s of the planet axis.

For both of these designs, however, the highest speed-ratios can be attained when the speed of the basic gear train is geared up in the first stage and then geared down in the second stage with an almost equal, reciprocal ratio, or vice versa, so that i_o approaches the value of 1 even closer.

The smallest gear diameters are obtained when the numbers of teeth of all four gears are almost equal, that is, when

$$z_1 = z_{p2} \; ; \qquad z_{p1} = z_1 \pm 1 \; ; \qquad z_2 = z_1 \mp 1 \; .$$

From worksheet 1, we then find that:

$$i_{s1} = \cfrac{1}{1 - \cfrac{(z_1 \pm 1)(z_1 \mp 1)}{z_1^2}} = \cfrac{1}{1 - \cfrac{z_1^2 - 1}{z_1^2}} = z_1^2$$

and $\qquad i_{s2} = 1 - i_{s1} = -(z_1^2 - 1).$

Thus, large speed-ratio steps with magnitudes of z_1^2 can be realized with a single planetary gear train; for instance, $i_{s1} = 40,000$ when $z_1 = 200$ or $i_{s1} = 40,401$ when $z_1 = 201$, etc. If, for extremely high speed-ratios, the number

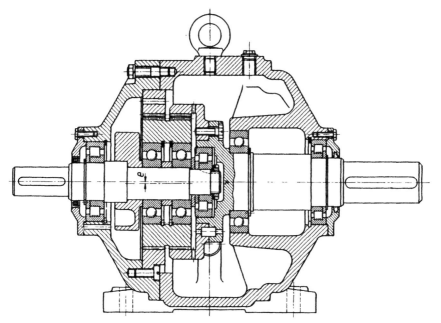

Fig. 76. Planetary transmission for extreme speed-ratios of the type shown in fig. 75. Manufactured by Prometheus Maschinenfabrik GmbH, Berlin, Series *UT,* for rated output torques of $T_1 \approx 5$ to $1,800$ Nm and $i_{s1} \approx 20$ to $10,000$.

of teeth becomes uneconomically high, bicoupled transmissions as described later in section 35k can be used, or several transmission stages can be connected in series.

Bicoupled transmissions according to section 35k also can be used when a very high speed-ratio, whose value lies between those of the z_1^2 stages, must be accurately realized.

If, in annular gear trains, according to fig. 21, the size of the planets approaches the size of the central gears, then obviously only one double-planet with a small axial eccentricity can be fitted inside of the annular gears as shown in fig. 76. Considering the high speed-ratio, the overall size of this transmission type is particularly small. However, it can be packaged even more densely when the two annular gear stages are nested as shown in fig. 77.

To assure trouble-free meshing in annular gear stages with standard involute gears, the difference between the number of teeth $\Delta z = z_1 - z_{p1}$, or $\Delta z = z_2 - z_{p2}$, must be equal or larger than 8. However, this minimum difference can be reduced to $\Delta z = 1$ when the tooth profiles are corrected as suggested by Clarenbach [32] (e.g., see fig. 17). This smallest difference $\Delta z = 1$ also can be realized with pin wheel transmissions such as the Cyclo transmission shown in fig. 78. Further investigations of involute gear stages with small differences between the number of planet and central gear teeth, have been conducted by T. Sunaga [24] and W. Schäfer [26].

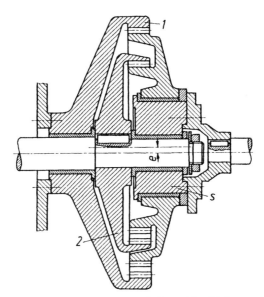

Fig. 77. Especially narrow planetary transmission with a high speed-ratio. Modification of the type shown in fig. 75, $i_o \approx +1$.

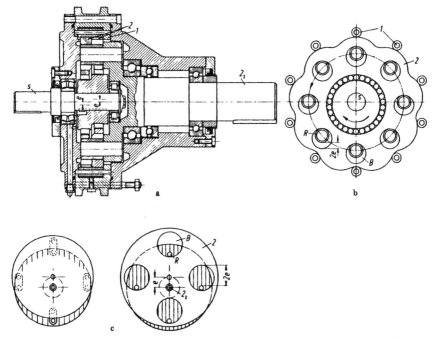

Fig. 78. Cyclo transmission, an open planetary transmission for single stage speed-ratios $i_{s2} = -9$ to -85. $T_{2z\,max} = 10$ to $34,000$ Nm: *a*, axial section; *b*, front view without housing; *c*, comparison between the output stage connected by pins (*right side*) and kinematically-equivalent output stage connected by parallel links (*left side*): *R*, roller, *B*, bore, *e*, eccentricity.

Unfortunately the *efficiencies* of these positive-ratio transmissions with basic speed-ratios $i_o \approx 1$ become the lower, the higher their speed-ratios are pushed up, which is contrary to their kinematical advantages. This is due to the fact that in this range the power-ratios P_R/P_{in} become equal to the speed-ratios i_{s1} and i_{s2}, as can be seen from figs. 66 and 67. The efficiencies themselves can be calculated with the aid of worksheet 5. With increasing speed-ratio they tend to zero when the carrier shaft is the input shaft. However, when the carrier shaft is the output shaft, η_{1s} and η_{2s} become negative when the speed-ratios i_{1s} and i_{2s} decrease towards zero which means that the value of the step-up speed-ratio is limited by the incidence of self-locking.

The Cyclo transmission of fig. 78 which has been developed by Siemens-Schuckert [6], initially as a reduction drive for electric motors, is a further example of an open positive-ratio transmission of the type shown in fig. 29b. Its annular gear *1* is a pin wheel and the planets *2* accordingly have cycloidal tooth profiles. The two planets, which are connected to the output, are arranged side by side to balance the carrier *s*, and at the same time provide two branches for the power-flow. In the Cyclo transmission, the rotation of the eccentrically positioned planets is transmitted with a speed-ratio of $i = +1$ to the central shaft 2_z through a parallel, pin type transmission, rather than a cardan shaft as shown in fig. 29. This part of the drive consists of rollers R which are supported by pins mounted on the face of an output flange. With a play of $2e$, which is equal to twice the eccentricity of the planet's orbit, they simultaneously engage the large holes B provided in the planet blanks and thus produce a motion which is analogous to the motion produced by the transmission shown in fig. 78c, where the parallel disks are coupled by a number of short links.

Alternately, the Cyclo transmission can be considered as a reverted planetary transmission of the type shown in fig. 21, where, however, input and output are reversed and the stage $p_1/1$ is replaced by the described pin-type transmission which has a speed-ratio of $i = +1$. Since $n_2 = n_{2z}$, the analysis of the transmission remains the same for both points of view and follows the procedure outlined earlier for simple planetary transmissions. With the gear sizes of the given example, the basic speed-ratio becomes

$$i_o = i_{12} = -\frac{z_2}{z_1} = -\frac{9}{-10} = 0.9 .$$

Since the annular pin wheel 1 is fixed to the housing, the speed-ratios are:

$$i_{s2} = \frac{i_o}{i_o - 1} = \frac{0.9}{0.9 - 1} = -9 \quad \text{and}$$

$$i_{2s} = \frac{1}{i_{s2}} = -\frac{1}{9} ,$$

as can be verified with worksheet 1, boxes 31 and 30 respectively. If η_o is known, the efficiencies of this transmission can be found from worksheet 5 where boxes 17 and 14 apply since the basic speed-ratio lies in the range $0 < i_o < 1$. Thus:

$$\eta_{s2} = \frac{i_o - 1}{i_o - 1/\eta_o} = \frac{-0.1}{0.9 - 1/\eta_o} = \frac{0.1}{1/\eta_o - 0.9} \quad \text{and}$$

$$\eta_{2s} = \frac{i_o - \eta_o}{i_o - 1} = \frac{0.9 - \eta_o}{-0.1} = 10\eta_o - 9 .$$

The latter equation indicates, that self-locking occurs when $\eta_o \leqq 0.9$, and the carrier shaft operates as the output shaft.

The actual basic efficiency of the Cyclo transmission, however, is higher than 0.9 since the rollers of the annular gear and the output pin-drive stage are free to rotate on their support studs and consequently the "tooth-friction losses" remain extremely low.

Fig. 15 indicates that positive-ratio transmissions with internal gears as shown in fig. 75, have substantially higher basic efficiencies than positive-ratio transmissions with external gears as shown in fig. 74. According to fig. 16, the geometrical loss factor f_L has a minimum around $\Delta z = 8$ and then increases as a consequence of the necessary profile correction, with a decreasing difference between the number of pinion and annular gear teeth. However, in the practically important speed-ratio range it remains always lower than for comparable transmissions with external gears.

Positive-ratio transmissions with high speed-ratios and an $i_o \approx 1$ are used only where small powers must be transmitted at unusually low speeds or in position and servo-drives. As power transmissions they are too uneconomical. Moreover, it would be difficult to dissipate the heat which is generated at higher loads.

Therefore, only those revolving or two-shaft drive trains which have a high positive or negative basic speed-ratio are suited as power transmissions for speed-ratios i of up to about 50. Preferred are the types shown in fig. 35 which offer good efficiency and the possibility of power branching, and fig. 21 whose two annular gear stages have an especially good efficiency.

If the required speed-ratio of a power transmission exceeds the capability of these simple planetary transmissions types, several of them can be connected in series to form multi-stage transmissions. Otherwise, bicoupled transmissions must be used, discussed in sections 32e, 34, and 35.

Section 23. Self-Locking Planetary Transmissions

Self-locking revolving drives with a basic speed-ratio i_o of approximately +1 can be used as two-shaft transmissions like self-locking worm gears when, for example, servo-drives for small loads are needed, and the input and output shafts must be coaxial. They can be used also in small hoists, for which a load brake is not specifically required.

The partial-locking property of a three-shaft transmission can be used also to satisfy the requirements of special drives such as shown in fig. 79 where the carrier of the self-locking positive-ratio gear train can be either locked by the brake B or can be rotated freely. In the first case, the speed-ratio between input 1 and output 2 is only slightly larger or smaller than 1. In the second case, the gear train operates in a partially-locked condition as a coupling with a speed-ratio i_{12} of precisely 1. This small difference between the speed-ratios could be used to obtain a phase shift—however, only in one direction—between two shafts.

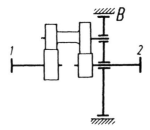

Fig. 79. Three-shaft transmission which is capable of self-locking for phase shift application in one direction. Brake B open: partial-locking with $i_{12} = 1$; brake B locked: $i_{12} = i_o \neq 1$.

The self-locking property of revolving drive trains has been used only seldom; perhaps because the conditions of self-locking, which only a few years ago were still disputed, are not widely understood, but certainly also because the efficiency in the unconstrained power-flow direction, that is, when the carrier shaft is an input shaft, is always less than 0.5. Another reason may lie in the fact that the tooth-friction losses cannot be accurately predicted. If, therefore, the design analysis is based on the tooth-friction losses as discussed in section 6, it is advisable to choose an operating point within the self-locking range which is sufficiently far from the self-locking limits.

At this point it is strongly emphasized that only the load-dependent rolling losses, that is, the tooth-friction losses and the load-dependent parts of the planet-bearing losses, influence the implementation of self-locking. The idling and speed-dependent losses do indeed influence the final overall efficiency, but in transmissions which do not yet operate in the self-locking range, they can always be compensated for by a corresponding increase of the input torque. Consequently, they must not be considered when the capability of a planetary transmission to become self-locking is to be investigated. Thus, for all efficiency considerations in the self-locking range, or close to it, a basic efficiency η_o, which has been calculated according to the recommendations given in section 6, must be substituted into the efficiency equations.

Section 24. Three-Shaft Transmissions as Superposition Drives for Power Division and Power Summation

a) Definitions

The terms "power division," describing the division of the input power of a single input shaft between two (or more) output shafts, and "power summation," describing the concentration of the power of two (or more)

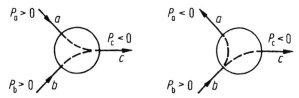

Fig. 80. Transition from power division (*right side*) to power summation (*left side*) when a shaft *a* changes its direction of rotation and thus the sign of its power.

input shafts in a single output shaft of *any* drive, do not characterize two different types of transmissions, but rather two different modes of operation which are possible in the same transmission. If, in a transmission with three connected shafts, one of these shafts, (e.g., shaft *a* in fig. 80) changes the sign of its speed or torque, then the sign of its power also changes. Thus, the mode of power division changes to the mode of power summation or vice versa. The summation of two quantities, independent of their signs, is frequently called superposition and, consequently, a planetary transmission which effects a power summation or superposition may also be called a superposition drive. Power division, of course, is the summation of two negative powers and thus is also a power superposition. The term superposition drive, therefore, is applicable to both modes of operation and quite generally characterizes a transmission with three or more connected shafts which receive positive or negative powers in varying changing proportions.

The simultaneous transmission of arbitrarily varying powers to several shafts is possible in three different ways:

1. The speed may be arbitrarily varied, while the shafts maintain a given constant torque ratio. This possibility exists for all revolving drives whose shaft torques, according to R 11 and eq. (46), bear a constant ratio which is dependent on the transmission configuration. In this case, however, the number of independent speeds, which can and must be specified, must be equal to the number of degrees of freedom of the transmission, that is, equal to two for a simple planetary transmission, or equal to four for each of the compound transmissions shown in figs. 97 and 111q. The dependent speed is then obtained as a sum to which each freely-chosen speed contributes an addend. For simple revolving drives we thus obtain, with eqs. (23) to (25), and eqs. (33) to (36),

$$n_1 = i_{12}n_2 + i_{1s}n_s \, ,$$

$$n_2 = i_{21}n_1 + i_{2s}n_s \, ,$$

$$n_s = i_{s1}n_1 + i_{s2}n_2 \, .$$

This summation is valid also for compound revolving drives with more than two degrees of freedom, for example, for transmissions as shown in fig. 97, where

$$n_e = i_{ea}n_a + i_{eb}n_b + i_{ec}n_c + i_{ed}n_d$$

and fig. 111q, where

$$n_B = i_{Ba}n_A + i_{BC}n_C + i_{BD}n_D + i_{BE}n_E \ .$$

In this context, i stands for the speed-ratio which exists between the shafts indicated by its subscripts when all other connected shafts are locked, that is, when the indicated shafts rotate positively constrained.

Revolving drives, consequently, are suited also for the superposition (summation) of speeds as will be discussed in more detail later in section 27.

2. The torques may be arbitrarily varied, while the given speed-ratios remain constant. This possibility exists for the usual, positively constrained transmissions with spatially fixed axes, whose more than two connected shafts bear a constant speed-ratio relative to each other which depends only on the transmission design (e.g., figs. 81 and 82).

For this type of transmission, the dependent torque is obtained as a sum, to which each independently given torque contributes an addend. If the losses are neglected, the input torque T_i for the power division drive of fig. 82 can be obtained from eq. (6) as:

$$T_i = -\left(\frac{1}{i_{ia}}T_a + \frac{1}{i_{ib}}T_b + \frac{1}{i_{ic}}T_c + \ldots + \frac{1}{i_{ih}}T_h\right) . \tag{66}$$

Therefore, positively constrained transmissions with more than two shafts can be used also when several torques must be superimposed (torque summation).

3. The speed may be arbitrarily varied at some of the connected shafts and, at the same time, arbitrarily given torques may be varied at others. This possibility exists, for example, for four-shaft bicoupled transmissions of the type shown schematically in fig. 111s.

Which of these three modes of operation is possible in a given transmission can be determined from its static, kinematic, and operational degree of freedom as will be discussed in the following section.

b) Static, Kinematic, and Operational Degree of Freedom

The number of the *independent* variables, that is, the number of the *independent* speeds and torques which completely specify the state of operation of a transmission, shall be called its operational degree of freedom. The degree of freedom, which specifies the number of independent speeds and which has already been discussed in section 3, henceforth shall be called the kinematic degree of freedom F_{kim}. The index "kim," short for "KIneMatic," is used to avoid a possible confusion with "kinetic." The degree of freedom which specifies the number of independent shaft torques that are needed to satisfy the static equilibrium conditions shall be called the static degree of freedom F_{stat}. Consequently, the operational degree of freedom is defined as

$$F_{op} = F_{kim} + F_{stat} .$$

Following general usage, this book shall always refer to the kinematical degree of freedom when the term "degree of freedom" is used without further specification and the symbol F carries no subscript.

According to section 3, a simple revolving drive train has a kinematic degree of freedom $F_{kim} = 2$. Only one of these torques can be arbitrarily chosen because, according to eq. (46), its three shaft torques bear a constant ratio with respect to each other. Thus, its static degree of freedom $F_{stat} = 1$ and its operational degree of freedom become

$$F_{op} = 2 + 1 = 3 = N_{cs}$$

where N_{cs} symbolizes the number of its connected shafts.

According to section 3, a conventional transmission has one kinematic degree of freedom $F_{kim} = 1$, even if it has more than two connected shafts as shown, for example, in figs. 81 and 82. Consequently, the speeds of all of its connected shafts must bear a constant ratio relative to each other. The output torques, on the contrary, can be freely chosen. In the drill head of fig. 82, for example, the drill loads produce eight arbitrary (output) torques T_a to T_h. Therefore, its input torque is determined by eq. (66) and can no longer be freely chosen. Consequently, the static degree of freedom is 1 less than the number of connected shafts, that is, $F_{stat} = N_{cs} - 1$ and the operational degree of freedom of this conventional transmission becomes

$$F_{op} = F_{kim} + F_{stat} = 1 + (N_{cs} - 1) = N_{cs}$$

as for a revolving drive.

If drive systems are investigated which are arbitrarily composed of conventional and revolving drives, the same result is obtained which is valid for any transmission. The operational degree of freedom F_{op} is equal to the number N_{cs} of the connected shafts:

$$F_{op} = F_{kim} + F_{stat} = N_{cs} . \tag{67}$$

If the kinematic degree of freedom has been determined as discussed in section 3, then the static degree of freedom obviously can be found from

$$F_{stat} = N_{cs} - F_{kim} \ . \qquad (68)$$

If a transmission is designed to transmit mechanical power, then it must, of course, be capable of bearing the required torque, that is, it must have a static degree of freedom of at least 1 which, with eq. (68), leads to the following guide rule:

> *The number of connected shafts of a power transmission* R 27
> *must exceed the number designating its kinematic degree of*
> *freedom by at least 1.*

Together with R 5, this general rule facilitates a quick examination of complex compound transmissions. It also points out that a constrained transmission must have at least two torque transmitting shafts, and a revolving drive with two degrees of mobility must have at least three.

c) Conditions for Stable Operation of Superposition Drives

A mechanical system which consists of three or more motors and machines can reach a steady-state point only when the operating characteristics of its components are matched to the characteristics of the superposition drive which connects them. This drive must receive a number of independent input speeds and torques equal to the number of its operational degrees of freedom F_{op}, and the characteristic curves of the motors and machines must yield stable points of intersection.

More specifically then, the described system is stable only when the transmission receives a number of shaft speeds, which equals the number designating its kinematic degree of freedom, and a number of torques equal to its static degree of freedom. However, for transmissions with $F_{kim} = 1$, or $F_{stat} = 1$, which includes all three-shaft transmissions, it is immaterial for which of the shafts the speeds are specified, and for which of the shafts the torques are specified.

If a transmission is connected to a motor or a machine whose interdependent torques and speeds are uniquely determined by its characteristic curve, then, as a consequence of this constraint, only its speed *or* its torque can be freely chosen. Therefore, each connected machine or motor with such a characteristic curve occupies one operational degree of freedom. This means that for each connected machine or motor whose speeds and torques are definitely related, the number of operational degrees of freedom of the superposition drive is reduced by 1. More specifically, the characteristic curve of a connected machine reduces the number of kinematic degrees of freedom when all static degrees of freedom of the transmission are occupied by prior constraints, and vice versa. Prior constraints are torques which are specified independently of the speeds, such as load torques of a hoist, and torque independent speeds, such as the speeds of a synchronous motor or generator. If all connected machines and motors have clearly defined operating characteristics, the machines operate with those transmission related speeds at which all their dependent torques are in equilibrium at their common operating point.

The second of these stability conditions, which specifies that the operating characteristic curves of the connected machines and motors must intersect in stable operating points, is met if, with increasing speed, the machine torques increase while the motor torques decrease.

It is understood that even in drive systems composed of several revolving and conventional gear trains possessing more than one kinematic degree of freedom and, at the same time, more than one static degree of freedom, only one speed can be freely chosen for a group of shafts which are positively connected internally. Likewise, only one torque can be freely chosen for the shafts of the revolving gear trains which already bear a constant torque ratio relative to each other.

Most practical drive systems consist of transmissions with a single kinematic degree of freedom and one or more static degrees of freedom, that is, conventional transmissions with spatially fixed axes, for which

$$n_a : n_b : n_c : \ldots = \text{const.}$$

Also used are simple revolving drive trains with a single static and two kinematic degrees of freedom. Transmissions with one static and more than two kinematic degrees of freedom are possible as compound planetary transmissions such as shown in fig. 12, whose torque ratio can be determined analogously from eq. (46):

$$T_a : T_b : T_c : T_d : T_e = \text{const.,} \quad \text{and thus}$$

$$F_{\text{stat}} = 1, \quad F_{\text{kim}} = 4, \quad \text{so that} \quad F_{\text{op}} = 5 .$$

Transmissions with three or more connected shafts usually have either one kinematic and several static degrees of freedom, or one static and several kinematic degrees of freedom, depending on whether their shaft speeds, or their shaft torques, bear a constant ratio relative to each other. In each of these cases, the connected machines and motors operate in fundamentally different modes which lead to fundamentally different operating characteristics for the overall drive system. These differences will be further explained by the following examples.

d) Examples of Transmissions with One Kinematic Degree of Freedom ($F_{kim} = 1$)

α) *Marine transmission for twin-engine propeller drive.* The marine transmission shown in fig. 81a has three connected shafts and, consequently, three operational degrees of freedom. As a conventional transmission it has one kinematic degree of freedom, so that according to eqs. (67) and (68), $F_{op} = 3$, $F_{kim} = 1$ and $F_{stat} = 2$. Of the three operational degrees of freedom, one is occupied on the output shaft by the operating characteristics of the propeller. The torques of the two diesel engines which, in the operating range, are essentially independent of the engine speeds, are adjusted separately for each engine by controlling the amount of fuel injected per working cycle. Thus, the two static degrees of freedom are occupied by the engine torques T_A and T_B. The torque of the propeller shaft is determined by the equation:

$$T_C = T_A' + T_B' = -(T_A i_{AC} + T_B i_{BC}) , \qquad (69)$$

where T_A' and T_B' are the propeller torque fractions due to the engine torques T_A and T_B. If the friction losses are neglected we obtain from eq. (6):

$$T_A' = -T_A i_{AC} \quad \text{and} \quad T_B' = -T_B i_{BC} .$$

For the given propeller torque T_C the propeller speed n_C is determined by the characteristic curve of the propeller as shown in fig. 81b, which, as previously mentioned, occupies the only kinematic degree of freedom. Consequently, the engine speeds become

$$n_A = n_C i_{AC} , \quad \text{and} \quad n_B = n_C i_{BC} .$$

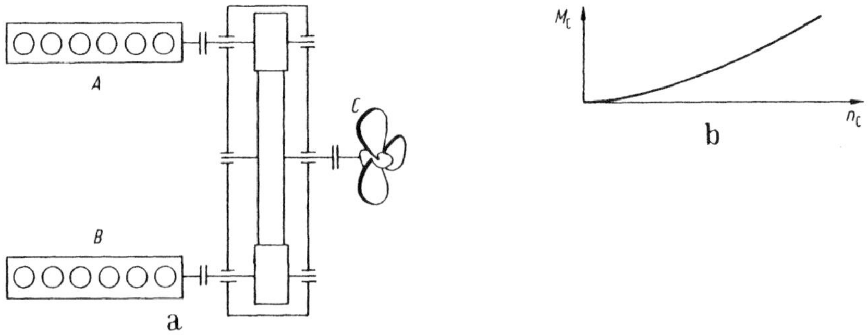

Fig. 81. Marine drive, *a*, with two input shafts and one kinematic degree of freedom; *b*, characteristic curve of the propeller.

Basically it is possible to choose the torque of one of the engines to be zero, that is, to disconnect one of the engines from the transmission. The propulsion system then remains operable as is well-known.

If, instead of controlling the engine torques, an attempt were made to individually control the engine speeds at arbitrarily chosen levels by means of governors, it would be found that it is impossible to predict the final state of operation since, because of $F_{kim} = 1$, only one of the speeds of the system can be freely chosen. Depending on the power delivered by the two motors, any one of the three following final states of operation could be reached.

1. The propeller rotates at the higher of the two chosen speeds provided the motor, which is governed at this speed, develops enough torque to drive the propeller and drag the other motor whose regulator then remains closed. In this case the kinematic degree of freedom is occupied by the governed higher speed, while the first static degree of freedom is occupied by the torque required to drag the other motor, and the second by the characteristic curve of the propeller.

2. The propeller rotates with a speed which lies between the two chosen speeds. The governors then operate one motor at the highest possible fuel injection rate and idle the other. This condition occurs when the maximum torque of the motor with the higher chosen speed is smaller than the propeller torque at this speed. In this case, the torque delivered by this motor at maximum fuel flow, and the torque required to drag the second motor, occupy the two static degrees of freedom. The propeller torque T_C can now be determined with eq. (69) and the propeller speed n_C, which occupies the only kinematic degree of freedom, from the characteristic curve of fig. 81b.

3. The propeller rotates at the lower speed which has been chosen for the second motor. This case occurs when the maximum torque of the first motor alone is just large enough to drive the propeller at a speed below the chosen speed of the second motor. The governor of the second motor then adjusts the fuel injection rate until the engine runs at its originally chosen speed. The only kinematic degree of freedom is thus occupied by the governed speed of the second engine, and the two static degrees of freedom by the peak torque of the first motor and the characteristic curve of the propeller.

Thus, in two of these cases, the characteristic curve of the propeller occupies one of the two static degrees of freedom. In the other case, it occupies the only kinematic degree of freedom.

β) *Multi-spindle drill head.* As a conventional transmission, the drive of the drill head shown in fig. 82 has one input and eight output shafts. According to R 6 and eqs. (67) and (68)

$$F_{op} = 9, \qquad F_{kim} = 1 \qquad \text{and thus} \qquad F_{stat} = 8 \ .$$

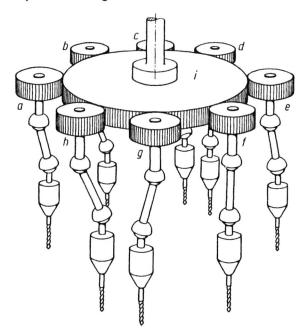

Fig. 82. Gear train with one degree of freedom for a multiple spindle drill head; input shaft *i*, output shafts *a–h*.

The torques needed to drive the eight drill bits are independent of the drill speeds and have priority in occupying the available eight static degrees of freedom. Since the kinematic degree of freedom is occupied by the characteristic curve of the drive motor, the system can operate at a steady state point.

γ) *Common drive for two centrifugal pumps.* Two identical centrifugal pumps A and B are to be driven with different speeds as shown in fig. 83a by a motor M which acts through a division drive.

The drive train has the following number of degrees of freedom: $F_{op} = 3$, $F_{kim} = 1$, $F_{stat} = 2$. Since $F_{kim} = 1$, all speeds bear a constant ratio relative to each other so that the two pump speeds n_A and n_B can be found easily for any motor speed n_M. The pump torques T'_A and T'_B, at n_A and n_B respectively, are then obtained from the common characteristic curve of the two pumps as shown in fig. 83c.

The diagram of fig. 83d shows the characteristic curve of the motor T_M and the two torque curves T'_A and T'_B of the pumps which have been transferred to the motor map. For example, the transfer of a point P_A of the characteristic curve of the pump A can be accomplished when the speed-ratio i_{MA}, and the efficiency η_{MA} of the gear train which is connected to the pump A, are known and the torques and speeds are converted separately with eqs. (6) and (1) as follows:

$$T'_A = -\frac{T_A}{\eta_{MA}} \cdot \frac{1}{i_{MA}} \quad \text{and} \quad n'_A = n_M = n_A \, i_{MA} \;.$$

Likewise for pump B:

$$T'_B = -\frac{T_B}{\eta_{MB}} \cdot \frac{1}{i_{MB}} \quad \text{and} \quad n'_B = n_M = n_B \, i_{MB} \;.$$

T'_A and T'_B represent the torques, which the motor shaft must transmit simultaneously at the speed $n'_A = n'_B = n_M$ in order to drive the pumps. During steady state operation, the sum of these torques must be equal to the motor torque.

The point of intersection between the characteristic curve of the motor and the summation curve $T'_A + T'_B$ determines the operating point on the motor map and thus the motor speed at which the motor torque and the pump torques are in a state of equilibrium.

For the chosen example and with the numerical values from fig. 83b, $T_M = T'_A + T'_B = 60 \, Nm$ and $n_M = 20s^{-1}$, so that the operating speeds of the pumps are found to be:

$$n_A = \frac{n_M}{i_{MA}} = \frac{20}{0.4} = 50 \sec^{-1} \quad \text{and}$$

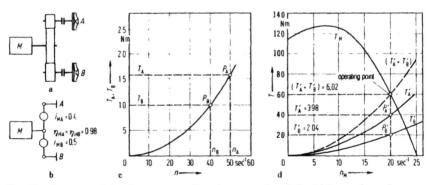

Fig. 83. Common drive for two centrifugal pumps A and B which consists of a conventional transmission with two output shafts and $F_{kim} = 1$: a, schematic of the drive; b, symbolic representation of the drive; c, characteristic curve of the two identical pumps A and B with their operating points P_A and P_B; d, map of the drive system with the characteristic curve of the motor T_M and the characteristic curves of the pumps T'_A and T'_B which have been transferred onto the motor map.

$$n_B = \frac{n_M}{i_{MB}} = \frac{20}{0.5} = 40 \sec^{-1} .$$

The corresponding operating points P_A and P_B can now be marked over n_A and n_B on the characteristic curve of the pumps shown in fig. 83c.

e) Examples of Revolving Drive Trains with One Static Degree of Freedom and Power Division

α) *Crusher with flushing water pump.* Fig. 84 schematically shows a crusher and an associated flushing water pump, which are simultaneously driven by a single motor. The speed n_1 of the motor M is assumed to be constant and independent of the required output torque. The speed of the crusher B should decrease as the size of the feed stock and, consequently, the required input torque increases, while the pressure p and the flow Q of the pump should increase simultaneously with increasing crusher torque. To satisfy these conditions, a planetary transmission I is used to split the motor output power between the crusher and the pump. A conventional step-down

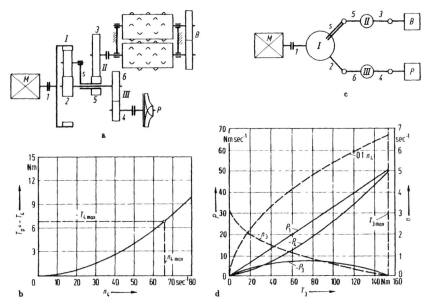

Fig. 84. Crusher and flushing water pump driven by a planetary transmission which operates in the power division mode: *a*, schematic of the system; *b*, characteristic curve of the pump showing the driven torque T_p; *c*, symbolic representation of the system; *d*, characteristic curves of the overall system as a function of the drive torque T_3 of the crusher B.

transmission is added in the drive line of the crusher to increase its input torque, and a conventional step-up transmission in the drive line of the centrifugal pump to match the system and pump characteristics. The parameters of the drive line may be given as:

$$i_o = \frac{n_1}{n_2} = -0.3, \qquad i_{II} = i_{53} = -4, \qquad i_{III} = i_{64} = -0.8 .$$

$$\eta_o = 0.98, \qquad \eta_{II} = \eta_{III} = 0.985, \qquad n_1 = \text{const} = 16\,s^{-1}$$

The pump torque T_4 depends on the speed of the pump as can be seen from its characteristic curve shown in fig. 84b, while the required drive torque T_3 of the crusher is a function of the momentary size and flow of the feed stock. T_3 should not exceed a given value T_{3max}.

The system has the following degrees of freedom:

$$F_{op} = 3 , \qquad F_{kim} = 2 , \qquad F_{stat} = 1 .$$

The only static degree of freedom is occupied by the continuously fluctuating torque T_3 at the crusher rolls, while the two kinematic degrees of freedom are occupied by the constant motor speed and the characteristic curve of the pump. Therefore, the system has a stable operating point.

For the given parameters, the torques can be calculated with eqs. (6), (44), and (45). Thus,

$$\frac{T_1}{T_3} = \frac{T_1}{T_s} \cdot \frac{-T_5}{T_3} = \frac{1}{i_o \eta_o^{r1} - 1} \left(\frac{-1}{\eta_{II} i_{53}} \right) \tag{70}$$

where according to R 4, $T_5 = -T_s$ (see fig. 84c).

The sign of $r1$ in eq. (43) can be determined even when the speeds are not precisely known by the following consideration: According to R 2 and R 3, we first define both the speed and torque of the input shaft of the transmission as positive, that is $n_1 > 0$ and $T_1 > 0$. Subsequently, we determine the other signs. According to worksheet 3, or R 13, shaft s is the summation shaft, so that according to R 8, $T_s < 0$. Since shaft s is also the output shaft, $P_s = T_s n_s < 0$, consequently $n_s > 0$. Also, since the input power P_1 must be larger than the partial output power P_s, $T_1 n_1 > -T_s n_s$, or $n_1/n_s > -T_s/T_1 > 1$, so that $n_1 > n_s$ and $T_1(n_1 - n_s) > 0$; that is, according to eq. (43): $r1 = +1$.

It is even easier to obtain $r1$ from worksheet 4, which shows that for negative-ratio transmissions and for the unequivocally given power-flow $1 <_s^2, r1 = +1$.

Thus, according to eq. (70)

$$\frac{T_1}{T_3} = \frac{1}{-0.3 \cdot 0.98 - 1} \cdot \frac{1}{0.985(-4)} = 0.196, \quad \text{and}$$

with eq. (45) and R 4

$$\frac{T_4}{T_3} = \frac{T_5}{T_3} \cdot \frac{-T_2}{-T_s} \cdot \frac{T_4}{T_6} = \frac{-1}{\eta_{11} i_{53}} \cdot \frac{i_o \eta_o^{r1}}{1 - i_o \eta_o^{r1}} (-\eta_{111} i_{64}), \tag{71}$$

that is,

$$\frac{T_4}{T_3} = \frac{1}{-0.985(-4)} \cdot \frac{-0.3 \cdot 0.98}{1 + 0.3 \cdot 0.98} (-0.985)(-0.8) = -0.0454 .$$

The speeds are found with eq. (23) and thus:

$$n_1 = i_o n_2 + (1 - i_o) n_s = i_o n_4 i_{64} + (1 - i_o) n_3 i_{53} ,$$

$$16 = (-0.3)(-0.8) n_4 + 1.3(-4) n_3 = 0.24 n_4 - 5.2 n_3 [s^{-1}] . \tag{72}$$

The overall efficiency of the transmission at a given operating point becomes:

$$\eta_{tot} = \frac{-T_3 n_3 - T_4 n_4}{T_1 n_1} .$$

If the speed of the crusher is lowered until it stops, that is, n_3 becomes 0, then, according to eq. (72), the pump reaches its highest speed,

$$n_{4max} = \frac{16}{0.24} = 66.7 \, sec^{-1}$$

and, simultaneously, its highest torque $T_p = -T_{4max} = 7 \, Nm$, as can be verified from the characteristic curve shown in fig. 84b. At this state of operation the crusher input torque T_3 and the motor output torque T_1 also reach their maximum values. According to eqs. (70) and (71) then:

$$T_{3max} = \frac{T_{4max}}{-0.0454} = -\frac{-7}{0.0454} = 154 \, Nm, \quad \text{and}$$

$$T_{1max} = T_{3max} \cdot 0.196 = 154 \cdot 0.196 = 30 \, Nm .$$

During stationary operation the torque cannot exceed these limits, so that the system possesses an inherent overload protection.

If the crusher draws an arbitrary average torque T_3, the pump torque becomes

$$T_4' = -0.0454 T_3' ,$$

and, consequently, the pump assumes a speed n_4' which is determined by its characteristic curve as shown in fig. 84d. With this speed, and the motor speed n_1, the crusher speed n_3' at this particular operating point can be obtained from eq. (72),

$$n_3' = \frac{0.24 n_4' - 16}{5.2} \; .$$

If we allow T_3' to assume a number of values between 0 and T_{3max}, then we obtain the parameters of the corresponding operating points and, consequently, the characteristic curves of the system, as shown in fig. 84d. The curves $P_3 = T_3 \cdot 2\pi \cdot n_3$ and $P_4 = T_4 \cdot 2\pi \cdot n_4$ of this map clearly show how the two transmission output powers vary with the changing crusher torque T_3.

β) *Automotive automatic transmission.* Similarly, planetary transmissions can be used to split the power-flow in automotive transmissions. In the Diwamatic transmission shown in fig. 85a, a planetary transmission I (input differential) is used to split the input power between the two power paths A and B. Through intermediate gear stages, A leads positively constrained to the output, while B connects to the pump P of the hydrodynamic torque converter W. The output power of the torque converter flows through the gear train II, which is a summation drive with one kinematic degree of freedom, and is subsequently added again to the power flowing along path A. The reunited total power is then transmitted to the output flange, either through the dog clutch K, or the intermediate gear stage III, which serves to reverse the direction of rotation of the output shaft.

The input differential I has the following degrees of freedom:

$$F_{I,op} = 3 \; , \qquad F_{I,kim} = 2 \; , \qquad F_{I,stat} = 1 \; .$$

One kinematic degree of freedom is preempted by the carrier speed n_s, which is directly proportional to the speed of the vehicle. The second kinematic and the static degree of freedom are occupied by the operating characteristic curves of the pump P and the motor.

If the vehicle is stopped, shaft A does not rotate and the total engine power flows along path B to the torque converter W. However, the carrier torque, which, according to eq. (44), depends on the engine torque, acts along path A so that the starting torque of the vehicle consists of the carrier torque and the large output torque of the stalled converter turbine which is added to it through the gear train II.

At medium vehicle speeds the planetary gear train I operates with power division as described earlier. The exact operating point, that is, the ratio between the two power-flows, can be obtained from the vehicle speed (pro-

134 / Simple Revolving Drive Trains

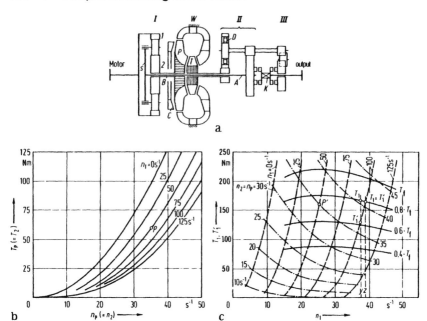

Fig. 85. Diwamatic transmission of the Voith-Getriebe KG showing an automotive transmission which operates in the power division mode. Schematic of the system, *a*: *I*, planetary gear train with power division (input differential); *II*, conventional transmission with one degree of freedom for the summation of the carrier and turbine power, overrunning clutch *D*; *III*, reversing gear stage with dog clutch *K*; *W*, torque converter with pump *P*, turbine *T*, and brake *C* which at high output speeds allows a purely mechanical power transmission; *A*, *B*, partial power paths which lead immediately to the output and to the pump of the converter. Characteristic curves, *b*, of the converter pump *P* corresponding to constant turbine speeds n_T, and thus to constant vehicle speeds. Characteristics map of the transmission, *c*: ———, assumed characteristic curves of the engine corresponding to constant positions of the throttle; T_f, full throttle torque; – – – –, characteristic curves of the pump for n_T = constant from fig. 85*b*, as mapped onto the characteristic curves of the engine; – · – · – · –, corresponding pump speeds $n_P = n_2$ = constant.

portional to n_s) and the characteristic curves of the converter pump and the engine.

Fig. 85b shows a family of characteristic curves for the converter pump where each of these curves corresponds to a constant turbine speed n_T and thus to a constant carrier speed n_s and a particular vehicle speed.

The solid lines in fig. 85c represent an assumed performance map of the engine where each of the characteristic torque curves corresponds to a constant throttle position.

Since, according to eq. (42), the torques of the engine and the pump bear a constant ratio relative to each other, the characteristic curves of the pump can be mapped, point by point, onto the performance map of the motor as follows: with the equation $n_s = n_T i_{sT}$, the constant speeds n_T which are the parameters of the characteristic curves of the pump, fig. 85, are expressed in terms of the carrier speed n_s. Each point P of these curves then represents a unique pump and carrier speed whose corresponding engine speed can be found with eq. (23):

$$n_1 = i_o n_2 + (1 - i_o) n_s \, .$$

The input torque $T_P = T_2$ of the pump at this operating point (n_1, n_2, n_s) can be read, over n_2, on the characteristic curve of the pump as shown in fig. 85b. A corresponding torque T_1, can then be calculated with eq. (42) which is written in the form

$$T_1' = -\frac{T_2}{i_o \eta_o^{r1}} \, .$$

Since the motor shaft is the sole input shaft and both shaft A and shaft 2 can only be output shafts, the power flow in this negative-ratio transmission clearly is $1 < \frac{1}{2}$, so that, according to worksheet 4, $r1 = +1$.

After mapping a series of points P of the characteristic curve of the pump into the corresponding points P' of the motor map, the dashed characteristic curves of the pump for the parameter n_T can be drawn so that the input torque T_1', which at a given operating point (n_2, n_s) must be transmitted to the pump, can be readily obtained. Stable operating conditions are possible only when, at a given turbine or vehicle speed, the torque T_1' required to drive the pump is equal to the motor torque T_1. On the motor map, this condition exists only at the points of intersection between a dashed curve, which represents a characteristic curve of the pump at a given constant turbine or vehicle speed, and a solid line which represents a constant position of the throttle. If T_1 and T_1' are not equal, (e.g., if $T_1 > T_1'$ as shown on line y in fig. 85c), then the speeds (n_1, n_2), increase until a stable operating point is reached which, for this example, lies at the point of intersection between the characteristic curves of pump and motor on line z.

As previously mentioned, together with n_1 and $n_T \sim n_s$, each point P' of the motor map also determines the speed $n_2 = n_P$ of the third shaft of the planetary transmission. Therefore, dash-dotted lines can be added on the map to represent the characteristic curves of the pump for the parameter n_P. With the aid of the converter map, and from the speeds and torques at the points of stable operation, the vehicle performance diagram could be constructed. However, the details of this procedure are beyond the scope of this book and the interested reader is referred to [8].

In the transmission of fig. 85, the pump impeller P is stopped by a brake

C when the vehicle travels at high velocities where the turbine torque approaches zero. Then the stationary turbine is disconnected from the output shaft by an overrunning (sprag) clutch D, so that the total engine power is transmitted to the output shaft without slip along the mechanical power path A.

γ) *Common drive for two centrifugal pumps.* The same two centrifugal pumps, introduced in example γ, section 24d, may again be driven simultaneously by the previously described motor, with the difference that a revolving drive is used now to divide the motor power between the two pumps. Fig. 86a shows a schematic of the drive system, fig. 86c the operating characteristic curve of the two identical pumps, and fig. 86d the characteristic curve of the motor. The parameters of the drive are given by: $i_0 = -0.2$, $i_{43} = -0.2$, $\eta_0 = 0.97$, $\eta_{43} = 0.98$. The two step-up gear trains are designed in such a way that the same constant speed-ratios are obtained as in the previous example, so that, according to worksheet 2, box 13, the speed-ratio

$$k_{MA} = k_{1s}i_{43} = \frac{1 - i_0}{1 - i_0/k_{12}} \cdot i_{43} = \frac{1 + 0.2}{1 - 0.2/0.5}(-0.2) = -0.4$$

when

$$k_{MB} = k_{12} = -0.5 \ .$$

As in the previous example, the common operating characteristic curve of the two pumps is transferred to the engine map in order to find the operating point, which is its point of intersection with the characteristic curve of the engine. The revolving drive has only one static degree of freedom. All torques, therefore, depend on each other, so that according to eqs. (44), (6), and (42), respectively:

$$T_3 = T_1 \frac{T_s}{T_1} \cdot \frac{T_3}{(-T_4)} = T_1(i_0\eta_0^{r1} - 1)(\eta_{43}i_{43}) \ ,$$

$$T_2 = -T_1 i_0 \eta_0^{r1} \ ,$$

where the exponent $r1 = +1$ can be determined from worksheet 4 for negative-ratio transmissions with the unequivocally given power-flow $1 <^s_:$. The output torques of the transmission are:

$$T_2 = -T_B = 0.194 T_1' \ , \quad \text{and} \quad T_3 = -T_A = 0.234 T_1' \ .$$

With these equations, the pump torques T_A and T_B can be calculated for a number of chosen values of the summation torque T_1'. The corresponding pump speeds $n_A = n_3$ and $n_B = n_2$ can then be found from the characteristic

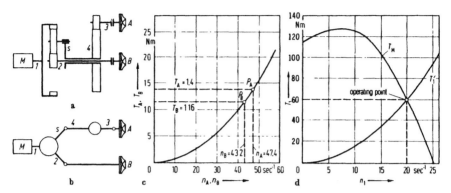

Fig. 86. Common drive for two centrifugal pumps containing a revolving drive train: *a*, schematic; *b*, symbolic representation of the system; *c*, characteristic curve of the two identical pumps; *d*, characteristic curve T_M of the motor and the common characteristic curve T_1' of the two pumps as functions of the motor speed n_1.

curve of the pumps which is given in fig. 86c. Under consideration of the intermediate speed ratio i_{43}, the corresponding motor speed n_1 can be calculated when the two pump speeds are introduced into eq. (23), so that

$$n_1 = i_o n_2 + (1 - i_o) n_3 i_{43} = -(0.2n_2 + 0.24n_3) \ .$$

The motor speed n_1 becomes negative when the direction of rotation of the pumps is assumed to be positive or vice versa, that is, the motor and the pumps rotate with an opposite sense of direction. If the corresponding values T_1' and n_1 are plotted on the motor map, fig. 86d, then the common operating characteristic curve T_1' of the two pumps is obtained, whose point of intersection with the characteristic curve T_M of the motor represents the stable operating point of the system: $T_M = T_1' = 60\,Nm$ and $n_1 = 20\,\sec^{-1}$. With T_1, the pump torques $-T_A = T_3 = 14\,Nm$ and $-T_B = T_2 = 11.4$ Nm, can be calculated, which determine the operating points P_A and P_B of the pumps. The corresponding speeds $n_A = -47.4\,\sec^{-1}$ and $n_B = -43.2$ $\sec^{-1}$ can be simply read from the abscissa of fig. 86c. A comparison between these operating points and the operating points of fig. 83c shows that the pump speeds at which the torques are in equilibrium are different when the transmission has two kinematic degrees of freedom, rather than one as shown in fig. 83.

f) Examples of Revolving Drive Trains with One Static Degree of Freedom and Power Summation

Two motors can simultaneously drive a machine through a three-shaft transmission, if, at any stable operating point of the system, the motors can adjust their output torques to the exact ratio required by the transmission,

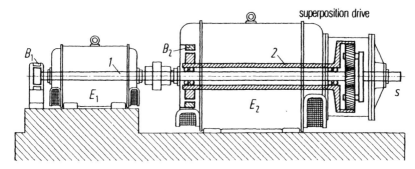

Fig. 87. Superposition drive for the summation of the powers of two electric motors E_1 and E_2 (the output speed can be varied by selective operation of the motors as shown in table 4); B_1, B_2, brakes (according to Voith, Heidenheim).

and thus adhere to R 12. However, because $F_{kim} = 2$, the speeds of the two motors are independent of each other and can assume the values required by their operating characteristics. If one of the motors is stopped, a torque from the other motor or the driven machine still acts on its output shaft and must be counteracted by a brake or stop. In this respect, the drive differs fundamentally from the summation drive with one kinematic degree of freedom as described in the previous example in section 24dα, fig. 81a.

Another condition for the operation of a summation drive is that its input torques act in accord with motor rotation, while the output torques of the motors oppose it (R 3, R 4). This point may appear to be self-evident, but is, nevertheless, useful for work with motors whose direction of rotation can be reversed, such as, for example, electric or hydraulic motors. When the direction of rotation of one of the connected motors is reversed, according to R 12 its torque remains unchanged, and thus its power $T\omega$ becomes positive, that is, it acts as a brake and absorbs power from the system. The reversal of one direction of rotation, therefore, changes the operation from power summation to power division, or vice versa (see R 18).

α) *Selective summation of the output powers of two electric motors.* As an example, fig. 87 shows a negative-ratio transmission as a superposition drive connected to two electric motors. By operating only one, or the other, or both motors with a common or opposite direction of rotation, a maximum of four different output speeds in either direction can be obtained at the output shaft s. If one of the two motors is switched off, the torque which acts on the stopped shaft must be counteracted by brakes B_1 or B_2, respectively. For a basic speed-ratio $i_0 = -4$ and the motor speeds n_1 and n_2, the four possible stable types of operation are summarized in table 4. The output speeds are calculated from eq. (25) and the types of operation

TABLE 4. POSSIBLE TYPES OF OPERATION OF THE SUPERPOSITION DRIVE
SHOWN IN FIG. 87

$n_1 \sec^{-1}$	$n_2 \sec^{-1}$	$n_s \sec^{-1}$	POWER-FLOW	TYPE OF OPERATION
+25	0	5	$1 \to s$	SR
−25	+25	15	$2 < {}^s_1$	PD
0	+25	20	$2 \to s$	SR
+25	+25	25	${}^1_2 > s$	PS

are power division *PD*, power summation *PS*, and simple speed reduction
SR. Each of the two motors drive one difference shaft of the superposition
drive and, therefore, their torques have the same signs regardless of their
directions of rotation (R 9). Therefore, the transmission operates with
power summation and shafts *1* and *2* are input shafts only when the directions of rotation of both motors are equal to the directions of the gear
torques, that is, also equal to each other (see line 4 in table 4). If the poles of
motor E_1 are reversed, only the sign of its speed changes. Its power, consequently, becomes negative, that is, it works as a brake. Motor E_2 now is the
only drive motor and its output power is split between the shafts *1* and *s* as
indicated by line 2 in table 4. Transmissions of this type are suited, for
example, for elevator drives, where only the two discrete operating conditions described by lines 1 and 4 are used for normal hoisting, and the slow
approach before stopping, respectively.

β) *Synchronizing drive for gantry crane.* A similar drive system is used
to synchronize the independent drive motor of gantry cranes. While one of
the two trucks is powered by a single electric motor, the other is driven by a
main motor and a synchronizing motor (figs. 88 and 89). Acting through a

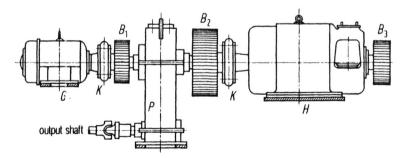

Fig. 88. Synchronizing drive for gantry cranes with individually powered trucks. *H*,
main motor; *G*, synchronizing motor; B_1, brake to counter the torque when the synchronizing motor is switched off; *P*, planetary transmission operating as superposition drive; B_2, B_3, main brakes; *K*, flexible coupling (according to Demag).

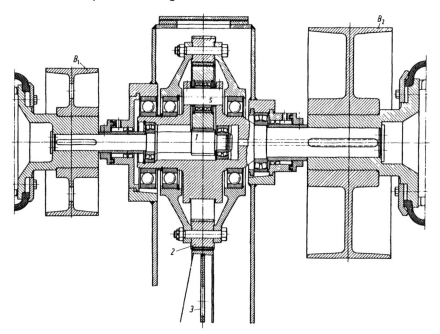

Fig. 89. Axial section of the superposition drive shown in fig. 88: *s*, carrier, driven by the main motor; *1*, central gear, driven by the synchronizing motor; *2*, annular gear whose additional external gearing meshes with the output gear, *3*.

planetary transmission, the latter can superimpose an additional speed component onto the basic speed of the main drive motor. During normal operation, however, the truck is driven only by the main motor. The trucks on either side of the crane bridge are always operated simultaneously and are designed to reach and maintain identical speeds. If, however, due to an uneven load distribution, the two trucks assume different speeds and the bridge is in danger of cocking in its tracks, then the synchronizing motor is switched on for a short time. By superposition, this slightly increases or decreases the speed of the truck and thus restores the synchronization between the two legs of the bridge. If the superposition drive is properly designed, the controlling synchronizing motor can be much smaller than the main motor. As in the previous example, power summation occurs only when the auxiliary motor increases the travel speed. If its direction of rotation is reversed to lower the travel velocity, it acts as a brake. The superposition drive then operates with power division. When the synchronizing motor is not operating, the reverse torque acting on its shaft must be counteracted by a brake.

Section 25. Three-Shaft Transmissions as Sensors for Torque Control and Overload Protection

If a three-shaft transmission is used as a simple reduction drive with a single input and output shaft, then, according to eq. (39) or (46), the torque of the third shaft must be counteracted by a suitable device. Since, according to eqs. (42), (44), or (45), this third torque bears a known ratio relative to each of the other two torques, it does at the same time indicate their value. Thus, such a transmission can simply and continuously monitor the torque load of the transmission and the coupled machinery through an external stationary measuring device. For this purpose it is only necessary to impose the supporting torque of the non-rotating third shaft on a force or torque transducer, through which it can be read-out, registered, or used as input to a torque control system.

Since eqs. (42), (44), and (45) contain the basic efficiency η_0^{r1}, the accuracy of the described method is limited by the accuracy with which η_0 is known for each operating condition. Moreover, the torque ratio which is obtained from each of the three equations depends on $r1$, that is, on the direction of the power-flow in the revolving drive. Thus, it may change whenever the direction of the power-flow changes during operation.

a) Torque Control

Fig. 90 schematically shows a planetary transmission as used in a torque control device. The input shaft 1 is driven in the direction indicated by the arrow; the carrier shaft is the output shaft and turns in the same direction as the input shaft 1. The output torque T_s, however, acts on the carrier shaft in

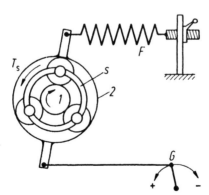

Fig. 90. Schematic of a speed-reducing revolving gear train with lever system for torque control: 1, input; s, output; 2, annular gear supported by the central spring F and anchor screw for the adjustment of the spring tension and thus the controlled torque; G, accelerator linkage of the engine.

the opposite direction, that is, counterclockwise. The annular gear 2 is supported in bearings, but is prevented from rotating by an adjustable spring and, thus, is restricted to limited oscillating motions. By a system of levers, this oscillation is transmitted to the accelerator of the engine which increases the engine torque T_1 when it turns in the positive direction, and vice versa. During operation, an equilibrium condition is reached between the spring tension, which corresponds to the torque T_2, the motor torque T_1, and the output torque T_s, which is thus determined. The output speed depends on the operating characteristic curves of the drive motor and the driven machinery whose torque should increase with increasing speed for the reasons discussed in section 24.

If the spring is moved to the right by turning the adjustable anchor screw, the accelerator moves to increase the flow of fuel to the engine, while the spring tension momentarily remains constant. Consequently, the engine torque and the torque T_2 increase simultaneously with the output torque. This, in turn, increases the spring tension and thus reduces the flow of fuel until a new equilibrium condition at a higher torque level is reached.

b) Overload Protection

To protect a drive system from excessive torque loads, a three-shaft gear train, which operates as a reduction gear, can be used. In this case the third shaft is held by a brake which is set for the maximum allowable system torque. Upon reaching the set torque limit, the third shaft breaks away and, thus, limits a further increase of all three shaft torques. According to eqs. (42), (44), and (45), the maximum input and output torques depend on the break-away torque of the brake.

Protection against shock loads can be obtained when the third shaft is held by a spring and damper system that can deflect and absorb the shock energy. In this case the third shaft goes through a damped oscillating motion whose peak amplitude depends on the compliance of the spring.

Fig. 91 shows a cross section of a transmission, which is simultaneously protected against shock loads by a series of compression springs, and against torque overloads by a hydraulic brake. Both the spring-loaded shock protection and the brake act on the carrier. The transmission, therefore, operates normally as a basic gear train. The break-away torque can be adjusted by altering the hydraulic pressure in the brake system. This transmission is used in the drive system of a generator which feeds the network of an electric railway system and experiences extremely high peak loads when the large traction motors start up and accelerate. Small shock loads are dissipated by the springs, large ones by a momentary slipping of the safety brake. The advantage offered by the planetary gear train essentially lies in the fact that the spring and brake system is stationary. For extreme loads,

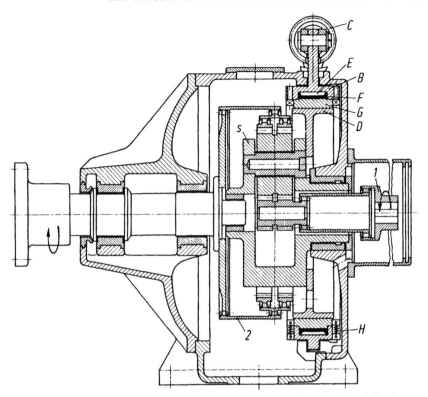

Fig. 91. Planetary transmission with herringbone gears and overload and shock protection as used in the drive system of a generator feeding the network of an electric railway system (according to BHS, Hüttenwerk Sonthhofen): *1*, input; *2*, output; *s*, carrier, connected to the hydraulic brake *B*, which is supported in the housing by several compression springs distributed around the circumference; *D*, brake drum; *E*, pressure side of hydraulic brake; *F*, seal; *G*, brake shoes; *H*, return springs.

such a device is easier to design and maintain than a slip clutch with rotating parts.

c) Four-Square Test Stand

In a similar arrangement, a three-shaft transmission can be used as a loading device for a four-square test stand. These test stands are used to test gear trains at or above their rated loads. Four-square test stands have a closed power-flow, that is, the output power of the test transmission is fed back to its input through a closed system of shafts and gear trains. Under stationary test conditions, therefore, only the relatively small power losses of the test transmission, and the other moving parts of the test stand, have to be made up continuously by the drive motor.

In fig. 92, a four-square test stand is shown schematically. If the gear ratio i_{56} of the intermediate gear train C is matched with the gear ratio i_{34} of the test transmission A, so that $i_{34}i_{56} = i_{2s}$ of the revolving transmission B, then the sun gear I does not rotate, since i_{2s} is the two-shaft speed-ratio of shafts 2 and s. As long as shaft I carries no load, the torques of shafts s and 2, as well as those of all other coupled shafts, must be zero. If, however, the motionless arm L on the sun gear shaft is loaded, a torque is imposed on shafts s and 2 which flows from the output shaft 2 through the gear trains A and C back to the input shaft s. For any chosen load torque T_1, the resulting shaft torques T_2 and T_s can be calculated from eqs. (42) and (44), while the torque loads carried by gear trains A and B can be obtained from eq. (6).

With the assumptions made for the example shown in fig. 92, the required motor torque T_M is obtained as a function of the load torque T_1. The torque T_6 is the output torque of gear train C, and T_s the input torque of gear train B, and thus $T_s = -T_6$. Provided the power flows through the system from B to A to C to B, the equilibrium conditions for shaft 6 require that:

$$T_M + T_6 + i_{56}\eta_{56}T_5 = 0$$

and thus, because $T_6 = -T_s$,

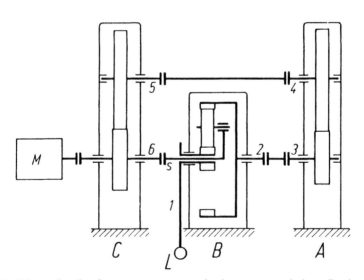

Fig. 92. Schematic of a four-square test stand: A, test transmission; B, planetary transmission with load lever L connected to the sun gear I; C, intermediate gear train for the recirculation of the power-flow; M, drive motor covering the power-losses. Assumed data: $i_o = -4$, $i_{56} = -5/8$, $i_{34} = -2$, $\eta_o = 0.98$, $\eta_A = \eta_C = 0.985$.

$$\frac{T_M}{T_1} = \frac{T_s}{T_1} - i_{56}\eta_{56} \cdot \frac{T_5}{T_1} \; .$$

From eqs. (44) and (7) and with

$$\frac{T_5}{T_1} = \frac{T_5}{T_4} \cdot \frac{T_4}{T_3} \cdot \frac{T_3}{T_2} \cdot \frac{T_2}{T_1} = (-1)(-i_{34}\eta_{34})(-1)(-i_o\eta_o^{rl})$$

we obtain

$$\frac{T_M}{T_1} = (i_o\eta_o^{rl} - 1) - i_{56}\eta_{56}(-i_{34}\eta_{34})(-i_o\eta_o^{rl}) \; .$$

For the power-flow $s2$ of the negative-ratio transmission B we find with worksheet 5, box 16, that $w1 = +1$ and thus:

$$\frac{T_M}{T_1} = -4 \cdot 0.98 - 1 + \frac{5}{8} \cdot 0.985 \cdot 2 \cdot 0.985 \cdot 4 \cdot 0.98 = -0.1659 \; .$$

If the direction of rotation of the test stand, or the direction of the torque acting on the arm L, is reversed, the power-flow in the system and in each gear train is also reversed. Gear trains A and C are loss-symmetrical, that is, they have the same losses in either direction of rotation. This, however, is not true for the revolving drive train B. Therefore, its efficiency and, consequently, the theoretically required motor output power change slightly when the direction of rotation is reversed. This problem does not arise when the transmission B has a positive speed-ratio and, therefore, operates in its basic mode where it is also loss-symmetrical. The test load, of course, must then be imposed on the carrier shaft.

Section 26. Three-Shaft Transmissions as Phase Adjusters

Drive systems, as shown in fig. 93, can be employed to vary the relative angular positions of two or more shafts rotating at the same speed. Therefore, they are used to change the position of the folding knife in rotary printing presses which are adjustable for various formats, and also to match the print rollers in multicolor printing presses. If the total speed-ratio between rolls A and B is $i_{AB} = 1$, while the arm C remains fixed, then the rolls rotate synchronously. If now the arm C is moved by an angle φ_C, roll B can be adjusted to lead or lag roll A by an angle φ_B during full operation. This process is understood more readily if it is assumed that the drive and, thus, roll A, does not rotate while this adjustment is made.

For a system as shown in fig. 93a which incorporates a negative-ratio transmission, worksheet 1 shows that:

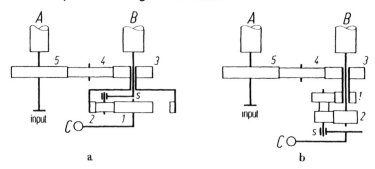

Fig. 93. Three-shaft transmission which permits shifting the relative phase of the rollers A and B, which operate at the same speed, by turning the lever C: a, negative-ratio transmission where the phase shift is effected by turning the sun gear; b, positive transmission where the phase shift is effected by turning the carrier.

Accordingly we have:

$$\frac{\varphi_B}{\varphi_C} = i_{BC} = i_{s1} = \frac{1}{1 - i_o} \, .$$

$$\frac{\varphi_B}{\varphi_C} = i_{BC} = i_{2s} = \frac{i_o - 1}{i_o}$$

for the drive shown in fig. 93b. The latter system is better suited for high-speed applications since the planet carrier is at rest and the planet bearings are not subject to the additional loads caused by the centrifugal forces.

As a further example, fig. 94 shows the principle of a propeller pitch-change mechanism. Acting through a planetary transmission and a worm gear C, its stationary electric motor M can turn the propeller blades E around their longitudinal axes during full flight. Through its sun gear I, the propeller shaft D drives the fixed-carier gear train A directly. Its gear 2, in turn, drives gear $2'$ of a second, completely identical positive-ratio gear train, whose carrier is stationary and whose sun gear I', therefore, rotates in phase with gear I and, thus, the propeller shaft. Gears 3 and 4 also rotate synchronously with the propeller shaft and, consequently, the worm gear shaft is stationary relative to the propeller. If, now, the torque motor M turns the carrier s of gear train A by an angle φ_s, then gears I' and 3 must turn through an angle φ_3 relative to the propeller shaft, where

$$\frac{\varphi_3}{\varphi_s} = i_{3s} = i_o' i_{2s} = i_o' \frac{i_o - 1}{i_o} = i_o - 1 \, .$$

Consequently, shaft 4 of the worm gear also rotates relative to the propeller and turns the propeller blade around its longitudinal axis, as pointed out

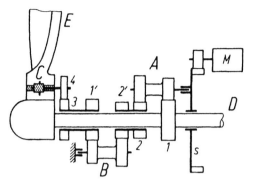

Fig. 94. Adjustable pitch propeller. The adjustment operates according to the phase shift principle and employs a three-shaft transmission: A and B, identical planetary gear trains whose i_0 is approximately $+1$; C, worm gear for the adjustment of an individual propeller blade E; D, propeller shaft; M, electrical servomotor to change the angular position of the carrier s (according to Vereinigte Deutsche Metallwerke AG [VDM].

earlier. In one practical design, the basic speed-ratio $i_0 = i_0'$ was chosen so close to $+1$ that the two gears of the double planets had the same number of teeth and, therefore, could be combined into a single wide-face gear while the central gears 1 and 2 of gear train A, and $1'$ and $2'$ of gear train B, with their slightly different numbers of teeth, were individually corrected to obtain equal center distances. As can be verified with the equation given above, this results in a very low speed-ratio i_{3s} which in combination with the worm drive allows a very precise propeller pitch adjustment.

Section 27. Three-Shaft Transmissions as Computing Mechanisms

a) Sum or Difference of Two-Shaft Angles

For recording or control purposes, it is sometimes necessary to form the sum or difference between two shaft speeds or shaft angles. This can be accomplished with revolving gear trains as can be seen from eqs. (23) to (25). However, one or more simple ratio gear trains must be connected to the input or output shafts of the revolving gear train in order to match the input speeds to the basic speed-ratio i_0. Basically, each of the three external shafts of the revolving gear train is suitable as the output shaft. It may be observed that in a three-shaft transmission the shaft whose speed equals the sum or difference of the *speeds* of the other two shafts is independent of the summation shaft (sec. 12), which transmits the *torques* of the other two shafts. Fig. 95 shows an example of a system of gear trains which can either add or subtract.

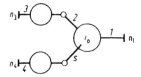

Fig. 95. Possible transmission arrangement for the formation of the sum or difference between two shaft angles or speeds according to eq. (23).

Since according to eq. (23)

$$n_1 = i_0 n_2 + (1 - i_0)n_s = i_0 i_{23} n_3 + (1 - i_0)i_{s4}n_4 \; ,$$

we obtain:

$$n_1 = n_3 + n_4, \quad \text{when} \quad i_{23} = \frac{1}{i_0} \quad \text{and} \quad i_{s4} = \frac{1}{1 - i_0} \; ,$$

$$n_1 = n_3 - n_4, \quad \text{when} \quad i_{23} = \frac{1}{i_0} \quad \text{and} \quad i_{s4} = -\frac{1}{1 - i_0} \; ,$$

$$n_1 = n_4 - n_3, \quad \text{when} \quad i_{23} = -\frac{1}{i_0} \quad \text{and} \quad i_{s4} = \frac{1}{1 - i_0} \; .$$

Fig. 96a shows another arrangement which yields the same result. If eq. (23) is divided by i_0, we obtain:

$$\frac{n_1}{i_0} = i_{15} \cdot \frac{n_5}{i_0} = n_2 + \frac{1 - i_0}{i_0} \cdot n_s = n_2 + \frac{1 - i_0}{i_0} \cdot i_{s4}n_4 \; .$$

Thus:

$$n_5 = n_2 + n_4 \quad \text{when} \quad i_{15} = i_0 \quad \text{and} \quad i_{s4} = \frac{i_0}{1 - i_0} \; ,$$

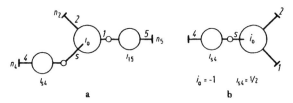

Fig. 96. Further possible transmission arrangement for the formation of the sum or difference between two shaft angles or speeds according to eq. (23): *a*, for an arbitrary i_0; *b*, for $i_0 = -1$.

$$n_5 = n_2 - n_4 \quad \text{when} \quad i_{15} = i_0 \quad \text{and} \quad i_{s4} = -\frac{i_0}{1 - i_0} \, ,$$

$$n_5 = n_4 - n_2 \quad \text{when} \quad i_{15} = -i_0 \quad \text{and} \quad i_{s4} = -\frac{i_0}{1 - i_0} \, .$$

If the difference in the last line is formed by a gear train with $i_0 = -1$, then $i_{15} = +1$ as in a coupling and $i_{s4} = 0.5$. In this case, the auxiliary gear train between shafts 1 and 5 no longer serves any useful purpose and can be omitted as shown in fig. 96b. The following sum or differences can then be formed by this arrangement as can be easily verified by simply transposing the first of these equations:

$$n_5 = n_4 - n_2 \, ; \qquad n_4 = n_2 + n_5 \, ; \qquad n_2 = n_4 - n_5 \, .$$

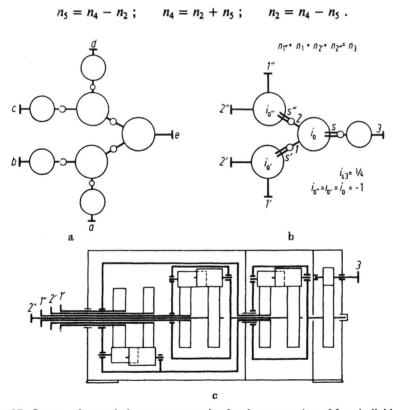

Fig. 97. System of several planetary gear trains for the summation of four individual shaft speeds or angles; $n_e = n_a + n_b + n_c + n_d$. The overall drive system has four degrees of freedom: a, system for arbitrary basic speed-ratios i_0 of the three planetary gear trains; b, $i_0 = -1$ for each planetary gear train, $i_{s3} = \frac{1}{4}$; c, schematic of a summation drive according to b.

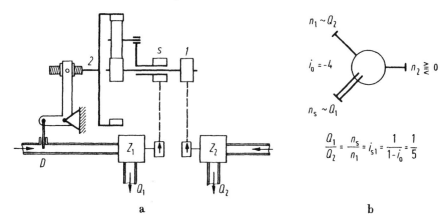

Fig. 98. Control system which maintains a constant mixing ratio $Q_1 : Q_2$ between two liquids; Z_1, Z_2, flow meters; D, control valve: a, schematic of the drive system; b, symbolic representation of the drive system with typical operating parameters.

Further gear systems of this type can be based on the analogous application of eqs. (24) and (25).

If, in a measuring device, an input speed must be weighted before it is summed, then it is necessary only to connect the corresponding input shaft to an auxiliary gear train with an appropriate speed-ratio. For instance, to add the masses of two liquid volumes whose densities differ by a factor of *1.2,* the corresponding input shaft must be connected to an auxiliary gear train which lowers its speed-ratio by a factor of *1/1.2.* Thus, with a gear system as shown in fig. 95, the desired output speed $n_1 = n_3 + 1.2n_4$ can be obtained when the auxiliary gear train *4* has a speed-ratio $i_{s4} = 1.2/(1 - i_0)$.

Frequently changing requirements of this type can be easily met when auxiliary transmissions with infinitely variable speed-ratios are used.

When more than two input speeds must be summed, several planetary gear trains can be connected as shown in fig. 97. Each of these contributes the partial sum which it forms from its two inputs. In general, all but one of these input shafts must be connected to an auxiliary gear train. If, however, as shown in fig. 97b, the summation drive consists of individual gear trains with basic speed-ratios equal to -1 as shown in fig. 96b, then only a single auxiliary gear train is needed.

b) Maintenance of a Constant Speed-Ratio

Three-shaft transmissions can also be used in control systems where a constant speed-*ratio* must be maintained. If, for example, the basic speed-ratio, or one of the other speed-ratios of a three-shaft transmission, equals the controlled speed-ratio, then the third shaft remains at rest as long as the other two shafts rotate with precisely the controlled speed-ratio. However, when the speed-ratio of these two shafts deviates from the desired value, then the third shaft starts to rotate in one or the other direction. Therefore, a control device can be connected to the third shaft, which, for example, increases or decreases the speed of the first shaft accordingly, so that the desired constant speed-ratio is maintained even when the speed of the second shaft fluctuates.

As an example, fig. 98 schematically shows a control system which maintains the flow rate of two liquids Q_1 and Q_2 at a constant ratio of *1:5*, independent of the total rate of flow. The meters Z_1 and Z_2 provide the input speeds n_s and n_1, or alternately, the angles φ_s and φ_1 for a planetary gear train with a speed-ratio $i_{s1} = 0.2$. If the ratio between the two input speeds deviates from the set point, then shaft *2* corrects the rate of flow Q_1 with the aid of a control valve D which is operated by an adjustable screw and bell crank mechanism.

Section 28. Three-Shaft Transmissions for Non-Uniform Rotary Motions: Pilgrim-Step Drives

Non-uniform rotary motions with periodically varying angular velocities can be generated with three-shaft transmissions when an oscillating motion is superimposed on a uniform rotation. As an example, fig. 99a shows a pilgrim-step drive used in spinning mills which is based on a transmission of the type shown in fig. 19. The central gear *1* is driven with a uniform speed through a gear *3*, while the other central gear *2* receives an oscillating motion from the rocker arm *S* of a cam drive, which is not shown in detail. The resulting motion of the carrier *s* is a non-uniform rotation whose momentary angular velocity ω_s can be calculated with eq. (25) when the speeds n are replaced by the angular velocities ω. The resulting expression can then be simplified with the help of worksheet 1, so that finally

$$\omega_s = \frac{i_o\omega_2 - \omega_1}{i_o - 1} = \frac{i_o\omega_2}{i_o - 1} + \frac{\omega_1}{1 - i_o} = \omega_2 i_{s2} + \omega_1 i_{s1} \ .$$

Thus, the angular velocity ω_s of the output shaft is obtained by the simple superposition of a uniform output speed $\omega_1 i_{s1}$ and a periodic output speed $\omega_2 i_{s2}$. Both speed components appear as if they were generated independently of each other, since the uniform speed component $\omega_1 i_{s1}$ is obtained when the input shaft is driven with a constant angular velocity ω_1 and the shaft *2* is locked; while the non-uniform speed component $\omega_2 i_{s2}$ results from the oscillating input speed ω_2 when the shaft *1* is locked. Fig. 99b shows an arbitrary example of the resulting motions and their super-

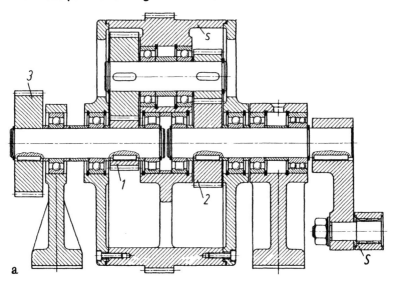

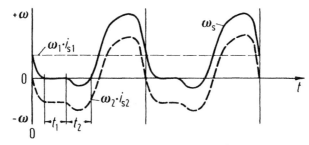

Fig. 99. Three-shaft transmission for the generation of a non-uniform output speed: *a*, schematic of the drive for a textile (combing) machine (according to Deutscher Spinnereimaschinenbau, Ingolstadt), with *3* and *1*, uniform input, *2* and *S*, oscillating input from the follower of a cam drive, and *s*, nonuniform output; *b*, resulting angular velocity ω_s at the output, due to the superimposed uniform input velocity ω_1 and an arbitrarily assumed oscillating input velocity ω_2.

position. It has been assumed that $i_o \omega_2 = \omega_1$ during a time period t_1 and $i_o \omega_2 < \omega_1$ during a period t_2 so that the output motion is stopped and then reversed (pilgrim step) for the time period t_2. These periods are compensated by an increased angular velocity during the remainder of the cycle, since, overall, the carrier has the average forward velocity $\omega_{s,AV} = \omega_1 i_{s1}$. At this point, it may be emphasized that for actual transmissions the assumption $\omega_1 = $ const. is never true, since all torques fluctuate due to the alternating acceleration and deceleration of the rotating masses. Thus, yet another non-uniformity is superimposed on the rotary motions shown in fig. 99b. Depending on the speed-ratio and the distribution of the mass moments of intertia, this non-uniformity influences the motions of the three shafts of the revolving drive to a larger or lesser degree.

Figs. 100a and 100b schematically show a pilgrim-step drive where the periodical change of the speed-ratio is accomplished through the use of a crank slot K. The revolving drive train is a negative-ratio transmission of the type described in fig. 34. Together with the annular gear 2, the crank slot is driven through gears 3 and 4 which have a speed-ratio $i_{34} = -1$. It imparts an oscillating motion with a fast and a slow half cycle on the carrier s. The transmission is constrained and has only one degree of freedom since the carrier is coupled to the input shaft 2 through the mating parts of the crank slot and the gears 3 and 4. Consequently, for each angular position φ_2 of the input shaft 2, a momentary speed-ratio ω_1/ω_2 can be determined which goes through one full cycle for each complete rotation of the crank and, because $i_{34} = -1$, for each complete rotation of the annular gear 2.

The periodical speed-ratio of this step drive can be derived as follows: starting from its median position as shown in fig. 100a, the angular position φ_s of the carrier s can be determined as a function of the angular position φ_4 of gear 4 and the length r of the crank. As shown in fig. 100c:

$$\sin \varphi_s = \frac{r \sin \varphi_4}{r_s} \quad \text{and} \quad \cos \varphi_s = \frac{1 - r \cos \varphi_4}{r_s}$$

where $r_s^2 = r^2 + l^2 - 2rl \cos \varphi_4$ (according to the law of cosines for the oblique triangle).

According to fig. 100d, the momentary tangential velocity at the radius r_s of the carrier is:

$$\varphi_s = \varphi_4 \cos (\varphi_4 + \varphi_s) \ ,$$

so that the corresponding momentary angular velocity becomes

$$\omega_s = -\frac{\omega_4 r}{r_s} \cos (\varphi_4 + \varphi_s) \ .$$

With $\cos (\varphi_4 + \varphi_s) = \cos \varphi_4 \cos \varphi_s - \sin \varphi_4 \sin \varphi_s$ and $\lambda = 1/r$, this leads to:

$$\frac{\omega_s}{\omega_2} = \frac{\lambda \cos \varphi_2 - 1}{1 + \lambda^2 - 2\lambda \cos \varphi_2} \ ,$$

where $\omega_4 = -\omega_2$ and $\varphi_2 = \varphi_2$ because $i_{24} = -1$.

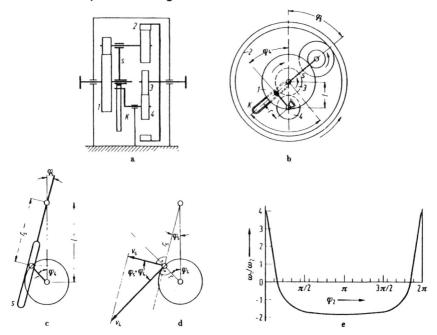

Fig. 100. Constrained pilgrim-step drive: a-b, schematic of the drive with crank slot K; c, schematic of the crank slot; d, determination of the velocities at the crank slot; e, momentary velocity ratio ω_1/ω_2 as a function of the angular position φ_2 of the gear 2.

Finally, with the aid of worksheet 2, the momentary speed-ratio of the overall transmission can be expressed in the form:

$$k_{12} = \frac{\omega_1}{\omega_2} = i_o + \frac{\omega_s}{\omega_2}(1 - i_o) = i_o + \frac{\lambda \cos \varphi_2 - 1}{1 + \lambda^2 - 2\lambda \cos \varphi_2}(1 - i_o) \ .$$

For the transmission of figs. 100a and 100b, which has a basic speed-ratio $i_o = -0.97$ and a crank-ratio $\lambda = 1.4$, this leads to

$$\frac{\omega_1}{\omega_2} = -0.97 + \frac{1.4 \cos \varphi_2 - 1}{2.96 - 2.8 \cos \varphi_2} \cdot 1.97 \ .$$

In fig. 100e this overall speed-ratio ω_1/ω_2 is plotted over the angular position φ_2 of the input shaft. For $\omega_1/\omega_2 = 0$, $\varphi_2 = \pm 28°$. At these points where the output velocity $\omega_1 = 0$, the output shaft I changes its direction or rotation (pilgrim-step). Consequently, this transmission cannot be driven through shaft I. Transmissions of this type are used to accelerate and decelerate the chain drive of the sheet-delivery mechanism in printing presses.

Section 29. Simple Three-Shaft Transmissions as Change-Gears

Simple revolving drive trains can be used as change-gears when they are equipped with clutches and brakes so that their shafts can be alternately connected to the input or the output, or can be locked by a brake. Since input and output can be interchanged, a revolving drive train can have a maximum of six, that is, three pairs of mutually reciprocal speed-ratios which for a known basic speed-ratio i_0 can be found directly from figs. 45 and 61. Because of the invariable relationship between the speed-ratios and the disproportionate expense for added shift mechanisms, clutches and brakes, such a six-speed transmission is without practical importance.

Economical designs, however, are possible for two forward speeds, or one forward and one reverse speed, when one of the three external shafts is rigidly connected, while the second is held by a brake, and the third is connected through an engaging clutch, or vice versa, as shown in fig. 101a. Thus, as can be seen from table 5, either two different positive speed-ratios or one positive and one negative speed-ratio can be realized, depending on whether the summation shaft or a difference shaft is the rigidly connected member.

If both clutches are engaged at the same time, an additional speed-ratio $i = +1$ is obtained. The other possible speed-ratios as given in table 5 can be verified with the help of fig. 101a. As an example, figs. 101b and 101c show two possible design solutions.

Transmissions with basic speed-ratios between 0 and 1 ($0 < i_0 < 1$) are identical to those with $i_0 > 1$. Only the symbols (1, 2) for the two central

TABLE 5. Possible Speed-Ratios of a Three-Shaft Change-Gear According to Fig. 101

TYPE OF SPEED-RATIOS	ROW	i_0	RIGIDLY-CONNECTED SHAFT	POSSIBLE SPEED-RATIOS		η
				i_{AB}	i_{BA}	
			a	i_{ab}, i_{ac}	i_{ba}, i_{ca}	
Two positive speed-ratios for a rigidly-connected summation shaft	1	<0	s	i_{s1}, i_{s2}	i_{1s}, i_{2s}	$> \eta_0$
	2	>1	2	i_{21}, i_{2s}	i_{12}, i_{s2}	$\gtrless \eta_0$
One positive and one negative speed-ratio (reversing gear) for a rigidly-connected difference shaft	3	<0	1	i_{12}, i_{1s}	i_{21}, i_{s1}	$> \eta_0$
	4	<0	2	i_{21}, i_{2s}	i_{12}, i_{s2}	$> \eta_0$
	5	>1	s	i_{s1}, i_{s2}	i_{1s}, i_{2s}	$\gtrless \eta_0$
	6	>1	1	i_{12}, i_{1s}	i_{21}, i_{s1}	$\leqq \eta_0$

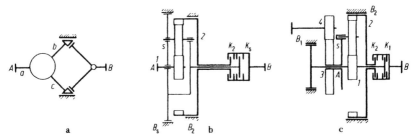

Fig. 101. Three-shaft change gear with two gear ratios and one direct speed: a, general symbolic representation; b, negative-ratio drive according to table 5, row 3, with clutches K_s and K_2 and brakes B_s and B_2, shaft I rigidly connected; c, negative-ratio drive according to table 5, row 1, with indirect input to $s \triangleq A$ through a series connection with the offset conventional gear train 43.

gears, whose selection is purely arbitrary, are interchanged. Therefore, these transmissions are implicitly included in the rows of table 5 where $i_o >$ 1 and need not be listed separately. According to R 25 a transmission cannot have two negative speed-ratios, since each revolving drive train has only one summation shaft.

As shown in fig. 101c, the listed design modifications whose carrier shafts are rigidly connected can only be built with offset connecting shafts, since, otherwise, shafts I and 2 would not be simultaneously accessible for each to be connected to one clutch and one brake.

One of the two speed-ratios of the transmissions, as described by rows 5 and 6 in table 5, is represented in fig. 45 by a dashed line which indicates that its associated efficiency is lower than the basic efficiency. Change-gears according to rows 5 and 6, therefore, are a second choice. The same is true for designs as described in row 2, provided their basic speed-ratios lie in the range $1 < i_o < 2$.

If one of two associated speed-ratios is given, then the other, as well as the basic speed-ratio i_o, is automatically determined and can be calculated with the aid of worksheet 1, or directly read from figs. 45 or 61.

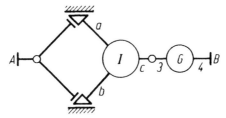

Fig. 102. Symbolic representation of a three-shaft change gear for two given speed-ratios with a conventional transmission G connected to the input or output shaft.

If an additional reduction gear G is connected to the input or output shaft as shown in fig. 102, both speed-ratios i_1 and i_2 can be arbitrarily chosen. In this case the planetary transmission must merely realize the speed-ratio step, that is, the ratio of the desired overall speed-ratios i_1/i_2. If both clutches are engaged simultaneously, a third speed-ratio $i_3 = i_{34}$ is obtained.

Thus, for a change-gear of this type:

$$i_1 = i_{ac}i_{c4}$$

$$\text{so that} \qquad \frac{i_1}{i_2} = \frac{i_{ac}}{i_{bc}}.$$

$$i_2 = i_{bc}i_{c4}$$

The basic speed-ratio i_0 for each of the six cases which are listed in table 5 can be found from i_{ac}/i_{bc} and worksheet 1. A suitable design can then be chosen from figs. 19 to 43.

Example: A transmission of the type shown in fig. 103a for the *two positive* ratios $i_1 = +1.8$ and $i_2 = +3.6$ can be found in rows 1 and 2 of table 5, which indicate that the summation shaft, that is, either the carrier shaft or shaft 2, must be rigidly connected. For the sake of an example, the case described in row 2 shall be investigated in more detail. Before a final choice could be made in reality, however, the second possibility as shown in row 1 also would have to be investigated.

Since shaft 2 is rigidly connected, we have the two possible solutions which either have the speed-ratios $i_1 = i_{s2}i_{34}$ and $i_2 = i_{12}i_{34}$ or $i_1 = i_{43}i_{2s}$ and $i_2 = i_{43}i_{21}$. For the first of these solutions we find from worksheet 1, boxes 31 and 1, that:

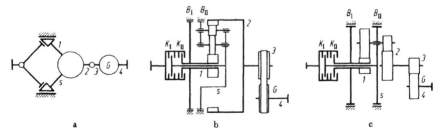

Fig. 103. Three-shaft change gear with the speed-ratios $i_1 = \pm 1.8$ and $i_2 = \pm 3.6$ and a conventional transmission G connected to the output: a, symbolic representation; b, realization by a positive-ratio gear train of the type shown in fig. 22 and the belt drive G; $i_1 = 1.8$ and $i_2 = 3.6$; c, realization by a positive gear train of the type shown in fig. 19; however, $i_1 = -1.8$ and $i_2 = -3.6$.

Solution α:
$$\frac{i_{s2}}{i_{12}} = \frac{i_{1.}}{i_{2.}} = \frac{1.8}{3.6} = \frac{i_o}{i_o - 1} \cdot \frac{1}{i_o} = \frac{1}{i_o - 1} ,$$

$$i_{12} = i_o = \frac{i_{2.}}{i_{1.}} + 1 , \qquad i_{s2} = \frac{i_o}{i_o - 1}$$

and, consequently,

$$i_o = \frac{3.6}{1.8} + 1 = 3.0 \quad \text{and} \quad i_{s2} = \frac{3}{3 - 1} = \frac{3}{2} , \qquad i_{12} = 3 .$$

For the second solution, we obtain from boxes 25 and 7 of worksheet 1 that

Solution β:
$$\frac{i_{2s}}{i_{21}} = \frac{i_{1.}}{i_{2.}} = \frac{1.8}{3.6} = \frac{i_o - 1}{i_o} \cdot i_o = i_o - 1 , \qquad \text{that is,}$$

$$i_o = \frac{1.8}{3.6} + 1 = 1.5 , \qquad i_{2s} = \frac{i_o - 1}{i_o} = \frac{1}{3} \quad \text{and} \quad i_{21} = \frac{1}{i_o} = \frac{2}{3} .$$

Consequently, the two auxiliary reduction drives which are coupled either to the input or output shaft have the speed-ratios:

Solution α:
$$i_G = i_{34} = \frac{i_{1.}}{i_{s2}} = \frac{1.8}{1.5} = 1.2 , \qquad \text{so that indeed,}$$

$$i_{1.} = i_{s2}i_G = 1.5 \cdot 1.2 = 1.8 \quad \text{and} \quad i_{2.} = i_{12}i_G = 3 \cdot 1.2 = 3.6 .$$

The direct gear, where the planetary gear train operates as a coupling, simply has the speed-ratio:

$$i_{3.} = i_G = 1.2$$

Solution β:
$$i_G = i_{43} = \frac{i_{1.}}{i_{2s}} = 1.8 \cdot 3 = 5.4 , \cdot$$

$$i_{1.} = i_{43}i_{2s} = 5.4 \cdot \frac{1}{3} = 1.8 ,$$

$$i_{2.} = i_{43}i_{21} = 5.4 \cdot \frac{2}{3} = 3.6 , \quad \text{and}$$

$$i_{3.} = i_{43} = 5.4 .$$

The two solutions differ in their efficiencies as well as in the speed-ratios of their direct gears i_3. If a good efficiency is essential, then solution α, where $i_o = 3$ is preferred. Here, at least in first gear, the efficiency $\eta_{s2} > \eta_o$, while for solution β, the efficiency in first gear $\eta_{2s} < \eta_o$ as indicated in fig. 45 by the dashed part of the corresponding speed-ratio curve.

On the other hand, solution β may be chosen if the speed-ratio $i_3 = 5.4$ of the direct gear is better suited for a particular application than the value $i_3 = 1.2$, provided of course that the following efficiency calculations yield a tolerable value.

If we assume that the efficiency of the reduction gear train G is $\eta_G = 0.985$, and the basic efficiency of the revolving gear train $\eta_o = 0.98$, then the overall efficiencies of the two design solutions can be obtained with the aid of worksheet 5 as follows:

Solution α:

$$\eta_{1.} = \eta_{s2}\eta_G = \frac{i_o - 1}{i_o - \eta_o} \cdot \eta_G = \frac{3 - 1}{3 - 0.98} \cdot 0.985 = 0.975 \ ,$$

$$\eta_{2.} = \eta_{12}\eta_G = \eta_o\eta_G = 0.98 \cdot 0.985 = 0.965 \ ,$$

$$\eta_{3.} = \eta_G = 0.985 \ .$$

Solution β:

$$\eta_{1.} = \eta_G\eta_{2s} = \eta_G\frac{i_o - 1/\eta_o}{i_o - 1} = 0.985 \cdot \frac{1.5 - 1/0.98}{1.5 - 1} = 0.945 \ ,$$

$$\eta_{2.} = \eta_G\eta_{21} = \eta_G\eta_o = 0.985 \cdot 0.98 = 0.965 \ ,$$

$$\eta_{3.} = \eta_G = 0.985 \ .$$

Fig. 103b shows a possible schematic layout for the solution α which incorporates a positive-ratio transmission as shown in fig. 22. Since $i_{12} = 3$, the rigidly connected shaft 2 of the basic transmission rotates with the lower speed and, therefore, is keyed to the annular gear. The positive speed-ratio i_{34} is realized with a belt drive.

Fig. 103c shows another possible design for solution α. In this case, a single gear stage G is connected to the output of a positive-ratio gear train of the type shown in fig. 19. Thus, $i_{34} = -1.2$, $i_{1.} = -1.8$ and $i_{2.} = -3.6$.

Section 30. Open Planetary Transmissions

Open planetary transmissions usually have two, and sometimes three, rotating shafts. If their central gear and their connected planets are arbitrarily labeled with the indices 1 and 2, then their speed-ratios, efficiencies, and torque-ratios can be calculated with the same equations and the same

worksheets as the simple reverted planetary transmissions. The open planetary transmissions which will be described in the following examples show
the simplicity of these calculations and, at the same time, may stimulate new
applications.

a) Twining-Machine Drive

Fig. 104 schematically shows an open, two-shaft, positive-ratio transmission of the type illustrated in fig. 29a. With each of its planets 2, two
bobbins of twine revolve about the planet shafts which at the same time
revolve with the driven carrier s. The annular gear 1 is held stationary. As
for reverted planetary transmissions we obtain:

$$i_o = \frac{n_1}{n_2} = -\frac{z_2}{z_1}, \qquad \text{so that} \qquad 0 < i_o < 1 .$$

According to worksheet 1, box 25,

$$i_{2s} = 1 - \frac{1}{i_o}, \qquad \text{and since}$$

$$\eta_o = \eta_{12}, \qquad \text{we find from worksheet 5, box 17, that}$$

$$\eta_{s2} = \frac{i_o - 1}{i_o - 1/\eta_o} .$$

If, for instance, $i_o = +1/3$, then the threads drawn from each pair of bobbins are twisted $i_{2s} = -2$ times (that is, against the input direction) per carrier revolution into cords, while at the same time the three cords are twisted
once in the input direction into rope.

b) Stirrer Drive

The stirrer drive shown in fig. 105 is an open, negative-ratio drive. If, for
instance

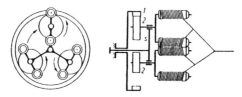

Fig. 104. Open positive-ratio gear train of a twining machine.

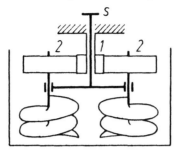

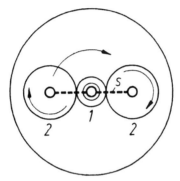

Fig. 105. Open negative-ratio gear train of a mixer.

$$i_o = \frac{n_1}{n_2} = -\frac{z_2}{z_1} = -2 \, ,$$

then, according to worksheet 1,

$$\frac{n_2}{n_s} = i_{2s} = \frac{i_o - 1}{i_o} = \frac{3}{2} \, .$$

With $\eta_o = \eta_{12}$, and worksheet 5, box 16, the overall efficiency becomes

$$\eta_{s2} = \frac{i_o - 1}{i_o - 1/\eta_o} \, .$$

According to the formula given above, the ratio between the absolute speed n_2 of the stirrers, and the speed n_s of the carrier, can be chosen not only within wide limits by an appropriate selection of i_o, but also can be infinitely varied during full operation when the central gear is not fixed but rather is powered by a separate, infinitely variable drive, or is coupled to the carrier through an infinitely variable transmission. The drive thus becomes an open planetary transmission with three rotating shafts. If the viscosities of the mixing stock vary, such a flexibility would offer substantial process-technological advantages.

c) Hoists

A hoist built by the Chisholme-Moore Manufacturing Co. is shown in fig. 106. It consists of an open, positive-ratio gear train of the type shown in fig. 31, which has only a single annular planet *1*. However, this planet cannot rotate about its own axis so that $n_1 = 0$. Rather it is driven in a circular path around the center of gear *2* by the two eccentrics *3* and *4*, which rotate synchronously. The eccentrics *3* and *4*, in turn, are driven by the gears *3* and *4*, which mesh with the input gear *5* as can be seen from fig. 106b. The inner central gear *2* serves as the output and, therefore, is connected to the load chain sprocket. Consequently, the speed reduction is:

$$i_o = \frac{n_1}{n_2} = -\frac{z_2}{z_1} = +\frac{16}{17} \, ,$$

$$i_{52} = i_{54}i_{s2} = -1\frac{i_o}{i_o - 1} = \frac{16 \cdot 17}{17 \cdot 1} = +16 \, .$$

With the assumption that $\eta_{54} = 0.98$ and $\eta_o = \eta_{12} = 0.99$, and according to worksheet 5, box 17, the efficiency becomes:

$$\eta_{52} = \eta_{54}\eta_{s2} = \eta_{54} \cdot \frac{i_o - 1}{i_o - 1/\eta_o} = 0.98 \cdot 0.853 = 0.836 \, .$$

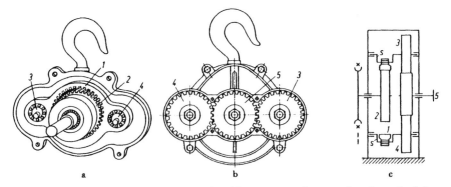

Fig. 106. Open, positive-ratio gear train with a non-rotating annular planet *1* of the Cyclone hoist manufactured by the Chisholme-Moore Mfg. Co. [4]: *a*, front view from the load chain side; *b*, front view from the input side; *c*, schematic of the gear train.

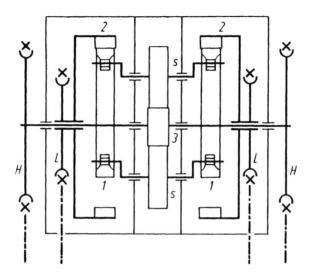

Fig. 107. Open positive-ratio gear train with non-rotating planet *1* as used in a hoist manufactured by the Chisholme-Moore Mfg. Co [4]: *H*, drive chain sprockets; *L*, load chain sprockets.

Fig. 107 shows a similar hoist, which is constructed from two symmetrically arranged, equal, open, positive-ratio transmissions of the type shown in fig. 29a. This hoist is equipped with two hand-chain sprockets *H* and two load-chain sprockets *L* which are coupled to the annular gears *2*. The eccentric driven planets *1* are designed as ring-type external spur gears. Thus:

$$i_o = -\frac{z_2}{z_1} > 1 \qquad \text{(see sec. 9)}$$

and, according to worksheet 1,

$$i_{32} = i_{3s}i_{s2} = i_{3s} \cdot \frac{i_o}{i_o - 1} \cdot$$

With worksheet 5, box 18, the efficiency becomes:

$$\eta_{32} = \eta_{3s}\eta_{s2} = \eta_{3s} \cdot \frac{i_o - 1}{i_o - \eta_o} \cdot$$

d) Self-Tensioning Belt Drive

Self-tensioning belt drives of the type shown in fig. 108 are open, negative-ratio transmissions, whose shafts *1* and *2* rotate while the carrier *s* is

164 / Simple Revolving Drive Trains

constrained by the drive belt. The tangential force exerted by the carrier, consequently, tensions the belt. Since the three torques of a revolving drive always bear the same ratio relative to each other, the belt tension is always proportional to the input torque. Therefore, even at the highest torques the belt will not slip, while, on the other hand, the device lowers the belt tension when the torque decreases, thus increasing the belt life.

Such a tensioning device can be located at the motor as shown in fig. 108a, or on the driven machine as shown in fig. 108b. In each of these cases,

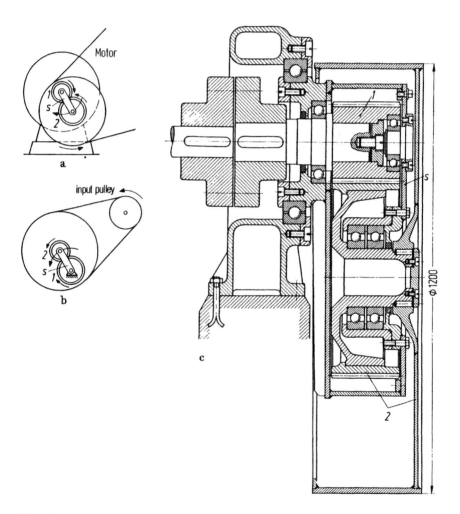

Fig. 108. Open negative-ratio gear train of a self-tensioning belt drive: *a*, attached to the motor; *b*, attached to the driven equipment; *c* section through an actual design [9].

it works only in the direction in which the carrier torque acts to tension the belt. The short arrows indicate the respective directions of rotation; the long arrow at the carrier indicates the direction in which the carrier would rotate if it were not constrained by the belt.

The carrier torque T_s of the drive shown in figs. 108a is obtained as for a reverted planetary transmission, either with the torque T_1 and eq. (44) as:

$$T_s = T_1 (i_o \eta_o^{r1} - 1) = T_1 (i_o \eta_o - 1) ,$$

or with the output torque T_2 and eq. (45) as:

$$T_s = T_2 \left(\frac{1}{i_o \eta_o^{r1}} - 1 \right) = T_2 \left(\frac{1}{i_o \eta_o} - 1 \right) ,$$

where according to eq. (43) $r1 = +1$, since R 2 stipulates that the signs of T_1 and n_1 are equal at the input shaft and $n_s = 0$.

In fig. 108b, shaft *1* is the output shaft. Therefore, an analogous consideration shows that $r1 = -1$. With the previously given equations T_s, consequently, becomes:

$$T_s = T_1 \left(\frac{i_o}{\eta_o} - 1 \right) = T_2 \left(\frac{\eta_o}{i_o} - 1 \right) .$$

The design of such a belt drive is treated in detail by Niemann [33].

III. COMPOUND EPICYCLIC
TRANSMISSIONS

To extend their range of application, epicyclic transmissions can be combined with other revolving or conventional drive trains to form larger drive systems. This leads to a large variety of compound transmissions whose operating characteristics at first sight often seem obscure, and whose analysis or synthesis frequently looks complicated. However, basically this is not true since the properties of these compound transmissions can always be reduced to the properties of the simple planetary transmissions. The bewildering multitude of combinations can be handled by grouping together drive systems with identical kinematic properties which can then be treated by a common method. In the following sections such a classification shall be established and the respective characteristic features shall be defined.

A. Symbols and Types

Section 31. Symbols and Modifying Subscripts or Superscripts

For compound drives, which consist of several epicyclic transmissions, the symbols 1, 2, s, and p, as defined in the previous part I (sec. 4), will be used to denote the gears or shafts of the first transmission. The corresponding gears or shafts of the second transmission are then denoted accordingly by the primed symbols, $1'$, $2'$, s', p' and the gears or shafts of a third revolving drive train by the double primed symbols, $1''$, $2''$, s'', p'', etc. Analogously, the basic speed-ratios of these transmissions are denoted by i_o, i_o', i_o'' and the basic efficiencies by η_o, η_o', η_o''. If, in general notation, the shafts of the individual revolving drives are symbolized by lowercase letters, we shall use a, b, c for the first transmission, a', b', c' for the second, etc. This type of notation permits us to analyze the individual component transmissions of a compound revolving drive, such as shown in fig. 109, by using directly, and without rethinking, the formulas which have been derived earlier for the simple epicyclic transmissions.

If conventional reduction drives with fixed axes are combined with revolving drives, their gears and shafts are denoted by a sequence of numbers starting with 3 as shown in fig. 10.

The shafts of two (or more) component transmissions which are rigidly coupled form a "coupling shaft." If a coupling shaft is also an external shaft, it is called a "connected coupling shaft"; if it is not connected to the outside, it is called "a free coupling shaft" (fig. 110). Shafts which are not coupled in this way, and either are the input or output shafts of a compound transmission, or are locked by a rigid connection with the housing, are called "connected monoshafts" and "fixed monoshafts," respectively (fig. 110).

The coupling shafts and the remaining monoshafts of the revolving compound drive trains are denoted by uppercase letters or Roman numerals ("uppercase subscription"), the shafts of the component transmissions, however, by lowercase letters or arabic numerals ("lowercase subscription") as described previously. In "simple bicoupled transmissions" (sec. 32), which consist of only two component transmissions, the connected-coupling shaft is always denoted with S and the free coupling shaft with F;

169

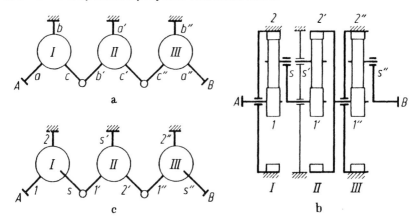

Fig. 109. Series-coupled revolving drive train consisting of three two-shaft transmissions which are connected in series: *a*, general symbolic representation; *b*, schematic representation of a typical transmission; *c*, symbolic representation of the transmission *b*.

the two monoshafts, however, are labeled with the Roman numerals *I* and *II* (see fig. 110).

If it is unknown or immaterial which of the shafts of the component transmissions of a simple bicoupled transmission should be the carrier shafts and which the central gear shafts, then these shafts are labeled with lowercase letters according to their functions in the bicoupled transmission. In fig. 110a, for example, *m*, *m'* are connected monoshafts, *c*, *c'* join to form the connected coupling shaft *S*, and *f*, *f'* join to form the free coupling shaft *F*. When the positions of the carrier shafts *s*, *s'*, and the central gear shafts *1*, *1'*, *2*, *2'* of the component transmissions are known, then these specific symbols, rather than lowercase letters, are used as shown in fig. 110b.

The different component transmissions are distinguished by Roman numerals as shown in fig. 110.

In the equations used to analyze compound or simple transmissions, the newly introduced symbols also are used as subscripts. This is consistent with our previous practice.

Section 32. Types

a) Derivation from Simple Revolving Drives

If a compound revolving drive is constructed from a number of simple revolving drives, each of the shafts of these component transmissions can be connected as an input or output shaft, can be fixed in the housing, or can be coupled to any shaft of another component transmission. The available simple transmissions, and the various types of compound epicyclic transmissions which result from their combinations, are shown in a series of schematic representations whose shafts are labeled with the chosen symbols (fig. 111).

If exclusively two-shaft epicyclic transmissions (which always have one stationary shaft and, hence, only one degree of freedom) are coupled in series or parallel, the resulting compound transmission likewise has only one degree of freedom. This is illustrated in figs. 111f and 111g. The single degree of freedom also carries over into various hookups of these two types.

If two simple three-shaft transmissions are connected at two of their shafts, a "simple bicoupled epicyclic transmission" or a "simple bicoupled planetary transmission" is obtained. The use of a special term such as "bicoupled transmission" for these compound transmissions must be credited to Wolf [13]. The term "simple" is used to characterize, unambiguously, the elementary bicoupled transmissions which consist of only two component transmissions. Where no misunderstanding is possible, this transmission will subsequently be called "simple bicoupled transmission" or, in short, "bicoupled transmission." Fig. 110 shows its symbolic representation and the designation of its shafts.

A simple bicoupled transmission has two degrees of mobility since its component transmissions together have four degrees of freedom, two of

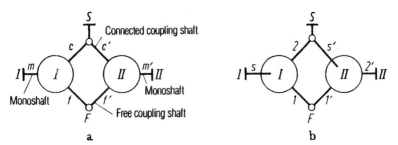

a   b

Fig. 110. Symbolic representation of a simple bicoupled transmission showing the designations of the shafts. The shafts of the bicoupled transmission have upper case symbols, those of the component transmissions lower case symbols. *I*, *II*, monoshafts of the bicoupled transmission; *S*, connected coupling shaft; *F*, free-coupling shaft. The shafts of the component transmissions have lower case subscripts. At *a*, the designations of the shafts of the component transmissions are chosen according to their position in the bicoupled transmission when the connections of the central gear shafts and the carrier shafts are unknown or inconsequential: *c*, *c'* form the connected coupling shaft *S*, *f*, *f'* the free coupling shaft *F*, and *m*, *m'* the monoshafts *I* and *II*. Example *b* shows the subscripts of the shafts of the component transmissions when the position of the carrier shafts *s*, *s'* and the central gear shafts *1*, *2*, *1'*, and *2'*, in the bicoupled transmission are known.

which are "bound" by the two internal connections which shall subsequently be called "couplings" (compare sec. 3). If, in addition, one of its three connected shafts is fixed by a rigid connection with the housing, a further degree of freedom is bound and it becomes a two-shaft bicoupled transmission. Three-shaft and two-shaft bicoupled transmissions differ in their operating characteristics in the same way as simple three-shaft and two-shaft revolving drives. Thus, they will be clearly distinguished by the introduction of the terms "simple bicoupled transmission with two degrees of mobility" and "simple constrained bicoupled transmission" or, in short, "constrained bicoupled transmission."

Two-shaft bicoupled transmissions which have a locked monoshaft (usually *II*) are commonly known as constrained bicoupled transmissions. A two-shaft bicoupled transmission "constrained" to one degree of mobility by a locked coupling shaft (instead of a locked monoshaft) is given the special designation: basic bicoupled transmission. Constrained bicoupled transmissions are frequently used and are being built in different configurations for a wide variety of applications. Thus, they will be treated separately and in detail. They will also be subdivided into special categories, namely, into simple constrained bicoupled transmissions, reduced bicoupled transmissions and variable bicoupled transmissions. A transmission type which is related to the simple bicoupled transmission is the symmetrical compound transmission. It is suited to accurately realize arbitrarily high speed-ratios and has been treated in detail by Willis [1].

If more than two revolving drive trains have multiple couplings, "higher bicoupled transmissions," rather than simple bicoupled transmissions, are formed. Thus far, transmissions of this type have seldom been built and their practical applications have not yet been investigated. Figs 111n to 111q show a few possible designs.

These compound transmissions will be defined subsequently and treated in more detail. As far as possible they will be classified by using notation which has been introduced earlier so that a minimum number of different types arise. These, however, are clearly distinguished through their operational characteristics and their mathematical treatment.

b) Series-Coupled Epicyclic Transmissions

As shown in fig. 109, these transmissions consist of several two-shaft transmissions which are connected in series; that is, they are queued up in the direction of the power-flow and are used to achieve high overall transmission ratios. One arbitrary shaft of each of the component transmissions is connected to the housing and thus becomes stationary. The output shaft of the first externally driven component transmission is connected to the input shaft of the next component transmission, and so on. Finally, the output shaft of the last component transmission serves at the same time as the

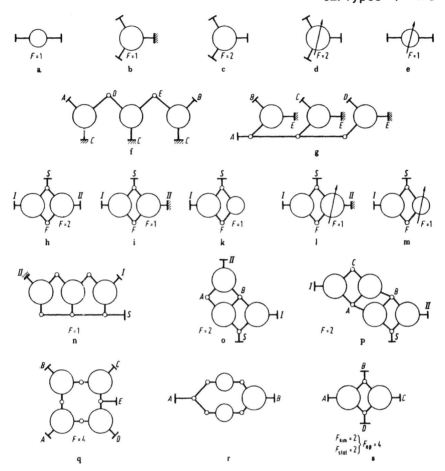

Fig. 111. Compound epicyclic transmission designs in symbolic representations. *A–e*, components of the compound epicyclic transmissions: *a*, conventional transmission with (spatially) fixed axes; *b*, simple revolving drive train with a fixed shaft; *c*, simple revolving drive train with two degrees of freedom; *d*, simple revolving drive train with an infinitely variable basic speed-ratio, e.g., a revolving traction drive; *e*, conventional variable-speed drive, e.g., V-belt or hydrostatic drive; *f* series-coupled epicyclic transmission; *g*, parallel-coupled epicyclic transmission. *H–m*, bicoupled transmissions: *h*, simple bicoupled transmission with two degrees of mobility; *i*, simple constrained bicoupled transmission with a revolving auxiliary component transmission (the symbols *h* and *i* are valid also for reduced bicoupled transmissions); *k*, simple constrained bicoupled transmission with a conventional auxiliary component transmission; *l–m*, variable bicoupled transmissions whose auxiliary transmissions have infinitely variable speed-ratios. *N–q*, higher bicoupled transmissions: *n*, constrained; *o–p*, with two degrees of mobility; *q*, with four degrees of mobility. *R*, symmetrical compound transmission; *s*, four-shaft bicoupled transmission.

output shaft of the compound transmission as, for instance, shaft B in fig. 109. Transmissions of this type have one degree of freedom and, thus, are constrained.

Kinematically, these series-coupled epicyclic transmissions behave like multi-stage conventional transmissions which consist of the stages I with a speed ratio i_1, II, with i_{II}, etc. With the symbols used in figs. 109a, b, and c, their overall transmission-ratio, consequently, can be expressed in the form:

$$i_{AB} = i_1 i_{II} i_{III} = i_{ac} i_{b'c'} i_{c''a''} = i_{1s} i_{o'} i_{1''s''} , \qquad (73)$$

where, according to worksheet 1, i_{1s} and $i_{1''s''}$ are functions of i_o and $i_{o''}$.

However, they are not loss-symmetrical since their component transmissions, as a rule, are not loss-symmetrical. Depending on the direction of the power-flow, their overall efficiency is

$$\left. \begin{array}{l} \eta_{AB} = \eta_{ac}\eta_{b'c'}\eta_{c''a''} = \eta_{1s}\eta_{o'}\eta_{1''s''} \\ \eta_{BA} = \eta_{a''c''}\eta_{c'b'}\eta_{ca} = \eta_{s''1''}\eta_{o'}\eta_{s1} , \end{array} \right\} \qquad (74)$$

or

where the individual efficiencies can be determined with the help of worksheet 5.

Series-coupled epicyclic transmissions are quite frequently used to generate high overall speed-ratios. Therefore, a guide rule which applies to a few special cases may be recorded at this point:

> *If at least one component transmission of a series-coupled*   R 28
> *compound drive train is self-locking in one of the power-flow*
> *directions, then the compound drive train itself is also self-*
> *locking in this power-flow direction.*

It can be readily understood that this also holds for an *even* number of component gear trains with negative efficiencies (i.e., which are self-locking)—although mathematically their overall efficiency, according to eq. (74), becomes positive.

c) Parallel-Coupled Epicyclic Transmissions

As shown in fig. 112, these transmissions consist of several two-shaft transmissions which have a common input shaft but separate output shafts, or, if the power-flow is reversed, several input shafts and a common output shaft. Also, some of the monoshafts can be input shafts, others can be output shafts. According to fig. 112, however, in all cases,

$$P_A + P_B + P_C + P_D = 0 ,$$

provided the losses can be neglected.

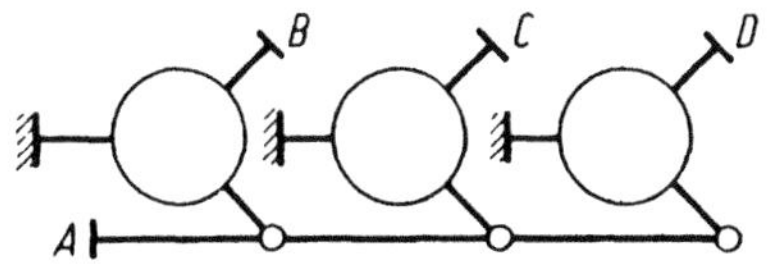

Fig. 112. Parallel-coupled revolving drive train with a common input at A and individual output shafts B, C, D, or vice versa.

At the common shaft A we may have either power division or power summation. To facilitate a mathematical analysis, the component transmissions are separated and then treated like conventional transmissions whose individual torques acting on shaft A are simply added. At this point, a detailed treatment of these easily comprehensible, constrained transmissions is not necessary, especially since they have little importance as revolving drive trains.

d) Bicoupled Transmissions with Two Degrees of Mobility

As shown in fig. 110, a simple bicoupled transmission with two degrees of mobility consists of two simple three-shaft transmissions which have two couplings. This is the basic type of the bicoupled transmission from which a multitude of constrained special designs can be derived by withholding one degree of freedom and varying the design. These special designs, however, evolved intuitively, and without reference to the basic transmission type which itself is seldom used.

This may be due to the fact that simple three-shaft transmissions can cover most of the drive problems which call for the application of transmissions with two degrees of mobility, and because, due to a lack of time, transmission designers usually do not attempt to analyze such a seemingly complicated compound transmission.

However, the widely held beliefs concerning the complexity of these compound transmissions are totally unfounded. In section 33, it will be shown that simple bicoupled transmissions can be analyzed and synthesized by the same methods that are used for simple epicyclic transmissions. Hopefully, this presentation will provide a better understanding of the particular operating characteristics of the bicoupled transmissions with two degrees of mobility, so that they will be utilized more often.

Basically, bicoupled transmissions with two degrees of mobility can be used wherever simple three-shaft transmissions may be applied (e.g., in the drive problems described in secs. 22 to 29). They will generally lead to more economical solutions than the simple three-shaft transmissions when they can be built smaller or lighter. They are especially suited where either very high basic speed-ratios or very accurate fractional speed-ratios are required which cannot economically be realized by a simple epicyclic transmission.

e) Constrained Bicoupled Transmissions

The constrained simple bicoupled epicyclic transmission (in shorter form, the constrained bicoupled transmission) is a special version of a bicoupled transmission with two degrees of mobility, where, as shown in fig. 113, one monoshaft (usually II) is connected to the housing and thus becomes fixed. As a compound transmission, it consists of a simple three-shaft transmission and a simple two-shaft transmission. The three-shaft transmission may be called the "main component transmission I," since it is always connected immediately to the input as well as to the output through its monoshaft and its connected coupling shaft. The two-shaft transmission may be called the "auxiliary component transmission II," since its only purpose is to maintain a constant speed-ratio between the input or output shaft, on the one hand, and the third shaft of the main component transmission on the other hand. The auxiliary component transmission can be a revolving gear train with a fixed central gear shaft as shown in fig. 113a, or a fixed-carrier gear train as shown in fig. 113b. Since, as far as their function is concerned, there is no difference between a fixed-carrier gear train and a conventional gear train with spatially-fixed axes, the latter can be used as an auxiliary gear train as shown in figs. 113c and 111k. When the fixed carrier shaft s' is considered equivalent to the stationary housing, then the transmission symbol used in fig. 113b stands for all auxiliary gear trains with spatially fixed axes. The symbolic representation of the auxiliary gear train shown in fig. 113c, therefore, will no longer be used in subsequent discussions.

f) Variable Bicoupled Transmissions

Constrained bicoupled revolving drive trains whose main or auxiliary component transmissions have an infinitely variable basic speed-ratio, also have an infinitely variable overall speed-ratio and, hence, are called "vari-

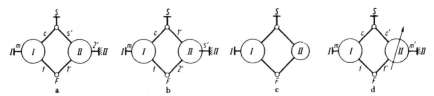

Fig. 113. Symbolic representation of the constrained bicoupled transmissions: I, main component transmission with connected monoshaft I; II, auxiliary component transmission with a constant (a, b, c) or infinitely variable (d) speed-ratio and a fixed monoshaft II; S, connected coupling shaft; F, free coupling shaft. A–d, constrained bicoupled transmissions whose auxiliary component transmission is: a, a simple epicyclic transmission; b, a basic transmission; c, a conventional transmission; d, a variable-speed drive ("variable bicoupled transmission").

able bicoupled transmissions." The component transmission whose speed-ratio is infinitely variable is called the "variable component transmission." Variable main component transmissions can be revolving traction drives, or revolving hydrostatic drives. Variable auxiliary component transmissions can be traction drives, belt or link-belt drives, or hydrostatic transmissions. Almost all of the variable bicoupled transmissions which have found practical applications are designed with a variable auxiliary component transmission as shown in fig. 113d. Transmissions of this type have been described in detail by Birkle [25].

g) Reduced Bicoupled Transmissions

Reduced bicoupled planetary transmissions are special designs of the simple bicoupled transmissions which have the following two unique features:

1. The combined carriers of the two component gear trains constitute the free coupling shaft and, therefore, can be reduced to a single unit.

2. The two central gears of the connected coupling shaft and their immediately meshing planets are respectively equal and, therefore, can be reduced to a single central gear and a single planet each. (The term "single-carrier bicoupled transmission" which Neussel [16] chose for this special transmission was not retained here, because it describes gear trains without the second of the characteristics mentioned above.)

At least one of the two component gear trains must have stepped planets as shown in figs. 19 and 34, or meshing planets as shown in figs. 22 and 36. Otherwise, the reduced bicoupled transmission becomes a simple planetary transmission. The combined carriers cannot constitute the connected coupling shaft since the two equal central gears, which then would have to constitute the free coupling shaft, would reduce to a single idler gear so that the

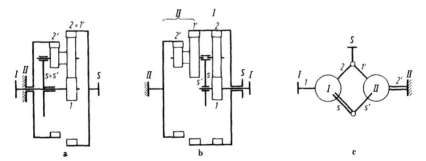

Fig. 114. Reduced bicoupled transmission, Minuteman Cover Drive [35]: *a*, schematic representation of the reduced bicoupled transmission; *b*, schematic representation of the functionally equivalent simple bicoupled transmission; *c*, symbolic representation of *a* and *b*.

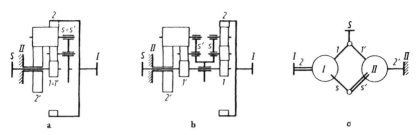

Fig. 115. Reduced bicoupled transmission, Ravigneaux gear set [34]: *a,* schematic representation of the reduced bicoupled transmission; *b,* schematic representation of the functionally equivalent simple bicoupled transmission; *c,* symbolic representation of *a* and *b.*

remaining bicoupled transmission would merely be a simple epicyclic transmission.

A reduced bicoupled transmission can be expanded into three different kinematically-equivalent, simple bicoupled transmissions. However, only one of them has the same power-flow and the same efficiency as the reduced bicoupled transmission and, thus, is functionally equivalent (sec. 37b). Therefore, the Wolf-symbol of this functionally equivalent bicoupled transmission can serve as a starting point for the analysis of the reduced bicoupled transmissions.

Figs. 114 and 115 show two well-known designs of reduced bicoupled transmissions, the Minuteman Cover Drive, and the Ravigneaux gear train, together with their functionally equivalent bicoupled gear trains and their Wolf-symbols.

h) Symmetrical Compound Transmissions

Willis [1] described a compound revolving drive train for very high and, at the same time, accurate speed-ratios, and also gave directions for its synthesis. As shown in fig. 116, it differs from a simple bicoupled transmission by having two fixed carrier, auxiliary component transmissions. One of its two connected shafts is the monoshaft of the main component transmission while the other is the connected coupling shaft between the two aux-

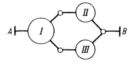

Fig. 116. Symmetrical compound transmission with the main component transmission *I* and the auxiliary component transmissions *II* and *III* which are symmetrically arranged.

iliary component transmissions. Because of their symmetrical construction these special transmissions shall be called "symmetrical compound transmissions."

i) Four-Shaft Bicoupled Transmissions

If the free coupling shaft of a simple bicoupled transmission with two degrees of mobility is also connected, a four-shaft bicoupled transmission is obtained as shown in fig. 111s. It, likewise, has two degrees of freedom. However, because of the additional external shaft the static degree of freedom increases to $F_{stat} = 2$.

To unequivocally describe the state of operation of such a drive, the torques of any two connected shafts, as well as the speeds of two arbitrary shafts, must be given.

To unequivocally describe the state of operation of such a drive, the torques of any two connected shafts, as well as the speeds of two arbitrary shafts, must be given.

k) Planetary Change-Gears

Transmissions which consist of two or more simple planetary gear trains, and whose overall speed-ratio can be altered from the outside in steps, are called "planetary change-gears." Thus, planetary change-gears are transmissions with several forward and/or reverse speeds which are used preferably as automotive transmissions. If put into gear, such a transmission can work as a simple planetary transmission, as a series-coupled epicyclic transmission, as a simple bicoupled or higher compound transmission, or as a reduced bicoupled transmission. A gear change can also be associated with a change between these transmission types, a condition which arises when the gear change causes one or more of the component gear trains to idle and thus alters the power-flow. According to their functions, two transmission types must be distinguished:

α) *Planetary change-gears with rigid couplings.* These transmissions consist of two or more rigidly-connected, simple planetary gear trains without a permanently fixed shaft as shown, for example, in figs. 117a, 117b, and 156, and having two degrees of freedom.

Two coupling shafts or monoshafts or one coupling shaft and one monoshaft are rigidly connected as input and output shafts. One of the other coupling shafts or monoshafts can be either locked by a brake, or two of them can be coupled together. This constrains the transmission and determines one of its gear-ratio steps. When two shafts are coupled together, all of the component gear trains of the transmission operate at the coupling

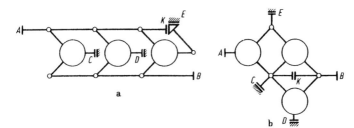

Fig. 117. Planetary change-gear with rigid couplings between its three component gear trains and 4 gear stages, one of which is the direct gear, $i = 1$. A, B, input and output; C, D, E, brakes for one gear stage each; K, clutch for the direct gear: a, in each gear the total power is transmitted by only one component gear train, while the other two are idling; b, in each gear the total power is transmitted by either two or all three component gear trains.

point and the speed-ratio becomes $i = 1$ which corresponds to the "direct gear."

The number of gear stages is equal to the number of component gear trains plus the direct speed-ratio. All speed-ratio steps can be freely chosen.

β) *Planetary change-gears with changeable couplings.* These transmissions consist of two or more simple planetary transmissions whose shafts can be coupled in several·different ways with the aid of several clutches or they can be connected to either the input or output. Alternately, they can be stopped by a number of brakes. As shown in fig. 118, the input or output can also be rigidly connected.

The selection of a particular gear stage is made by coupling the appropriate shafts with the aid of their clutches and locking one of the shafts. How-

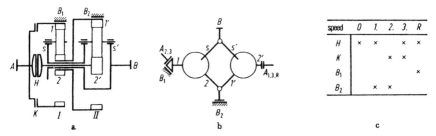

Fig. 118. Hydra-Matic transmission [40]: B_1, B_2, brakes; K, mechanical clutch; A, input; B, output; H, hydrodynamic clutch which can be operated by changing its fluid volume. A, schematic representation of the transmission; b, symbolic representation; c, shift-schematic (in each speed, the marked clutches or brakes are engaged).

ever, the transmission can be put into direct gear merely by using the clutches.

l) Higher Bicoupled Transmissions

Higher bicoupled transmissions are compound transmissions which consist of more than two simple three-shaft transmissions. Each of these is coupled through at least two of its shafts with one or two other component transmissions as shown in figs. 111n to 111q. They will be called "higher bicoupled transmissions" to distinguish them from simple bicoupled transmissions which consist of only two, three-shaft, component transmissions. Higher bicoupled transmissions have as many degrees of mobility as non-locked monoshafts, and as many coupling shafts as component transmissions, where a coupling shaft can connect two or more component transmissions. Any coupling shaft can become a connected coupling shaft and, according to R 27, at least one of the coupling shafts must be connected.

When a connected coupling shaft is locked, the component transmissions which are affected by this measure become constrained. They operate either as reduction gears which are connected at either end of the transmission or as series-coupled intermediate gears. This reduces the degree of mobility of the higher bicoupled transmission. If, (e.g., in fig. 111o) the coupling shaft B, instead of shaft S, were connected and locked, the transmission would be reduced to a series-coupled revolving drive train with a power-flow $II \rightarrow A \rightarrow S \rightarrow I$. Similarly, if in fig. 111p the coupling shaft A instead of shaft S is chosen as the connected coupling shaft, and is locked at the same time, the transmission reduces to a simple constrained bicoupled transmission with the connected monoshaft II, the free coupling shaft S, the locked "monoshaft" A, the connected coupling shaft B, and a series-coupled revolving drive $I \rightarrow C \rightarrow B$ which is connected at the input B.

Higher bicoupled transmissions can increase the range of applications of the simple bicoupled transmissions. To date, however, they have been used rarely and their operating characteristics and advantages have not been systematically investigated.

Fig. 111n shows the symbolic representation of an executed design. Figs. 111o, 111p, and 111q, however, show arbitrarily chosen examples. The transmissions shown in figs. 111n, 111o, and 111p can operate as either (higher) bicoupled transmissions with two degrees of mobility or (higher) constrained bicoupled transmissions, depending on whether one of their two monoshafts is locked. A further example is the automotive transmission operating with the brake C set (figs. 156g and 157).

B. Operating Characteristics of Compound Epicyclic Transmissions

Section 33. Perfect Analogy between Simple Bicoupled Transmissions and Simple Epicyclic Transmissions

a) Development of the Analogy

Simple bicoupled transmissions with two degrees of mobility have two degrees of freedom and three connected shafts. In this respect they are completely similar to the simple epicyclic transmissions as already pointed out by Wolf [13]. The utilization of this analogy leads to a substantial simplification of the analysis and, especially, the synthesis of these bicoupled transmissions as well as of the widely used transmission designs derived from them.

It can be readily understood that it is possible to perform the same mental experiment with a bicoupled transmission as was described in section 11d for a simple epicyclic transmission. Consequently, the speeds of its shafts can also be determined with the aid of the generalized Willis equation. If, unlike section 11d, we do not assume that the interior construction of the transmission is unknown, but rather proceed from a bicoupled transmission whose components are given and whose coupling shafts and monoshafts we know, then we can calculate the speed-ratios between any two external shafts when the third is locked. However, given a transmission as shown in fig. 119, we cannot readily determine from its schematic representation (fig. 119a) which of the three shafts I, II, or S should be locked in order to find the speed-ratio between the other two shafts. However, if we consider the symbolic representation of the transmission, fig. 119b, we will realize that the bicoupled transmission becomes a series-coupled transmission whose speed-ratio can be most readily determined if we lock the connected coupling shaft S. Then, according to section 32b, and worksheet 1, box 13:

$$i_{\mathrm{I}\,\mathrm{II}} = i_{\mathrm{mf}}i_{\mathrm{f'm'}} = i_{\mathrm{1s}}i_{\mathrm{1'2'}} = (1 - i_{\mathrm{o}})i_{\mathrm{o'}} \ .$$

This indicates that the connected coupling shaft S corresponds to the carrier shaft s of a simple revolving drive for which we likewise obtain the simplest

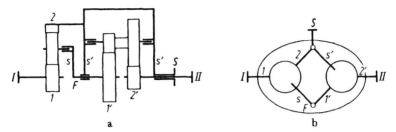

Fig. 119. Example for a simple bicoupled transmission: *a*, schematic representation; *b*, symbolic representation.

state of motion, namely, that of the basic transmission when the carrier shaft is locked.

For our analogy we could basically assume that an arbitrary external shaft of the bicoupled transmission corresponds to an arbitrary shaft of a simple planetary transmission. That this is true can be proven easily with the help of the generalized Willis equation as discussed in section 11d. These assumptions would also yield the correct efficiency values as given in section 15g. However, the specific correspondence between the shafts which has been chosen for the subsequently discussed analogy will require the least computational effort and, therefore, we will use it exclusively to simplify the analysis of simple bicoupled transmissions.

b) Characteristics of the Analogy

If the analogy between the bicoupled transmissions and the simple epicyclic transmissions is investigated in more detail, a complete correspondence, not only between the speed characteristics, but all other typical parameters such as the torque characteristics, the efficiencies, and power-flows will be found.

This perfect analogy with the equations developed earlier for the simple revolving drive trains makes it possible to analyze bicoupled transmissions with two degrees of mobility, as well as constrained bicoupled transmis-

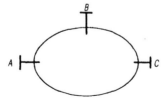

Fig. 120. Wolf symbol for a bicoupled transmission with unspecified internal design (analogous to fig. 46).

TABLE 6. ANALOGOUS DESIGNATIONS FOR SIMPLE REVOLVING DRIVE TRAINS
AND SIMPLE BICOUPLED DRIVE TRAINS

SIMPLE REVOLVING DRIVE TRAINS	SIMPLE BICOUPLED DRIVE TRAINS
Carrier shaft (spider) s	S connected coupling-shaft
Basic transmission (fixed-carrier shaft)	Basic bicoupled transmission (fixed connected coupling-shaft)
Connected shafts of the basic transmission } 1, 2	I, II monoshafts { connected shafts of the basic bicoupled transmission
Basic speed-ratio $i_o = i_{12}$	$i_{I\,II}$ series speed-ratio
Basic efficiency $\eta_o = \eta_{12}$ $\eta_o^{-1} = 1/\eta_{21}$	$\eta_{I\,II}$ series efficiency $\eta_{I\,II}^{-1} = 1/\eta_{III}$ [see defining equations (84) and (85)]
Three-shaft drive with two degrees of mobility	Three-shaft drive with two degrees of mobility
Two-shaft drive with one degree of mobility (constrained)	Two-shaft drive with one degree of mobility (constrained)
One central gear locked	One monoshaft locked
Spider locked: reverts to the basic transmission	Connected coupling-shaft locked: reverts to the basic bicoupled transmission, made up of two series-coupled simple revolving drive trains
Rolling-power P_R	P_R series power
Coupling-power P_S	P_S coupling-power

sions, reduced bicoupled transmissions, and variable bicoupled transmissions. It is necessary only to observe the analogous designations of symbols and subscripts given in the upper part of table 6, where "lowercase" applies to the simple revolving drive trains and "uppercase" to the simple bicoupled transmissions. In this connection it should be noted that when the internal design of a simple bicoupled transmission, like that of fig. 119, is unknown or irrelevant, the transmission may be represented by the simple Wolf-symbol of fig. 120 with uppercase designations for its three shafts.

The series speed-ratio $i_{I\,II}$ is defined as the speed-ratio which is obtained when the connected coupling shaft is locked, since the bicoupled transmission then works as a series-coupled transmission, in accordance with section 32b. With the notation used in fig. 110,

$$i_{I\,II} = \left[\frac{n_I}{n_{II}}\right]_{(n_S = 0)} = i_{mf} i_{f'm'} \, . \tag{75}$$

If shaft I is the input shaft, then the "series-efficiencies" of these series-coupled transmissions become:

$$\eta_{III} = \eta_{mf}\eta_{f'm'} \cdot \tag{76}$$

Accordingly,

$$\eta_{III}^{-1} = \frac{1}{\eta_{III}} = \frac{1}{\eta_{m'f'}\eta_{fm}} \tag{77}$$

when shaft II is the input shaft.

In series-coupled revolving drive trains the dissymmetry of the losses can become quite significant because both component transmissions are revolving gear trains which, as a rule, are not symmetrical with respect to their losses. The simplified assumption that basic transmissions are practically loss-symmetrical (which led to eq. (14) of section 5) is no longer applicable to series-coupled revolving trains. Therefore, we must distinguish rigorously between eqs. (76) and (77). If the position of all shafts and the basic efficiencies η_o and $\eta_{o'}$ of the component transmissions are known, then we can obtain η_{mf} and $\eta_{f'm'}$ from worksheet 5.

For the transmission shown in fig. 119 we obtain (with the help of worksheets 1 and 5):

$$i_{III} = i_{1s}i_{1'2'} = (1 - i_o)i_{o'} \quad \text{and}$$

$$\eta_{III} = \eta_{1s}\eta_{1'2'} = \frac{1 - i_o\eta_o}{1 - i_o}\eta_{o'} \,, \quad \text{since} \quad i_o < 0 \,.$$

However,

$$\eta_{III} = \eta_{2'1'}\eta_{s1} = \eta_{o'} \cdot \frac{i_o - 1}{\dfrac{i_o}{\eta_o} - 1} \,, \quad \text{for } i_o < 0 \,.$$

c) Speeds

If a bicoupled transmission works as a series-coupled revolving drive with a locked connected coupling shaft, then its state of motion corresponds to that of a basic transmission which has been described by Willis as the first partial motion (see sec. 11), and

$$\frac{n_I'}{n_{II}'} = i_{III} \,.$$

If, now, an equal "coupling speed" n_S is superimposed on all three external shafts, then the relative motions between the three shafts do not change. Analogous to eq. (22), the basic speed-ratio becomes

$$\frac{n_I - n_S}{n_{II} - n_S} = i_{III}$$

or
$$n_1 - i_{1\text{II}}n_{\text{II}} - (1 - i_{1\text{II}})n_S = 0 .$$ (78)

Because of the analogy between eqs. (22) and (78), eqs. (23) to (36), which all have been derived from eq. (22), as well as worksheets 1 and 2, remain valid for simple bicoupled transmissions when the lowercase subscripts are replaced by uppercase subscripts. Thus, the worksheets can be used immediately for all simple bicoupled transmissions and the transmission types derived from them, that is, the reduced bicoupled transmissions and variable bicoupled transmissions. With uppercase subscripts, eqs. (23) to (25) become:

$$n_1 = i_{1\text{II}}n_{\text{II}} + (1 - i_{1\text{II}})n_S ,$$ (79)

$$n_{\text{II}} = \frac{n_1 + (i_{1\text{II}} - 1)n_S}{i_{1\text{II}}}$$ (80)

$$n_S = \frac{i_{1\text{II}}n_{\text{II}} - n_1}{i_{1\text{II}} - 1} .$$ (81)

d) Torques

Since the superposition of a coupling speed evidently does not change the torque ratios of either a simple or bicoupled epicyclic transmission, eqs. (40) and (41) remain valid when written with analogous subscripts. However, it must be reemphasized that the series-coupled epicyclic transmissions cannot be treated as being loss-symmetrical. Therefore, if shaft I is the input shaft,

$$\frac{T_{\text{II}}}{T_1} = -i_{1\text{II}}\eta_{1\text{II}} ,$$

and if shaft II is the input shaft,

$$\frac{T_{\text{II}}}{T_1} = -\frac{i_{1\text{II}}}{\eta_{1\text{II}}} .$$

However, even if the loss-dissymmetry of the series-coupled epicyclic transmissions is considered, we can write eqs. (42) and (43) analogously as

$$\frac{T_{\text{II}}}{T_1} = -i_{1\text{II}}\eta_{1\text{II}}^{r1}$$ (82)

and

$$r1 = \frac{T_1(n_1 - n_S)}{|T_1(n_1 - n_S)|} = \pm 1 ,$$ (83)

if we *define* the series efficiency of the simple bicoupled transmissions as:

$$\eta_{\mathrm{III}}^{+1} = \eta_{\mathrm{III}} , \quad \text{when} \quad r\mathrm{I} = +1 \tag{84}$$

and $\quad \eta_{\mathrm{III}}^{-1} = \dfrac{1}{\eta_{\mathrm{III}}} , \quad \text{when} \quad r\mathrm{I} = -1 .$ (85)

Analogous to eqs. (44) and (45), the torque ratios then become:

$$\frac{T_{\mathrm{S}}}{T_{\mathrm{I}}} = i_{\mathrm{III}}\eta_{\mathrm{III}}^{r\mathrm{I}} - 1 , \tag{86}$$

$$\frac{T_{\mathrm{S}}}{T_{\mathrm{II}}} = \frac{1}{i_{\mathrm{III}}\eta_{\mathrm{III}}^{r\mathrm{I}}} - 1 . \tag{87}$$

e) Efficiencies

Also, the efficiencies of the three-shaft and two-shaft *bicoupled* transmissions can be calculated like the efficiencies of the *simple* three-shaft and two-shaft transmissions. However, the dissymmetry of the losses, which can be quite appreciable in series-coupled transmissions, can no longer be neglected.

Because of the perfect analogy between simple bicoupled and simple revolving drives, worksheets 4 and 5 are valid without any reservation for three- and two-shaft bicoupled transmissions respectively, if the indices *1, 2,* and *s* are replaced by *I, II,* and *S.*

f) Series Power and Coupling-Power

If a bicoupled transmission operates solely with Willis' first partial motion, using the series gear train furnished by a locked connected coupling shaft, the total power flows from *I* to *II* through the free coupling shaft. This "series power" P_{R} thus corresponds to the rolling-power P_{R} of the simple revolving drive. (Because of this analogy and the coincidence of the corresponding equations, the same subscript R is used to denote rolling-power and series power.) With the analogous indices we can write,

$$\text{series power of shaft I} : P_{\mathrm{RI}} = T_{\mathrm{I}}(n_{\mathrm{I}} - n_{\mathrm{S}}) , \tag{88}$$

$$\text{series power of shaft II:} \ P_{\mathrm{RII}} = T_{\mathrm{II}}(n_{\mathrm{II}} - n_{\mathrm{S}}) . \tag{89}$$

With the second of Willis' partial motions the following coupling-powers are then superimposed on the three shafts,

$$\text{coupling-power of shaft } I : P_{\mathrm{SI}} = T_{\mathrm{I}}n_{\mathrm{S}} ,$$

$$\text{coupling-power of shaft } II: P_{\mathrm{SII}} = T_{\mathrm{II}}n_{\mathrm{S}} , \tag{90}$$

coupling-power of shaft $S : P_{SS} = T_S n_S$.

Thus, if we use the uppercase subscripts I instead of 1, II instead of 2, and S instead of s, table 1 is equally valid for simple bicoupled transmissions.

g) Internal Power-Flows

The internal power-flows in simple bicoupled transmissions are likewise analogous to those in simple revolving drives and figs. 49 and 49 remain valid for simple bicoupled transmissions when upper case subscripts are used. While negative-ratio bicoupled transmissions (i.e., transmissions with a negative series speed-ratio) operate without opposing series and coupling futile power-flows, such power-flows may occur in positive-ratio transmissions as shown in figs. 49. Entirely independent of these power-flows, however, futile coupling-power and futile rolling-power can flow in each component gear train of a bicoupled transmission if this component gear train has a positive basic speed-ratio.

The opposing futile coupling-power and futile series power, which—analogous to the futile coupling-power and futile rolling-power in simple positive-ratio transmissions—can occur in bicoupled transmissions with a positive series speed-ratio, are not identical with the circulating powers that will be described later in section 35f.

If the series power becomes larger than the total external power, then the total efficiency of a bicoupled transmission becomes smaller than its series efficiency and vice versa, since only the series power, not the coupling-power, is associated with losses. If uppercase subscripts are used and the losses are neglected, the ratio P_R/P_{in} of the series power to the external power in bicoupled transmissions can be obtained immediately from figs. 66 and 67 for constrained bicoupled transmissions, and from figs. 68, 69 and 70 for bicoupled transmissions with two degrees of mobility.

h) Summation Shaft and Total-Power Shaft

Like a simple epicyclic transmission, a simple bicoupled transmission has a summation shaft and two difference shafts and, according to worksheet 3, the positions of these shafts depend on whether its series speed-ratio is $i_{III} < 0$, or $0 < i_{III} < 1$, or $i_{III} > 1$. A simple bicoupled transmission has a total-power shaft whose position, according to figs. 63 to 65, depends on the speed-ratios of its shafts I, II and S and its series speed-ratio i_{III}. To prove these statements, procedures can be followed which are analogous to those outlined in section 15a for the simple epicyclic transmissions.

i) Self-Locking and Partial-Locking

Like a simple positive-ratio transmission, a bicoupled transmission with a positive series speed-ratio is capable of self-locking when

$$\eta_{III} < i_{III} < \frac{1}{\eta_{III}} . \tag{91}$$

However, self-locking occurs only when the connected coupling shaft is the sole output shaft. Self-locking, then, is possible over a wider range of speed-ratios i_{III} than in simple, positive-ratio transmissions, since, in

$$\eta_{III} = \eta_{mf}\eta_{f'm'} , \quad \text{and} \quad \frac{1}{\eta_{III}} = \frac{1}{\eta_{m'f'}\eta_{fm}} ,$$

η_{III} can be the product of two rather low revolving-carrier efficiencies. This occurs when the component transmissions are positive-ratio gear trains, which, in a series-coupled transmission, operate with a low revolving-carrier efficiency, that is, with a high rolling power-ratio as can be verified with figs. 66 and 67. On the contrary the basic efficiency of a simple revolving drive is always close to 1.

As long as its series efficiency $\eta_{III} > 0$, and eq. (91) is not satisfied, a bi-coupled transmission is not capable of self-locking. This has been pointed out by Birkle [25]. A drastic illustration is the bicoupled transmission shown in fig. 121, which has the basic speed-ratios $i_o = i_{o'} = 1.02$, and the basic efficiencies $\eta_o = \eta_{o'} = 0.97$. As a series-coupled revolving drive it cannot operate when the shaft S is locked and, at the same time, shaft I is the input shaft, since each of the two series-coupled gear trains is self-locking when the power-flow is in the indicated direction (R 28). According to worksheet 5, box 9 and 12:

$$\eta_{ls} = \eta_{1's'} = \frac{i_o\eta_o - 1}{i_o - 1} = \frac{1.02 \cdot 0.97 - 1}{1.02 - 1} = -0.53 ,$$

$$\eta_{sl} = \eta_{s'1'} = \frac{i_o - 1}{\dfrac{i_o}{\eta_o} - 1} = \frac{1.02 - 1}{\dfrac{1.02}{0.97} - 1} = 0.388 .$$

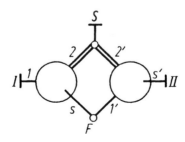

Fig. 121. Simple bicoupled transmission whose component gear trains are capable of self-locking: $i_o = i_{o'} = 1.02$ and $\eta_o = \eta_{o'} = 0.97$.

Mathematically, the series efficiency becomes positive, that is,

$$\eta_{III} = \eta_{1s}\eta_{1's'} = (-0.53)(-0.53) = 0.281 \, ,$$

and

$$\eta_{III} = \eta_{s'1'}\eta_{s1} = 0.388 \cdot 0.388 = 0.15 \, .$$

The series speed-ratio becomes:

$$i_{III} = i_{1s}i_{1's'} = (1 - i_o)(1 - i_{o'}) = (-0.02)(-0.02) = 0.0004 \, .$$

Since now $1/\eta_{III} > \eta_{III} > i_{III}$ and not as required by eq. (91) $1/\eta_{III} > i_{III} > \eta_{III}$, the condition under which the bicoupled transmission can become self-locking is no longer satisfied. The speed-ratio of the bicoupled transmission can be obtained with the help of worksheet 1:

$$i_{IS} = 1 - i_{III} = 1 - 0.0004 = 0.9996 \approx 1 \, ,$$

and the efficiency with the help of worksheet 5, box 8:

$$\eta_{IS} = \frac{\dfrac{i_{III}}{\eta_{III}} - 1}{i_{III} - 1} = \frac{\dfrac{0.0004}{0.15} - 1}{0.0004 - 1} = 0.9977 \, .$$

Despite the capability of both component gear trains to be self-locking, the calculated total efficiency of the bicoupled transmission is almost equal to 1 and, therefore, it is not self-locking. This, at first sight surprising result, becomes clear when the speeds and torques are considered: since $i_{IS} \approx 1$, the gear train I on the left-hand side operates at the coupling point without losses. Shafts $1'$ and $2'$ of gear train II on the right-hand side transmit only approximately 2% of the external torque T_S, so that for all practical purposes the gear train II idles and its tooth-friction losses can be

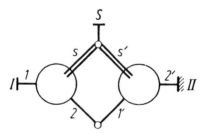

Fig. 122. Constrained bicoupled transmission which is capable of self-locking but whose component gear trains are not capable of self-locking: $i_{12} = -0.25$; $i_{1'2'} = -4.2$; $\eta_{12} = \eta_{1'2'} = 0.97$. Thus: $i_{III} = 1.05$; $\eta_{III} = 0.941$.

neglected. However, as discussed in section 6, in a practical transmission of this type the idling losses which have not been considered in this calculation would have to be taken into account and the final efficiency would assume a somewhat smaller value.

According to eq. (91), however, a bicoupled transmission can be self-locking even if its two component gear trains are not self-locking because, for example, they have negative basic speed-ratios. The constrained transmission shown in fig. 122 may serve as an example. Its parameters have been chosen as follows:

$$i_0 = i_{12} = -0.25 \, , \qquad i_{0'} = i_{1'2'} = -4.2 \, , \qquad \eta_0 = \eta_{0'} = 0.97 \, ,$$

$$i_{III} = i_{12}i_{1'2'} = (-0.25)(-4.2) = 1.05 \, ,$$

$$\eta_{III} = \eta_{12}\eta_{1'2'} = 0.97 \cdot 0.97 = 0.941 \, .$$

For the bicoupled transmission we thus obtain, with worksheet 1,

$$i_{IS} = 1 - i_{III} = 1 - 1.05 = -0.05 \, ,$$

and with worksheet 5, box 9,

$$\eta_{IS} = \frac{i_{III}\eta_{III} - 1}{i_{III} - 1} = \frac{1.05 \cdot 0.941 - 1}{1.05 - 1} = -0.24 \, .$$

Since this efficiency is negative, the bicoupled transmission is self-locking for the power-flow direction *IS*.

In bicoupled transmissions with two degrees of mobility, which are self-locking according to the conditions expressed by eq. (91), partial self-locking can occur in the same way as has been described in section 15f for simple three-shaft transmissions.

To summarize what has been said, thus far, about the use of the analogy between simple and bicoupled epicyclic transmissions, we can formulate the following guide rules:

All equations derived for simple epicyclic transmissions are R 29
equally valid for simple bicoupled transmissions when shafts
1 and 2 are replaced by the analogous monoshafts I and II,
the carrier-shaft s by the analogous connected coupling shaft
S, and the fixed-carrier efficiencies η_{12} and η_{21} by the anal-
ogous series efficiencies η_{III} and η_{III}.

For the majority of its possible couplings, series-coupled epi- R 30 *cyclic transmissions are not loss-symmetrical and, therefore, as a rule $\eta_{1\text{II}} \neq \eta_{\text{III}}$.*

k) Application of the Analogy to Revolving Hydrostatic Drives

Worksheets 1 to 5, which can be used to determine the speeds and efficiencies of two-shaft and three-shaft transmissions, are fully valid also for revolving hydrostatic drives such as shown in fig. 44, [52].

The subscripts, which are analogous to those of the simple epicyclic transmissions, can be found as follows: Each of the two hydrostatic components of the transmission possesses a *monoshaft* and one of these is arbitrarily denoted by *1*, the other by *2*. Both components are rigidly connected through the housings of the hydraulic components, that is, pump and motor. Together these housings constitute the *carrier* of the transmission and, therefore, they are denoted by the symbol *s*. It would, however, be possible to connect the cylinder block of the left-hand side component to the housing of the right-hand side component. Since the connected parts would again form the carrier of the transmission, they would be denoted by the letter *s*, while the housing of the left-hand side unit would become a monoshaft and as such would be denoted by the symbol *1* or *2*.

If the transmission has a *closed* hydraulic system, its *basic speed-ratio i* is defined by the ratio of the displacement volumes V_1 and V_2 of pump and motor:

$$i_{\text{o}} = i_{12} = \frac{n_1}{n_2} = -\frac{V_2}{V_1} \cdot \frac{1}{\eta_{\text{V}}^{\text{rI}}},$$

where η_{V} is the *volumetric* efficiency of the transmission. It depends, of course, on the leakage losses of the hydraulic components and thus the basic speed-ratio i_{o} is influenced by pressure, speed, and temperature (viscosity). Only for a first estimate may we assume that η_{V} has a value of 1. At the same time, the equation above determines the sign of the displacement ratio V_2/V_1. If the displacement volumes of pump and motor are infinitely variable, and either of them is changed through zero, its sign and, consequently, the sign of the speed-ratio i_{o} change, so that the output speed of the drive changes its direction of rotation.

If the hydraulic and mechanical friction losses are considered through the introduction of the mechanical efficiency factor η_{m} then the *torque ratio* becomes:

$$\frac{T_2}{T_1} = \frac{V_2}{V_1} \cdot \eta_{\text{m}}^{\text{rI}}.$$

Thus, according to the definition given in eq. (13), the basic efficiency η_{o} becomes:

$$\eta_o = \eta_{12} = -\frac{n_2 \, T_2}{n_1 \, T_1} = \frac{V_1 \, V_2}{V_2 \, V_1} \cdot \eta_{V_{12}}\eta_{m_{12}} = \eta_{V_{12}}\eta_{m_{12}} \; .$$

If such a revolving hydrostatic drive is loss-symmetrical (see sec. 5), that is, when $\eta_{12} \approx \eta_{21}$, then its two-shaft and three-shaft efficiencies can be determined with the aid of worksheets 5 and 4 respectively.

Completely analogous to the rolling-power in planetary transmissions a hydrostatic power $P_h \triangleq P_R$ occurs in hydrostatic transmissions. In the fixed-carrier mode this hydrostatic power equals the external power. The direction of the power-flow is determined by the exponent $r1$ which has been introduced earlier and can either be calculated from eq. (43) in section 12 or can be found directly from worksheets 4 and 5. Because of the perfect analogy, the characteristic differences between positive- and negative-ratio transmissions, as well as the conditions for self-locking and partial self-locking (sec. 15) are valid also for revolving hydrostatic drives.

Revolving hydrostatic drives are frequently used as component transmissions in variable bicoupled drive trains (sec. 36). Because of the analogy, these hydrostatic main or auxiliary component transmissions can be treated like simple planetaries. In the analysis of the bicoupled transmission they are, therefore, considered as revolving drive trains, as is any simple planetary transmission. The analogy as described in sections 33 to 36 is thus directly applicable to bicoupled transmissions which contain a hydrostatic main or auxiliary component transmission. The frequently encountered use of additional intermediate drive trains will be investigated later in section 36d.

Section 34. Bicoupled Transmission with Two Degrees of Mobility

In section 32d, the simple bicoupled transmission with two degrees of mobility has been defined as the fundamental type of the bicoupled transmissions. Its symbolic representation and the designation of its shafts are shown in fig. 110. The relationships between the speeds, torques, and powers of its three external shafts can be derived with the help of the analogy from the simple epicyclic transmission as described in section 33. The subsequent discussions, therefore, deal with the internal construction of these transmissions and the relationships which determine their analysis and design.

a) Number and Designations of Possible Couplings

Each of the three shafts *1*, *s*, or *2* of a component transmission can be a monoshaft of the bicoupled transmission. Once the monoshaft is deter-

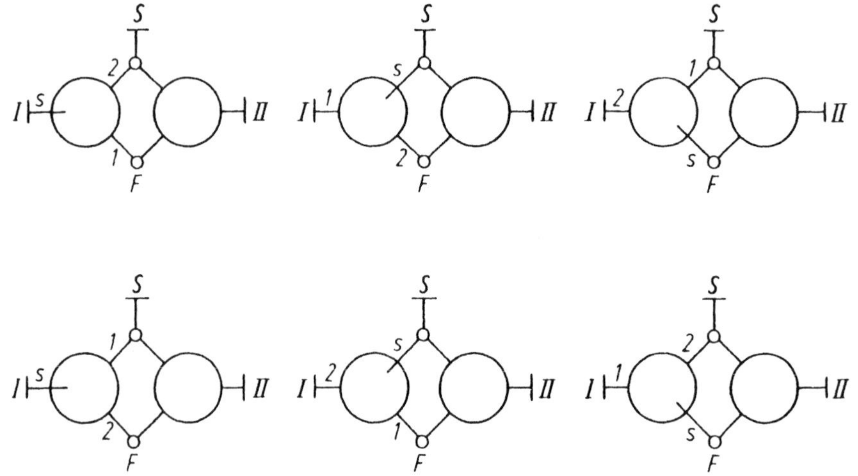

Fig. 123. Possible distribution of the three shafts of a main component transmission between the monoshaft I, the connected coupling shaft S, and the free coupling shaft F. The different $c-f$ shaft pairs (ref. fig. 113) in vertical alignment result from interchanging central gear shafts I and 2.

mined, either one of the other two shafts can become part of the connected coupling shaft. Consequently, as shown in fig. 123, the monoshaft and the connected and free coupling shafts of each component transmission can be arranged in $3 \times 2 = 6$ different ways so that $6 \times 6 = 36$ different couplings are possible between the two component transmissions of a bicoupled transmission.

To reduce this multitude of possibilities, we can make use of the fact that the six possible shaft combinations of each simple transmission can be grouped in three pairs where the component transmissions of each pair differ only in that its central gear shafts I and 2 change places. Both transmissions of each pair are thus identical when their basic speed-ratios are reciprocal with respect to each other (fig. 123). For each component transmission, therefore, the pair of modifications which has the same position of the carrier shaft needs to be considered only once. This reduces the number of possible couplings of the bicoupled transmission to $3 \times 3 = 9$.

These couplings are indicated by an ordered pair of the symbols c, f, or m and f', c' or m' respectively of those shafts of the bicoupled transmission which are the carrier shafts of the two component transmissions. The nine possible coupling cases and their designations are shown in fig. 124. Each of these has four variations which can be obtained by exchanging the designations of the central gear shafts I and 2, or I' and $2'$. If, at the same time, however, the basic speed-ratio $i_0 = x$ is replaced by $i_0 = 1/x$ and/or $i_{0'} = y$

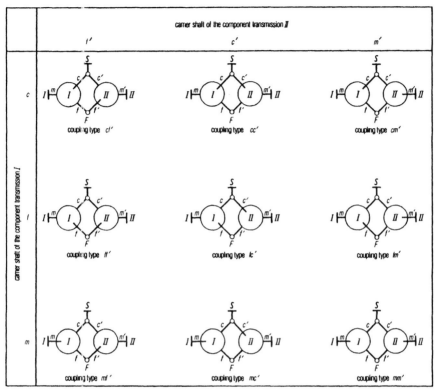

Fig. 124. The possible structural couplings of the simple bicoupled transmissions are denoted according to the positions of the carrier shafts of the component transmissions, e.g., the coupling case *cm'* in the third column of the first row is characterized by $s = c$ and $s' = m'$. Each coupling case has four possible identical arrangements which are obtained by interchanging the arbitrarily selected positions of the central gear shafts *1* and *2* and/or *1'* and *2'*.

by $i_{o'} = 1/y$, then the four variations reduce to four identical transmissions.

The nine possible coupling cases make it possible to design nine structurally different simple bicoupled transmissions. Therefore, we shall refer to them more accurately as "structural coupling cases."

However, it is possible to find bicoupled transmissions which belong to different structural coupling cases but are kinematically equivalent. These bicoupled transmissions are characterized by the fact that in each of their two component transmissions the summation shafts occupy the same position. Since, in each of the two component transmissions, the shafts *c, m,* or *f,* and *c', m',* or *f',* can be the summation shafts, they can again be coupled in 3 × 3 = 9 different ways which we shall call the nine "kinematical coupling cases." Like the nine coupling cases shown in fig. 124, they shall be indicated by an ordered pair of the symbols m, c, or f, and m', c', or f' respectively of those shafts of the bicoupled transmission which are the *summation shafts* of the two component transmissions.

According to R 13, in all negative-ratio transmissions the carrier shaft is also the summation shaft. If, therefore, a bicoupled transmission consists of two negative-ratio component transmissions, then the designations of the structural and kinematical coupling cases are identical.

b) Shaft Speeds

All speeds of a bicoupled transmission with two degrees of mobility are unambiguously determined when the speeds of two of the four shafts I, II, S, and F are known.

For the analysis of the speed characteristics of a bicoupled transmission we must know its coupling case according to fig. 124, the basic speed-ratios of its component transmissions, and two of its speeds. If the known speeds belong to the same component transmission, then the speed of the third shaft and, subsequently, the speeds of the other component transmission can be found with the aid of worksheet 2. If the speeds of the two mono-shafts of the bicoupled transmission are known, then we must first determine its series speed-ratio $i_{\text{I II}}$ according to section 33c from which we can subsequently find its third speed n_s. The speed of the free coupling shaft which is the only unknown speed, can again be found with the help of worksheet 2.

For example, an analysis of the speed characteristics of the transmission shown in fig. 125a: To begin with, every shaft of the transmission is labeled.

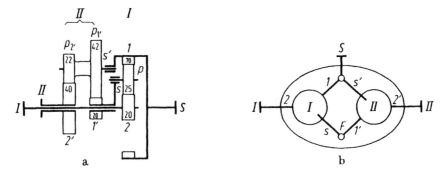

Fig. 125. Simple bicoupled transmission with two degrees of mobility illustrating the transmission analysis in secs. 34b–d: *a*, schematic representation of the gear train showing the numbers of teeth; *b*, Wolf symbol whose shaft designations are transcribed from *a*.

Then their speed-ratios are determined from the number of teeth of their gears. The sequence of the necessary steps is as follows:

1. The two component gear trains shown in fig. 125a are arbitrarily denoted as *I* and *II*. The shafts of gear train *I* are then labeled *1*, *2* and *s*, and its connected monoshaft with *I*. Likewise, the shafts of gear train *II* are labeled with *1'*, *2'*, *s'*, and *II*.

2. The designations are then transferred to the Wolf-symbol. As shown in fig. 125b, shafts *I*, *II* and *S* are the connected shafts:

Gear train *I:* *s* is a part of the free coupling shaft; *2* is the connected monoshaft *I;* *1* is a part of the connected coupling shaft.

Gear train *II:* *s'* is a part of the connected coupling shaft; *2'* is the connected monoshaft *II;* *1'* is a part of the free coupling shaft.

3. Determination of the basic speed-ratios according to section 9. The number of teeth of the individual gears are given in fig. 125a:

$$\text{Gear train } I: i_o = \frac{n_1}{n_2} = \left(-\frac{z_p}{z_1}\right)\left(-\frac{z_2}{z_p}\right) = \frac{20}{-70} = -\frac{2}{7}$$

$$\text{Gear train } II: i_{o'} = \frac{n_{1'}}{n_{2'}} = \left(-\frac{z_{p1'}}{z_{1'}}\right)\left(-\frac{z_{2'}}{z_{p2'}}\right) = \frac{42}{20}\cdot\frac{40}{22} = \frac{42}{11}.$$

4. Determination of the series speed-ratio according to eq. (75), section 33 and worksheet 1:

$$i_{III} = \frac{n_I}{n_{II}} = i_{mf}i_{f'm'} = i_{2s}i_{1'2'} = \frac{i_o - 1}{i_o}\cdot i_{o'}$$

$$= \frac{-9/7}{-2/7}\cdot\frac{42}{11} = \frac{9\cdot 21}{11} = \frac{189}{11}.$$

5. Determination of the shaft speeds:
Given $n_I = 25s^{-1}$; $n_{II} = -12s^{-1}$. Sought are all other shaft speeds:

$$n_S = n_I\cdot\frac{n_S}{n_I} = n_I k_{SI}.$$

With k_{SI} from worksheet 2, box 19 and

$$k_{III} = \frac{n_I}{n_{II}} = -\frac{25}{12},$$

we obtain:

$$k_{SI} = \frac{1 - \dfrac{i_{III}}{k_{III}}}{1 - i} = \frac{1 + \dfrac{189 \cdot 12}{11 \cdot 25}}{1 - \dfrac{189}{11}} = -0.5715 ,$$

so that $n_S = n_1 k_{SI} = 25(-0.5715) = -14.29 s^{-1}$.

Finally,

$$n_F = n_{1'} = n_s = n_2 \cdot \frac{n_s}{n_2} = n_2 k_{s2} .$$

With

$$n_2 = n_I \quad \text{and} \quad k_{s2} = \frac{k_{12} - i_o}{1 - i_o} \quad \text{from worksheet 2, where}$$

$$k_{12} = k_{SI} = -0.5715 ,$$

$$n_F = n_I \cdot \frac{k_{12} - i_o}{1 - i_o} = 25 \frac{-0.5715 - \left(-\dfrac{2}{7}\right)}{1 - \left(-\dfrac{2}{7}\right)} = -5.556 s^{-1} .$$

With the previously calculated series speed-ratio $i_{III} = 189/11 \approx 17.2$, and the given speed-ratio $k_{III} = -(12/25) = -0.48$, we can determine $k_{SI} = -0.58$ faster by using fig. 64 if only approximate values of the shaft speeds are sought. It is then

$$n_S = n_1 k_{SI} \approx -25 \cdot 0.58 = -14.5 s^{-1} ,$$

or with $k_{IIS} \approx 0.83$ from fig. 63,

$$n_S = \frac{n_{II}}{k_{IIS}} \approx \frac{-12}{0.83} = -14.4 s^{-1} .$$

The speed of the free coupling shaft can be found from the known speeds of gear train I:

$$n_F = n_s = n_2 \cdot \frac{n_s}{n_2} = n_1 k_{s2} .$$

With $k_{12} = k_{SI} = -0.58$ and $i_o = -(2/7) \approx -0.29$, $k_{s2} \approx -0.23$ can be found from fig. 63. Then

$$n_F \approx -25 \cdot 0.23 = -5.7 \sec^{-1} .$$

According to section 11c, the speeds of the planets relative to the carriers become:

$$(n_p - n_s) = (n_1 - n_s)\left(-\frac{z_1}{z_p}\right) = (-14.29 + 5.56)\frac{70}{25} = -24.4s^{-1} ,$$

$$(n_{p2'} - n_{s'}) = (n_{2'} - n_{s'})\left(-\frac{z_{2'}}{z_{p2'}}\right) = (-12 + 14.29)\left(-\frac{40}{20}\right) = -4.16s^{-1} .$$

This analysis could also be performed graphically with the aid of a Kutzbach diagram, as will be illustrated by another example in section 40, figs. 156b, c, d.

c) Torques

Each of the three shaft torques of each component transmission bears a constant ratio relative to each of the others which, according to eqs. (42), (44) and (45), is determined for each of these transmissions by its basic speed-ratio i_o or $i_{o'}$ and its basic efficiency η_o or $\eta_{o'}$. Moreover, it is obvious from fig. 2, and also is expressed in R 4, that the torques of the two shafts which together constitute the free coupling shaft are equal and opposed to each other. On the basis of these two rules, we can calculate all the torque ratios of not only a simple bicoupled transmission, but of any arbitrary compound epicyclic transmission as well.

For a simple bicoupled transmission the torques of the two shafts, which together constitute the *connected* coupling shaft, thus are also determined. As a rule, these torques are different and depending on the basic speed-ratios i_o and $i_{o'}$ of the component transmissions they may have equal or opposite signs. Their sum is equal to the external torque of the connected coupling shaft.

If the input and output shafts of a bicoupled transmission are interchanged, the power-flow in the component transmissions is reversed and the exponents $r1$ and $r1'$ change their signs. Thus, the torque ratios of a bicoupled transmission depend on the position of the input and output shafts.

For example: Determination of the torques of the transmission shown in fig. 125: The torque $T_1 = 50$ Nm may be given. The speeds are as calculated in the previous section. Since we found that $n_1 > 0$, shaft I is an input shaft:

1. Determination of the basic efficiencies η_o and $\eta_{o'}$ as described in section 6. The friction coefficient $\mu_1 = 0.07$ and the coefficient f_L for standard gears can be found from fig. 15:

Gear train I:

$$i_{2p} = -\frac{25}{20} = -1.25 , \qquad z_{pinion} = z_2 = 20 , \qquad f_L = 0.19$$

$$i_{p1} = \frac{70}{25} = 2.8 \ , \qquad z_{\text{pinion}} = z_p = 25 \ , \qquad f_L = 0.068$$

$$\zeta_o \approx \mu_t \Sigma f_L = 0.07 \ (0.19 + 0.068) = 0.0181 \ ,$$

$$\eta_o = 1 - \zeta_o = 1 - 0.0181 \approx 0.982 \ .$$

Gear train *II:*

$$i_{p\acute{2}2'} = -\frac{40}{22} = -1.82 \ ; \qquad z_{\text{pinion}} = z_{p\acute{2}} = 22 \ , \qquad f_L = 0.160 \ ,$$

$$i_{1'p\acute{1}} = -\frac{42}{20} = -2.1 \ ; \qquad z_{\text{pinion}} = z_1 = 20 \ , \qquad f_L = 0.166 \ ,$$

$$\zeta_{o'} = 0.07 \ (0.160 + 0.166) = 0.0228 \ ,$$

$$\eta_{o'} = 1 - 0.0228 = 0.977 \ .$$

2. Determination of the appropriate exponent $r1$. When T_1, n_1, and n_s are given, the exponents $r1$ and $r1'$ can be determined from eq. (43) in section 12. However, as will be explained subsequently, the exponent can be found faster with the aid of worksheet 4 or from figs. 63 to 65.

Gear train *I*:
Since $i_o < 0$, the first four rows of worksheet 4 are applicable. Since $k_{12} = -0.57 < i_o$, shaft *1* is the total-power shaft and since shaft $I = 2$ is the input shaft, the power-flow is $\frac{5}{2} > 1$ according to R 14. Thus we find from worksheet 4 that

$$r1 = -1 \ .$$

Gear train *II*:

With $i_{o'} = 42/11 > 1$, the third and last group of rows on worksheet 4 must be used. Since $k_{1'2'} = n_F/n_{II} = 0.463, 0 < k_{1'2'} < 1$ and thus shaft $2'$ is the total-power shaft. Since shaft s is an input shaft (see power-flow of gear train *I*), shaft $1'$ must be an output shaft. Thus the power-flow is $2' <_{s'}^{1'}$, for which we find from worksheet 4 that:

$$r1' = +1 \ .$$

We also could have obtained the total-power shafts of the two component gear trains with i_o and k_{12} from fig. 63 or 64 and with $i_{o'}$ and $k_{2's'} = n_{II}/n_S$ from figs. 63 or 65.

3. Determination of the torques from eqs. (42), (44), or (45), or worksheet 6. With the input torque $T_1 = T_2 = 50$ Nm and $r1 = -1$ and $r1' = +1$,

$$T_1 = -\frac{T_2}{i_o \eta_o^{r1}} = -\frac{50 \cdot 0.982}{-2/7} = 171.9 \text{ Nm} ,$$

$$T_s = T_2 \left(\frac{1}{i_o \eta_o^{r1}} - 1 \right) = 50 \left(\frac{0.982}{-2/7} - 1 \right) = -221.9 \text{ Nm} ,$$

$$T_{1'} = -T_s = 221.9 \text{ Nm} ,$$

$$T_{II} = T_{2'} = -T_{1'} i_{o'} \eta_{o'}^{r1} = -221.9 \cdot \frac{42}{11} \cdot 0.977 = -827.7 \text{ Nm} ,$$

$$T_{s'} = T_{1'} (i_{o'} \eta_{o'}^{r1} - 1) = 221.9 \left(\frac{42}{11} \cdot 0.977 - 1 \right) = 605.8 \text{ Nm} .$$

Consequently, the torque of the connected coupling shaft becomes:

$$T_S = T_1 + T_{s'} = 171.9 + 605.8 = 777.7 \text{ Nm} .$$

4. Simplified approximate determination of the torques. The torques can be determined approximately when the losses are neglected, that is, we assume that $\eta_o = \eta_{o'} = 1$. This simplification, however, is permissible only when the efficiencies of the component gear trains are close to 1, that is, when the rolling-power is not substantially larger than the total external power (e.g., when it is no more than twice as large). We shall investigate the validity of this assumption with the aid of figs. 68, 69, or 70.

Gear train I:

Since $i_o = -2/7 = -0.286$, gear train I is a negative-ratio transmission and fig. 70 is applicable. In section 2 we found that $k_{12} = k_{SI} = -0.57$ so that $P_R/P_{tot} \approx 0.62$. For gear train I we could have omitted this check since, according to section 14g, and R 15, $P_R < P_{tot}$ in negative-ratio gear trains.

Gear train II:

$$i_{o'} = \frac{42}{11} = 3.82 > 1 \qquad \text{so that fig. 69 applies:}$$

$$k_{1'2'} = \frac{n_{1'}}{n_{2'}} = \frac{-5.556}{-12} = 0.463 , \qquad \text{which leads to} \qquad \frac{P_R}{P_{tot}} = 0.2 .$$

Thus, for both component gear trains $P_R < P_{tot}$, so that $\eta > \eta_0$ and for the purpose of an approximation the losses can indeed be neglected. Consequently, if we let $\eta_0 = 1$, and omit the determination of $r1$ and $r1'$, we can calculate the torques from eqs. (42), (44), and (45) as follows:

$$T_1 \approx -\frac{T_2}{i_0} = -\frac{50}{-2/7} = 175\,\text{Nm} ,$$

$$T_s \approx T_2\left(\frac{1}{i_0} - 1\right) = 50\left(-\frac{7}{2} - 1\right) = -225\,\text{Nm} ,$$

$$T_{1'} = -T_s \approx 225\,\text{Nm} ,$$

$$T_{2'} \approx -T_{1'}i_{0'} = -225 \cdot 3.82 = -860\,\text{Nm} ,$$

$$T_{s'} \approx T_{1'}(i_{0'} - 1) = 225\,(3.82 - 1) = 635\,\text{Nm} ,$$

$$T_S = T_1 + T_{s'} \approx 175 + 635 = 810\,\text{Nm} .$$

Check where the power-flow mode is not considered because $\eta = 1$. According to eqs. (82) and (86),

$$T_{2'} = T_{II} = -T_I i_{III} = -50 \cdot \frac{189}{11} = -860\,\text{Nm} ,$$

$$T_S = T_I(i_{III} - 1) = -50 \cdot \frac{189 - 11}{11} = 810\,\text{Nm} \qquad \text{and}$$

$$T_I + T_{II} + T_S = 50 - 860 + 810 = 0 .$$

A comparison of these approximate torque values with the accurate values calculated in section 3 shows that the difference for all shafts is less than 5%.

A check of the power-flow mode as conducted in section 4 is required only for the individual component gear trains. If it shows that the power-ratio P_R/P_{tot} is small, then an approximate calculation of the shaft torques on the basis of the assumption that $\eta_0 = 1$ is sufficiently accurate even if the bicoupled transmission is self-locking as, for example, the transmission shown in fig. 122, section 33i. If, however P_R/P_{tot} is large, considerable *percentage* deviations occur for some of the shaft torques even when the bicoupled transmission has a very high efficiency, such as 0.9977, as does the transmission shown in fig. 121.

d) Efficiencies

The efficiencies of simple bicoupled transmissions can be determined by the same methods discussed in section 15 for simple revolving drive trains. The necessary equations are tabulated in worksheet 4. However, the loss dissymmetry of series-coupled compound transmissions can no longer be neglected since it may be substantially larger than in simple revolving drive trains.

To simplify the time-consuming determination of the efficiencies and the necessary prior determination of the exponent rI, the efficiency equations for simple bicoupled transmissions have been compiled, ready for use, in worksheet 4.

Analogous to fig. 52 the position of the total-power shaft for each of the three ranges of the series speed-ratio $i_{\mathrm{III}} < 0$; $0 < i_{\mathrm{III}} < 1$ and $i_{\mathrm{III}} > 1$ is unambiguously characterized by the speed-ratio k_{III}. Therefore, worksheet 4 lists the efficiency equations for each power-flow in terms of k_{III}. If the efficiency of a bicoupled transmission for a particular state of motion is sought and its series-ratio i_{III}, and an arbitrary speed-ratio k, are known, then k_{III} can be determined with the help of worksheet 2. Subsequently, the two applicable power-flows with the same total-power shaft can be found from worksheet 4. Which of these applies can be decided if, in addition, any of the shafts is known to be the input or output shaft. At the same time this identifies the correct efficiency equation.

Worksheet 4 also lists the values of the exponent rI which are definitely associated with each power-flow. This value is needed for an accurate determination of the torques using eqs. (82), (86), or (87).

Example: Determination of the efficiency of the transmission shown in fig. 125. Its specifications are as determined in the previous sections:

$$i_{\mathrm{III}} = 17.2 \; ; \qquad \eta_0 = 0.982 \; ; \qquad \eta_{0'} = 0.977 \; ; \qquad k_{\mathrm{III}} = -2.08 \; ;$$

$$i_0 = -0.286 \; ; \qquad i_{0'} = 3.82 \; ; \qquad \text{shaft } I \text{ is the input shaft.}$$

1. Series efficiencies:
For $i_0 < 1$, we find from worksheet 5:

$$\eta_{\mathrm{III}} = \eta_{2s}\eta_{1'2'} = \frac{\eta_0 - i_0}{1 - i_0}\eta_{0'} = \frac{0.982 + 0.286}{1 + 0.286}0.977$$
$$= 0.963 \; ,$$

$$\eta_{\mathrm{III}} = \eta_{2'1'}\eta_{s2} = \eta_{0'}\frac{i_0 - 1}{i_0 - 1/\eta_0} = 0.977 \cdot \frac{-1.286}{-0.286 - 1/0.982}$$
$$= 0.963 \; .$$

2. Overall efficiencies:

For $i_{III} > 1$, the lower group of rows on worksheet 6 must be used. Since $k_{III} = -2.08 < 0$, shaft S is the total-power shaft. Thus, with shaft I as input shaft, the power-flow is $\overset{I}{II} > S$ and the efficiency becomes

$$\eta_{\overset{I}{II}>S} = \frac{(k_{III} - i_{III})(1 - i_{III}\eta_{III})}{(k_{III} - i_{III}\eta_{III})(1 - i_{III})}$$

$$= \frac{(-2.08 - 17.2)(1 - 17.2 \cdot 0.963)}{(-2.08 - 17.2 \cdot 0.963)(1 - 17.2)} = 0.994 .$$

3. Total efficiency of the same transmission at different speeds:

For $\quad n_I = 25$, $\quad n_S = -1s^{-1}$, $\quad k_{IS} = 25/-1 = -25$,

and with worksheet 2:

$$k_{III} = \frac{k_{IS}i_{III}}{k_{IS} - 1 + i_{III}} = \frac{-25 \cdot 17.2}{-25 - 1 + 17.2} = 48.86 .$$

For the same transmission, the same lower group of rows on worksheet 4 is applicable and since $k_{III} = 48.86 > 17.2$, that is, since $k_{III} > i_{III}$, the input shaft has now become the total-power shaft. Thus the power-flow is $I < \overset{II}{S}$ and the efficiency

$$\eta_{I<\overset{II}{S}} = \frac{k_{III} - i_{III} + i_{III}\eta_{III}(1 - k_{III})}{k_{III}(1 - i_{III})}$$

$$= \frac{48.86 - 17.2 + 17.2 \cdot 0.963(1 - 48.86)}{48.86(1 - 17.2)} = 0.962 .$$

Section 35. Constrained Bicoupled Transmission

a) Treatment as a Special Case of the Simple Bicoupled Transmission with Two Degrees of Mobility

According to the definition given in section 32e, the constrained bicoupled transmission is merely a special case of the bicoupled transmission with two degrees of mobility where the monoshaft of component transmission II is always connected to the housing so that its speed $n_{II} = 0$. Thus, the constrained bicoupled transmission can be analyzed as shown in section 34 for the bicoupled transmission with two degrees of mobility.

It has operating characteristics identical to those of the simple two-shaft transmissions, and can be analyzed like these transmissions using the analogy between the two transmission types (see sec. 33).

Nevertheless, a special treatment of the constrained bicoupled transmissions is justified since the locked monoshaft simplifies its analysis. In view of the many different applications of this transmission type, this advantage should be used, particularly in special purpose designs of reduced and infinitely variable bicoupled transmissions.

b) Main and Auxiliary Component Transmissions

The analysis and synthesis of a constrained bicoupled transmission becomes very clear when, as shown in fig. 113 and discussed in section 32e, the transmission is resolved into a main component transmission I and an auxiliary component transmission II. In fig. 113, the main component transmission is a simple three-shaft transmission to which the input and output of a bicoupled transmission are connected. Its speed characteristics are unequivocally determined by the fact that the speed-ratio k_{cf} of its shafts c and f must equal the speed-ratio $i_{c'f'}$ of the constrained auxiliary component gear train. The speed characteristics, however, also determine the overall operating characteristics of the main component transmission which can be analyzed completely with the rules given in part II of this book. Its speeds n_m and n_c are also the speeds of the bicoupled transmission, and its torque T_m is the torque of the monoshaft of the bicoupled transmission:

$$n_m = n_I \qquad n_c = n_S \qquad T_m = T_I \ .$$

c) Speeds

When the basic speed-ratios of the main and auxiliary component transmissions, and the structural coupling case as described in fig. 124, are known, then the speed-ratios of the auxiliary component transmission can be determined with the aid of worksheet 1. Subsequently, the speed-ratio of the bicoupled transmission can be determined with the aid of worksheet 2, or the diagrams in figs. 63 to 65, since according to fig. 110,

$$k_{cf} = i_{c'f'} \ .$$

In section 35h, the equations needed for a layout of such a bicoupled transmission are derived and the results plotted in the form of a family of characteristic curves. Independent of the structure of the respective main and auxiliary component transmissions, these equations are generally valid for constrained bicoupled transmissions.

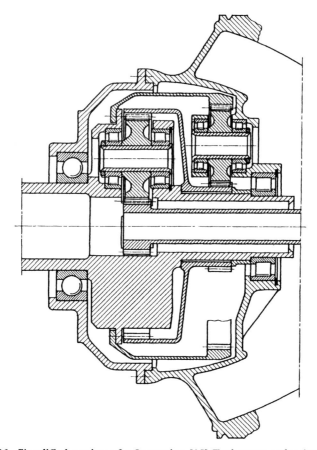

Fig. 126. Simplified section of a Lycoming [15] Turboprop-reduction drive.

For example, the speed analysis of the turboprop reduction drive shown in fig. 126: Before an analysis can begin, the Wolf-symbol of the transmission must be drawn. The proper labels for this symbol have been discussed in section 31 and must be carefully chosen. It is expedient to start with a schematic sketch of the transmission from which the Wolf-symbol can be derived. In general, the determination of the speeds proceeds according to the following sequence.

1. A schematic sketch of the transmission as shown in fig. 127a is drawn.

2. The component gear train which has three rotating shafts is designated as the main gear train I, the gear train with one locked shaft as the auxiliary gear train II.

3. The carrier shafts of the main gear train and the auxiliary gear train are denoted by s and s', respectively.

4. The central gear shafts of the main gear train are arbitrarily denoted by *1* and *2* and those of the auxiliary gear train by *1'* and *2'*.

5. A symbolic representation of the bicoupled transmission, as shown in fig. 127b, is drawn. The component gear trains are denoted by *I* and *II*, the external shafts by *I*, *S* and *II*, where shaft *II* is locked; see also fig. 113.

6. The labels of the shafts are transcribed from fig. 127a to fig. 127b, where *1* becomes the monoshaft *I* and *s'* becomes the fixed monoshaft *II*; *s* and *1'* form the connected coupling shaft; *2* and *2'* constitute the free coupling shaft.

7. According to section 9, the basic speed-ratios of the component gear trains are obtained from the ratio of the pitch circle diameters:

$$i_o = \frac{n_1}{n_2} = \left(-\frac{d_p}{d_1}\right)\left(-\frac{d_2}{d_p}\right) = \frac{d_2}{d_1} = -4.3$$

$$i_{o'} = \frac{n_{1'}}{n_{2'}} = \left(-\frac{d_{p'}}{d_{1'}}\right)\left(-\frac{d_{2'}}{d_{p'}}\right) = \frac{d_{2'}}{d_{1'}} = -0.36 \ .$$

8. Speed-ratio of the auxiliary gear train:

$$i_{c'f'} = i_{1'2'} = i_{o'} = -0.36 \ .$$

9. Speed-ratio of the bicoupled transmission. From worksheet 2, box 18, it is found that:

$$i_{IS} = k_{1s} = (1 - i_o) + \frac{i_o}{k_{s2}} \ . \quad \text{With} \quad k_{s2} = i_{1'2'} = -0.36 \ ,$$

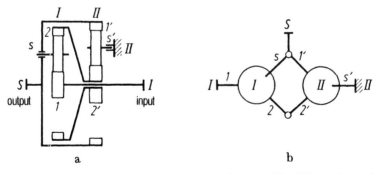

a   b

Fig. 127. Analysis of the simple bicoupled transmission of fig. 125: *a*, schematic representation of the gear train; *b*, symbolic representation according to Wolf with specific shaft designations as in *a*. The transmission corresponds to the structural coupling case *cm'* of fig. 123.

$$i_{IS} = 1 + 4.3 + \frac{-4.3}{-0.36} = 17.24 \ .$$

Check with the aid of the analogy:

$$i_{III} = i_o i_{2's'} = i_o \frac{i_{o'} - 1}{i_{o'}} = -4.3 \frac{-1.36}{-0.36} = -16.24 \ ,$$

$$i_{IS} = 1 - i_{III} = 17.24 \ .$$

Check with the aid of fig. 65: For $k_{s2} = -0.36$ and $i_o = -4.3$, we find that

$$k_{s1} \approx 0.06 \quad \text{or} \quad k_{1s} \approx \frac{1}{0.06} \approx 17 \ .$$

d) Torques

As far as the torques are concerned, there is no difference between a constrained bicoupled transmission and a bicoupled transmission with two degrees of mobility, since, according to R 11, the torque of the locked monoshaft *II* is independent of its speed. Therefore, the simplest method to determine the external torques T_I, T_{II}, and T_S of a bicoupled transmission is to use the analogy and eqs. (82) to (87) where the appropriate exponent r_I can be found on worksheet 5. The shaft-torques of the simple gear trains can be calculated with eqs. (42) to (45) as described in section 34c.

Example: Determination of the torques of the bicoupled transmission shown in figs. 126 and 127. The input torque $T_I = 350$ Nm and the basic efficiencies $\eta_o = \eta_{o'} = 0.985$ may be given. According to the results obtained in the previous sections, $i_{III} = -16.24$; $i_o = -4.3$; $i_{o'} = 0.36$. With worksheet 5 the efficiency is found to be:

$$\eta_{III} = \eta_{12}\eta_{2's'} = \eta \frac{i_o - \eta_{o'}}{i_{o'} - 1} = 0.985 \frac{-1.345}{-0.36 - 1} = 0.974$$

With eqs. (82) and (86) and $r_I = +1$ from worksheet 5, box 7, the values of the torques become:

$$T_{II} = -T_I i_{III} \eta_{III}^{r_I} = +350 \cdot 16.24 \cdot 0.974 = 5540 \, \text{Nm} \ ,$$

$$T_S = T_I (i_{III} \eta_{III}^{r_I} - 1) = 350(-16.24 \cdot 0.974 - 1) = -5890 \, \text{Nm} \ .$$

Starting with $T_1 = T_I$, where $T_1 > 0$, $n_1 > 0$, $n_s < 0$, and with $r_1 = +1$ from eq. (43), the torques of the simple gear trains become

$$T_2 = -T_1 i_o \eta_o^{r1} = -350(-4.3)0.985 = 1480\,\text{Nm} ,$$

$$T_s = T_1(i_o \eta_o^{r1} - 1) = 350(-4.3 \cdot 0.985 - 1) = -1830\,\text{Nm} ,$$

$$T_{2'} = -T_2 = -1480\,\text{Nm} ,$$

$$T_{1'} = T_s - T_s = -5890 + 1830 = -4060\,\text{Nm} .$$

Check: For the bicoupled transmission, as well as for each of the component gear trains, the sum of the torques must equal zero, $\Sigma T = 0$.

e) Efficiency and Self-Locking

According to the analogy described in section 33e, the efficiency of a bicoupled transmission can be found from the equations compiled in worksheet 5 which are ready for use. The conditions under which self-locking can occur are discussed exhaustively in section 33i.

Example: Determination of the efficiencies of the transmission shown in figs. 126 and 127 using the results obtained in the previous two sections: $i_{1\text{II}} = -16.24$; $\eta_{1\text{II}} = 0.974$. Since the bicoupled gear train has a negative speed-ratio, box 7 on worksheet 5 applies and

$$\eta_{\text{IS}} = \frac{i_{1\text{II}}\eta_{1\text{II}} - 1}{i_{1\text{II}} - 1} = \frac{-16.24 \cdot 0.974 - 1}{-16.24 - 1} = 0.976 .$$

If we would interchange the input and output, then we would find from box 10 in the same column of worksheet 5 that:

$$\eta_{\text{S1}} = \frac{i_{1\text{II}} - 1}{\dfrac{i_{1\text{II}}}{\eta_{1\text{II}}} - 1} = \frac{-17.24}{\dfrac{16.24}{0.974} - 1} = 0.975 , \qquad \text{where}$$

$$\eta_{1\text{II}} = \eta_{\text{S'2'}}\eta_{21} = \frac{i_{o'} - 1}{i_{o'} - \dfrac{1}{\eta_{o'}}}\,\eta_o = \frac{-1.36}{-1.375} \cdot 0.985 = 0.974 .$$

f) Power-Flow in Constrained Bicoupled Transmissions

According to fig. 52, the total-power shaft of the main component transmission is determined when its basic speed-ratio i_o and the speed-ratio of its two coupling shafts, which is equal to the speed-ratio of the auxiliary component transmission, have been chosen. If, in addition, one of its shafts is known to be an input or output shaft, which usually is true for the monoshaft I, then, according to R 14, the external power-flow of the main component transmission is completely known, that is, it is known for each of its shafts, whether it is an input or output shaft (see sec. 15c).

Since each of the three shafts of the main component transmission can be the total-power shaft, the power-flow can generally take three different paths through the main component transmission and, thus, through the bicoupled transmission. With shaft I as input shaft, the following power-flows are possible.

1. As shown in fig. 128a, the input shaft $I = m$ is also the total-power shaft. Consequently, the other two shafts c and f are output shafts. The part of the output power which is transmitted by shaft f becomes the input power of the auxiliary component transmission. It flows through the auxiliary component transmission and is then transmitted as its output power to the connected coupling shaft S. There it combines with that part of the output power of the main component transmission which is transmitted directly to S by shaft c. Together, the two power components constitute the total effective output power of the bicoupled transmission. This type of power-flow is shown in fig. 128b and is defined as power division (PD). The torque transmitted by the connected coupling shaft equals the sum of the torques transmitted by the two connected shafts: $|T_S| = |T_c| + |T_{c'}|$.

2. As shown in fig. 128c, shaft c, which is a part of the connected coupling shaft, is the total-power shaft. Since $I = m$ is an input shaft, the second partial-power shaft f must also be an input shaft. Consequently, shaft c is the sole output shaft and as the total-power shaft it carries a larger power than is introduced at I by the partial-power shaft m. The difference flows through c' and f' and back to the main component transmission through the free coupling shaft f. Thus, it flows in a closed path through

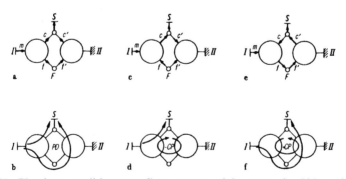

Fig. 128. The three possible power-flow patterns of the constrained bicoupled transmission, depending on the position of the total-power shaft of the main component transmission: a–b, the total-power shaft at m results in power division; c–d, the total-power shaft at c causes a negative circulating power; e–f, the total-power shaft at f causes a positive circulating power.

both component transmissions and is therefore called the *"circulating power,"* which, of course, is not an effective power. Birkle [25] calls it *"negative"* circulating power since it flows through the auxiliary component transmission in a direction which is opposite to the direction of the partial power in the power division mode. This type of power-flow is illustrated in fig. 128d. Following Birkle we shall define it as negative circulating power $(-CP)$. In this power-flow mode the circulating power is the *only* power which flows through the auxiliary component transmission and thus the connected coupling shaft S transmits only the *difference* of the torques carried by the two coupled shafts: $|T_S| = |T_c| - |T_{c'}|$.

3. As shown in fig. 128e, shaft f, which is connected to the free coupling shaft F, is the total-power shaft. Like shaft c in the previous example, shaft f is the sole output shaft. It transmits a larger power than introduced at $I = m$. The difference is again the circulating power which flows through both component transmissions in a closed path. In this case, however, its direction in the auxiliary component transmission is identical with the direction of the partial power-flow in the power division mode (PD).

This power-flow mode is illustrated in fig. 128f. It shall be defined as positive circulating power and it is characterized by the fact that the effective power *and* the circulating power simultaneously flow through the auxiliary component transmission. However, only the difference of the torques which are transmitted by the two connected shafts acts on the connected coupling shaft: $|T_S| = |T_{c'}| - |T_c|$.

If the signs of the external torques of the bicoupled transmission which act at I and S change, while their direction of rotation remains the same, that is, if input and output are exchanged, then the signs of *all* of the powers in the bicoupled transmission are reversed. However, the power-flow modes—power division and circulating power—are unaffected because they depend only on the position of the total-power shaft in the main component transmission.

These considerations are expressed in the following guide rule:

A constrained bicoupled transmission operates with internal   R 31
power division when the speed-ratio of the auxiliary com-
ponent transmission is chosen in such a way that the mono-
shaft m of the main component transmission becomes the lat-
ter's total-power shaft. If the total-power shaft of the main
component transmission is part of the free coupling shaft,
then it operates with positive circulating power; if part of the
connected coupling shaft it operates with negative circulating
power.

Which of the shafts, in any particular case, acts as the total-power shaft can be found with the aid of the figs. 63, 64, or 65, provided the basic speed-ratio i_0 of the main component transmission and one of its speed-ratios k are known.

Whether a constrained bicoupled transmission operates with power division or with circulating power depends on the mutual positions of the summation shafts of its main and auxiliary component transmissions. This criterion for its state of motion has been introduced by Wolf [13]. If, in a constrained bicoupled transmission which operates with power division, the total-power shaft m simultaneously is the summation shaft of the main component transmission, then the two coupled partial-power shafts are difference shafts and their torques have the same sign. Consequently, the two shafts must also have the same direction of rotation. The auxiliary component transmission, therefore, has a positive speed-ratio, and, according to R 25, its fixed monoshaft $II = m'$ must be a difference shaft as shown, for example, in fig. 129a.

If, however, the transmission operates with power division and the total-power shaft m is a difference shaft, then the two partial-power shafts, that is, the summation shaft and one of the difference shafts, rotate in opposite directions. The auxiliary component transmission then has a negative speed-ratio so that according to R 25, its fixed monoshaft must be the summation shaft, as shown in fig. 129b. If these conditions for the auxiliary component transmission are not satisfied, then a positive or negative circulating power will arise in the bicoupled transmission, as indicated in the captions for figs. 129c and 129d. This leads to R 32:

> *A constrained bicoupled transmission operates with power* R 32
> *division when its two monoshafts are a summation shaft and*
> *a difference shaft. If both monoshafts are summation shafts*
> *or difference shafts, a circulating power flows in the*
> *transmission.*

These circulating powers are not identical with the futile series powers and the futile powers flowing at the coupling speed which have been described in section 33g. They have been discussed in this section merely to

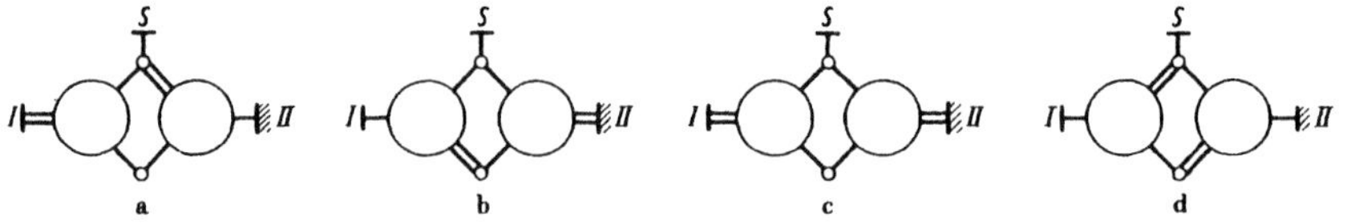

Fig. 129. Examples showing the effect of summation-shaft positioning on the power-flow in constrained bicoupled transmissions: *a–b*, power division because monoshafts *I* and *II* serve different functions; *c–d*, circulating power arises because monoshafts *I* and *II* have identical functions.

develop a better understanding of the behavior of compound revolving drive trains. A quantitative consideration of these power-flows, however, is not required for the analysis and layout of these transmissions.

g) Power-Flow in Auxiliary Component Transmissions

It is frequently of interest to know what percentage of the input power of a bicoupled transmission flows through the auxiliary component transmission. An investigation of this problem by Birkle [25] led to a number of very simple relationships when the assumption was made that the bicoupled transmission incurs no losses, so that $\eta_{IS} = \eta_{SI} = 1$. Birkle defined a power-ratio which he further qualified by the subscript o to indicate the assumption of a loss-free transmission.

If, with the symbols introduced in fig. 110a, for a loss-free transmission

$$\epsilon_o = \frac{\text{input power of the aux. component trans.}}{\text{input power of the constrained bicoupled trans.}} = \frac{P_{f'}}{P_I} \quad (92)$$

$$= -\frac{P_f}{P_I} = -\frac{n_f T_f}{n_m T_m} ,$$

then, according to eq. (6) and with $\eta = 1$, that is, for a loss-free transmission:

$$\frac{T_f}{T_m} = -i_{mf} .$$

With the generalized Willis-formula and an arbitrarily chosen subscript key $c \triangleq 1, m \triangleq 2, f \triangleq s,$ the speed-ratios

$$k_{fm} = \frac{k_{cm} - i_{cm}}{1 - i_{cm}} \quad \text{and} \quad i_{mf} = \frac{i_{cm} - 1}{i_{cm}}$$

can be obtained with the aid of worksheets 2 and 1. Then:

$$\epsilon_o = k_{fm} i_{mf} = \frac{k_{cm} - i_{cm}}{1 - i_{cm}} \cdot \frac{i_{cm} - 1}{i_{cm}} = 1 - \frac{k_{cm}}{i_{cm}} .$$

With $k_{cm} = i_{SI}$, the power-ratio becomes: $\quad \epsilon_o = 1 - \frac{i_{SI}}{i_{cm}} ,$ \quad (93)

where i_{SI} is the overall speed-ratio of the bicoupled transmission, and i_{cm} the basic or revolving-carrier speed-ratio between those shafts c and m of the main component transmission which at the moment form the input and output shafts of the bicoupled transmission.

For the synthesis of bicoupled transmissions where the power-flow in the auxiliary component transmission must be considered, the following equation expresses the power-ratio exclusively in terms of the characteristic parameters, that is, the speed-ratios of the bicoupled and auxiliary component transmissions.

If the next-to-last equation above is solved for i_{cm}, it becomes

$$i_{cm} = \frac{k_{cm}}{1 - \epsilon_0}.$$

If now the equation $k_{12} = f(k_{1s})$ from worksheet 2 is transcribed with the previously given subscript key, it takes the form

$$k_{cm} = \frac{k_{cf} i_{cm}}{k_{cf} - 1 + i_{cm}}.$$

This equation can likewise be solved for i_{cm} and then becomes:

$$i_{cm} = \frac{k_{cm} k_{cf} - k_{cm}}{k_{cf} - k_{cm}}.$$

Since the two equations in i_{cm} must be equal, we can write

$$\frac{k_{cm} k_{cf} - k_{cm}}{k_{cf} - k_{cm}} = \frac{k_{cm}}{1 - \epsilon_0} \quad \text{and solve for } \epsilon_0, \text{ so that}$$

$$\epsilon_0 = \frac{k_{cm} - 1}{k_{cf} - 1}.$$

With $k_{cm} = i_{SI}$ and $k_{cf} = i_{c'f'}$, the desired form of the power-ratio finally becomes

$$\epsilon_0 = \frac{i_{SI} - 1}{i_{c'f'} - 1}. \tag{94}$$

This equation has been derived earlier by Laughlin, Holowenko and Hall [10]. Plotted for $-\infty < i_{c'f'} < \infty$ and $-\infty < i_{SI} < \infty$, it is given in fig. 130. Because it has been assumed that $\eta = 1$, eq. (94) and fig. 130 are independent of the power-flow. Therefore they are valid, whether shaft I or shaft S is the input shaft. If a bicoupled transmission operates with power division, the power-ratio $0 < \epsilon_0 < 1$; if a positive or negative circulating power flows through the auxiliary transmission, $\epsilon_0 > 1$ or $\epsilon_0 < 0$ respectively, as can be expected from fig. 128.

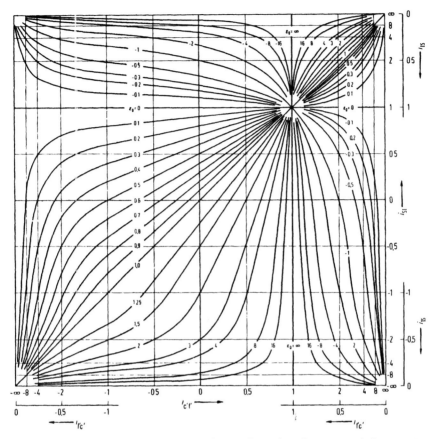

Fig. 130. Dependence of the power-ratio ϵ_0 of the loss-free transmission on the speed-ratios i_{S1} of the constrained bicoupled transmission and $i_{c'f'}$ of the auxiliary component transmission.

$$\epsilon_0 = \frac{\text{input power of the auxiliary component transmission}}{\text{input power of the constrained bicoupled transmission}} = P_{c'}/P_S = T_{c'}/T_S$$

Eq. (94) shows that the power-ratio ϵ_0 depends only on the speed-ratio of the bicoupled and auxiliary component transmissions but is independent of the type of coupling between the component transmissions, which are considered loss free. This finding is very useful for the practical synthesis of bicoupled transmissions.

The theoretical power-ratio ϵ_0 corresponds well with the power-ratio ϵ of a real transmission as long as the transmission losses are small, that is, as long as a calculation with the aid of worksheet 5 confirms that the chosen coupling case has a high efficiency.

The power-ratio of a given real transmission can be found when the speeds and torques which have been calculated as described in section 34c and section 35c, are substituted into eq. (92). Thus, if shaft I is the input shaft:

$$\epsilon = - \frac{n_f T_f}{n_m T_m} ,$$

and if shaft S is the input shaft:

$$\epsilon = \frac{n_{c'} T_{c'}}{n_S T_S} = \frac{T_{c'}}{T_S} . \tag{95}$$

In [25] the equations defining ϵ are tabulated in general form for all of the coupling cases shown in fig. 124 where the carrier of the auxiliary component transmission is fixed, that is, cases cm', mm' and fm'.

It may be pointed out that the power-ratio ϵ or ϵ_0 relative to the connected coupling shaft S equals the ratio between the torque T_c of the auxiliary component transmission and the total torque T_S. This is expressed by eq. (95) and is a consequence of the fact that $n_{c'} = n_S$. Thus, the power-ratio permits a quick estimate of the torque load and consequently the dimensions of the auxiliary component transmission and, therefore, is an important tool for the design of bicoupled transmissions. The power-ratio especially influences the relative expense for the auxiliary transmission when its design calls for a traction or hydrostatic drive (see sec. 36, variable bicoupled transmissions). Since even modern traction drives can transmit only limited torques, their loads have to be kept as low as possible by choosing a small power-ratio ϵ_0. To facilitate this task, a map of speed-characteristic curves has been provided which will be described in the following section. It permits the consideration of ϵ_0 while the constrained bicoupled transmission is still in the layout stages.

h) Speed Characteristics and Speed-Ratio Equations; Application to Design of Bicoupled Transmissions

Since, in a constrained bicoupled transmission, the speeds of all shafts are unequivocally associated, a map of speed characteristic curves, as shown in fig. 131, can be established with the aid of the generalized Willis equation (sec. 11d). This map is independent of the type of auxiliary transmission used. The speed-ratios i_{iS} of the bicoupled transmission and $i_{f'c'}$ of the auxiliary component transmission, which are to be determined, have been chosen as coordinates. Parameters are the speed-ratio i_{mf}, which characterizes the main component transmission, and the power-ratio ϵ_0, as obtained from eq. (94). The latter determines the power fraction flowing in the auxiliary com-

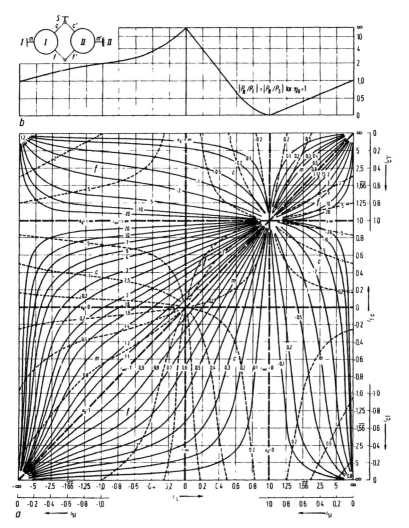

Fig. 131. Generally valid map of speed characteristics for constrained bicoupled transmissions, variable bicoupled transmissions, and reduced bicoupled transmissions: i_{IS}, speed-ratio of the bicoupled transmission; $i_{f'c'}$, speed-ratio of the auxiliary component transmission II; i_{mf}, basic or revolving-carrier speed-ratio between the shafts m and f of the main component transmission I; ϵ_o, power-ratio (ratio between the input power of the auxiliary component transmission II and the total input power) of the loss-free transmission; c, m, f, total-power shaft of the main component transmission in the respective operating ranges, which are bordered by heavily dashed lines. *Above:* P_R/P_1 is the "series power" as a function of the external (input) power (the series power is analogous to the rolling-power in simple epicyclic transmissions).

ponent transmission and at the same time the torque distribution in the connected coupling shaft. An associated diagram, which is located above the map, shows the ratio of the series-power P_R to the external power P_1. Because of the analogy, and according to eq. (64) and worksheet 1, this ratio depends on the overall speed-ratio of the bicoupled transmission but is independent of the speed-ratio of the auxiliary component transmission and the design of the main component transmission. A similar diagram where

$$\frac{P_R}{P_1} = i_{s2} = 1 - \frac{1}{i_{1s}}$$

has been given by Laughlin, Holowenko and Hall [11]. Analogously,

$$\frac{P_R}{P_1} = 1 - \frac{1}{i_{IS}} \ .$$

If P_R/P_1 is known, the efficiency of the transmission can be roughly estimated, since

$P_R/P_1 < 1$ means that $\eta_{IS}, \eta_{SI} > \eta_{III}, \eta_{III}$,

$P_R/P_1 > 1$ means that $\eta_{IS}, \eta_{SI} < \eta_{III}, \eta_{III}$,

$P_R/P_1 \rightarrow \infty$ means that the transmission operates in the range where self-locking may occur.

The map of the speed characteristics itself is nothing more than a map of characteristic curves of the main component transmission with the coordinates $i_{IS} = k_{mc}$ and $i_{f'c'} = k_{fc}$ which has been constructed from the identical diagram of fig. 65 by using the subscript key $m \triangleq 1$, $c \triangleq s$ and $f \triangleq 2$. According to this subscript key, the basic speed-ratio $i_o = i_{12}$ becomes the speed-ratio i_{mf} which may be the basic, or a rotating-carrier, speed-ratio of the chosen main component transmission.

For an exact determination of the relationships between i_{IS}, $i_{f'c'}$ and i_{mf}, we must transcribe the indices of eq. (27) with the previously introduced subscript key. Thus we obtain the following equations which are valid for all constrained bicoupled transmissions as well as for the reduced and variable bicoupled transmissions derived from them:

$$i_{IS} = 1 - i_{mf} + i_{mf}i_{f'c'} \ , \tag{96}$$

$$i_{f'c'} = \frac{i_{IS} - 1 + i_{mf}}{i_{mf}} = \frac{i_{IS} - 1}{i_{mf}} + 1 \ , \tag{97}$$

$$i_{mf} = \frac{i_{lS} - 1}{i_{f'c'} - 1} .$$ (98)

If, during the layout phase of a bicoupled transmission, its overall speed-ratio i_{lS} and the speed-ratio $i_{f'c'}$ of its auxiliary component transmission have been chosen under consideration of P_R/P_l and ϵ_0, then nine different bicoupled transmissions can be found, each of which satisfies the given speed-ratio conditions. It is only necessary to enter the corresponding subscripts of the nine possible couplings between the main and auxiliary component transmission shown in fig. 124 into fig. 131, one at a time. Before this can be done, however, a subscript key which is characterized by the carrier position, must be chosen for each component transmission. The subscripts 1 and 2 or $1'$ and $2'$ can then be arbitrarily assigned to the two remaining shafts. For instance, for the coupling case cc': $c \triangleq s$ and $c' \triangleq s'$. The other subscripts can be arbitrarily assigned, for example, $m \triangleq 1, f \triangleq 2$ in the main section and $m' \triangleq 2', f' \triangleq 1'$ in the auxiliary section. According to this key: $i_{mf} = i_{12} = i_o$ for the main component transmission and $i_{f'c'} = i_{1's'}$ for the auxiliary component transmission, so that with worksheet 1: $i_{o'} = 1 - i_{1's'} = 1 - i_{f'c'}$.

Out of the nine possible main and auxiliary component transmission combinations, that one is finally chosen which offers the most suitable design (compare figs. 19 to 43) and has the required basic speed-ratio and the best possible efficiency. The efficiencies of the best transmission designs, which structurally may be quite different, can be determined by the methods discussed in sections 33e and 35e.

An additional nine bicoupled transmissions can be obtained when the input and output are exchanged, that is, when the desired speed-ratio is entered as $i_{SI} = 1/i_{lS}$ instead of i_{lS}.

The special advantage offered to the transmission synthesis by the map shown in fig. 131 and the eqs. (96) to (98) which it represents, lies in the fact that the kinematic characteristics of the desired transmission can be found quickly and without working through confusing design details. Then the most suitable design can be chosen from the possible configurations.

Example: Find the efficiency of a bicoupled transmission as shown in figs. 126 and 127, whose speed-ratios $i_{lS} = 17.24$ and $i_{f'c'} = -1/0.36 = -2.778$ have been determined earlier in section 35c.

Fig. 131b shows that $P_R/P_l < 1$, so that the transmission lies in a range of good efficiencies, while fig. 131a indicates that the monoshaft m is the total-power shaft and, therefore, the transmission operates with power division. It further shows that approximately 70% of the power and thus also approximately 70% of the torque transmitted by the output shaft S, flow through the auxiliary gear train. The transmission corresponds to the coupling case cm' of fig. 124.

220 / Compound Epicyclic Transmissions

The possible kinematically-equivalent modifications of this bicoupled transmission have been compiled in table 7. It has been assumed that $\eta_o =$ $\eta_{o'} = 0.985$. The speed-ratio $i_{mf} = 16.24/-3.778 = -4.3$ has been determined with eq. (98).

TABLE 7. Kinematically-Equivalent Component Gear Trains
of the Bicoupled Transmission Shown in Fig. 127

MAIN GEAR TRAIN KEY			$i_{IS} = k_{mc}$ $= 17.24$	$i_{f'c'} = k_{fc}$ $= -2.778$			
m	c	f			i_o	i_{mf}	η_{mf}
s	1	2	k_{s1}	k_{21}	0.811	$i_{s2} = -4.30$	$\eta_{s2} = 0.925$
1	s	2	k_{1s}	k_{2s}	-4.30	$i_{12} = -4.30$	$\eta_{12} = 0.985$
1	2	s	k_{12}	k_{s2}	5.3	$i_{1s} = -4.30$	$\eta_{1s} = 0.982$

AUX. GEAR TRAIN KEY			$i_{f'c'} = -2.778$			
m'	c'	f'		$i_{o'}$	$i_{f'm'}$	$\eta_{f'm'}$
s'	$1'$	$2'$	$i_{2'1'}$	-0.36	$i_{2's'} = 3.778$	$\eta_{2's'} = 0.989$
$1'$	s'	$2'$	$i_{2's'}$	0.265	$i_{2'1'} = 3.778$	$\eta_{2'1'} = 0.985$
$1'$	$2'$	s'	$i_{s'2'}$	0.735	$i_{s'1'} = 3.778$	$\eta_{s'1'} = 0.960$

Thus, the series speed-ratio of the bicoupled transmission $i_{I\,II} = i_{mf}i_{f'm'} = -4.3 \cdot 3.778 = -16.24$ and the series efficiency $\eta_{I\,II} = \eta_{mf}\eta_{f'm'}$, so that with the aid of worksheet 5 the overall efficiency can be obtained from the equation:

$$\eta_{IS} = \frac{i_{I\,II}\eta_{I\,II} - 1}{i_{I\,II} - 1} .$$

TABLE 8. Efficiencies of Kinematically-Equivalent Modifications
of the Bicoupled Transmission Shown in Fig. 127

COUPLING CASE	cf'	cc'	cm'	ff'	fc'	fm'	mf'	mc'	mm'
$\eta_{I\,II}$	0.946	0.970	0.974	0.943	0.967	0.971	0.888	0.911	0.915
η_{IS}	0.949	0.972	0.976	0.946	0.969	0.973	0.894	0.916	0.920

A comparison between the overall efficiencies, η_{IS}, of all nine coupling cases in table 8, shows that the coupling case cm', which actually has been used, has the highest possible efficiency. This result is understandable since the coupling case cm' is the only possible combination of two negative-ratio transmissions for which, according to R 26, higher individual efficiencies can be expected than for positive-ratio transmissions.

If we now examine the other nine possible transmissions with reversed input and output shafts, that is with the speed-ratio $i_{IS} = 1/17.24 = 0.058$, then fig. 131b immediately shows that their power-ratio P_R/P_I becomes rather large. Consequently, considerably smaller efficiencies must be expected. A further investigation of these transmissions is therefore omitted.

l) Optimum Bicoupled Transmissions

A constrained bicoupled transmission for a given speed-ratio can be realized with any one of the nine structural coupling cases shown in fig. 124. For each of these coupling cases the basic speed-ratio of one of the two component transmissions can be freely chosen within a wide range to achieve the required speed-ratio. Finally, each of the basic speed-ratios can again be realized with a number of different transmission types. Thus the most suitable or optimum transmission modification which satisfies the given speed-ratio requirement must be chosen from a multitude of possible designs.

An optimum design is chosen primarily by considering the following three criteria: the size or volume of the transmission, its efficiency and, because of the centrifugal forces which act on the planet bearings, the carrier speed.

As stated in R 26, the rolling-powers of positive-ratio transmissions are larger and, consequently, their efficiencies smaller than those of kinematically-equivalent negative-ratio transmissions. Positive-ratio component transmissions should therefore be considered only in cases where the required speed-ratio i_{IS} is so high that it cannot be realized with negative-ratio component transmissions. As a rule, positive-ratio transmissions are also more expensive.

A further reduction in the number of possible modifications occurs because a given overall speed-ratio cannot be realized with each of the nine kinematical coupling cases (sec. 34a). For instance, the kinematical coupling case fm' can be used only for negative speed-ratios i_{IS} and i_{SI}, coupling cases mc', mf', cm', cf', and fc' only for positive speed-ratios i_{IS} and i_{SI}, but the coupling cases mm', cc', and ff' can be used for positive as well as negative speed-ratios i_{IS} and i_{SI}. This claim can be verified without calculation if we use R 4, R 8 and R 9 to determine whether for a given kinematical coupling case (that is, for a given position of the two summation shafts of the bicoupled transmission), the torques of shafts I and S can only have equal or opposite signs, or whether, depending on i_o and $i_{o'}$, both equal and opposite signs are possible. According to R 2 equal signs of the torques signify a negative speed-ratio i_{IS} or i_{SI} and vice versa.

Schnetz [37] can assign only a limited range of speed-ratios i_{IS} to each of the nine coupling cases shown in fig. 132 since for bicoupled transmissions which consist only of negative-ratio component transmissions, the kinematical and structural coupling cases are identical (sec. 34a). The speed-ratio ranges shown in fig. 132 are valid when only the simplest negative-ratio gear trains as shown in figs. 32 and 33, are used as component gear trains.

According to fig. 132, at most four different coupling cases remain to cover a given speed-ratio $i = i_{IS}$. However, a desired speed-ratio can also be realized as $i = i_{SI} = 1/i_{IS}$ and the coupling cases which are suited for this

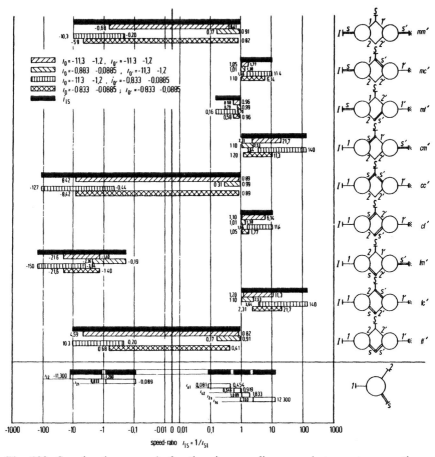

Fig. 132. Speed-ratio ranges i_{IS} for the nine coupling cases between two negative-ratio transmissions according to fig. 32. Below each overall range the partial ranges are shown which are covered by the four variations of each coupling case (as obtained by interchanging the sun and annular gears in both component gear trains). For comparison, the six possible speed-ratio ranges of a simple negative-ratio drive train according to fig. 32 are shown below.

approach can be found in fig. 132 above the reciprocal value of i_{IS}. Thus, a given speed-ratio can be realized with a total of between three and eight coupling cases. However, there are four design modifications to each of these coupling cases which are obtained by interchanging the sun and annular gears of each of its two component gear trains. These design modifications cover different and sometimes overlapping parts of the overall speed ratio range, i_{IS}, which are shown in fig. 132 and are based on the possible speed-ratio ranges i_o and $i_{o'}$ of their component gear trains. If a desired speed-ratio i_{IS} can be realized with one or more modifications of the three to eight coupling cases, then each of these which are kinematically equivalent, must be considered as a possible choice for the optimum transmission.

Poppinga [43] and later Jensen [49] described a method to optimize bicoupled transmissions with respect to their outside diameters and calculated their efficiencies. From the possible combinations of the basic speed-ratios i_o and $i_{o'}$ which yield the same overall speed-ratio i_{IS}, that one is chosen as the optimum solution whose annular gears have the smallest outside diameters D relative to the diameter d_{min} (the pitch circle diameter of the equal-sized smallest gears) of either the sun gears or planet gears. The results, which are plotted in fig. 133, show that for a given speed-ratio only one of the suitable coupling cases has a smallest relative outside diameter. Moreover, each of the coupling cases has its smallest relative outside diameter at a different overall speed-ratio i_{IS} which occurs when both sun gears and the planet gears *simultaneously* assume the smallest diameter d_{min}. The equal diameters of the annular gears then become $D = 3d_{min}$ and the associated basic speed-ratios $i_o = i_{o'} = -3$ or $i_o = i_{o'} = -1/3$. If all the possible combinations of these two basic speed-ratio values are considered, then four transmission variations are obtained for each coupling case. Because of the reciprocity of the two basic speed-ratios, however, these variations form one or two identical pairs, depending on the coupling case, so that a total of twenty-four transmissions occur which have the same minimum outside diameter but different speed-ratios i_{IS} and i_{SI}. As R 25 explains, two-thirds of these speed-ratios are positive, one-third are negative.

The described optimization process, however, yields only an approximation of the smallest practically possible transmission. Therefore, another optimization process which is better suited to the requirements of transmission engineering is based on the assumption that the smallest gears have an equal minimum number of teeth z_{min}, an equal input torque T^* and equal bending stresses σ_b at the roots of the teeth, rather than equal diameters. If, as a condition for geometrical similarity, it is further required that the product $b \cdot 1/m$ (symbols according to ISO–701) of the face width and the diametral pitch ($P = 1/module = 1/m$) remains constant, then the ratio between the diameters of those two gears x and y in each component gear train which have the same minimum number of teeth z_{min} and transmit a

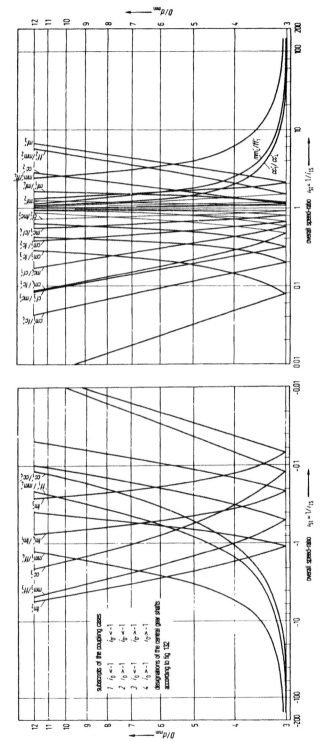

Fig. 133. The smallest relative diametes D/d_{min} for the four variations of each of the nine structural coupling cases as obtained by the purely geometrical optimization procedure described by Poppinga [43] and Jensen [49]: $d_{min} \hat{=}$ pitch circle diameters of the equal-sized smallest gears (sun gear and/or planets) of each component gear train: $D \hat{=}$ pitch circle diameters of the equal-sized annular gears. Component gear trains are according to fig. 32.

tangential force F, can be calculated according to the theory of models ([45, 46] and sec. 43a) as follows:

$$\frac{T_x}{T_y} = \frac{d_x F_x}{d_y F_y} = \frac{m_x z_{min} F_x}{m_y z_{min} F_y}, \qquad \left[m = \frac{d}{z} = \text{module.} \right]$$

Since the bending moment in the root section of a tooth $M_r \sim F h_t$, where h_t is the whole depth of the tooth, $h_t \sim m$ and the section modulus $I/c = t_r^2 b/6$, where t_r is the chordal thickness at the root, $t_r \sim m$ and $b \sim m$, the root bending stress becomes:

$$\sigma_b \sim \frac{Fm}{bm^2} \sim \frac{F}{m^2} = \frac{F_x}{m_x^2} = \frac{F_y}{m_y^2},$$

$$\frac{F_x}{F_y} = \frac{m_x^2}{m_y^2},$$

so that $$\frac{T_x}{T_y} = \frac{m_x^3}{m_y^3} \quad \text{or} \quad \frac{d_x}{d_y} = \sqrt[3]{\frac{T_x}{T_y}}.$$

This last equation shows that the ratio between the diameters of the smallest gears in each component gear train equals the cube root of the ratio of their torques. As a rule, therefore, they cannot have the same size when the transmission has been designed according to the presented considerations.

Another difference between this and the optimization process described by Poppinga results from the fact that all gear diameters are referenced to an equal input torque T^*. If this input torque T^* acts on the annular gear of the main gear train, then the pitch diameter d and the torque T of the latter's smallest gear become smaller than when T^* acts directly on the sun gear. Thus the transmission builds smaller when it is driven through the annular gear rather than through the sun gear of the main gear train. In the optimization procedure described by Schnetz [37], these conditions are considered by referencing all outside diameters of the transmission, that is, the diameters of the annular gears, to the diameter d^* of a fictitious pinion gear which has a minimum number of teeth $z = z_{min}$ and transmits the input torque T^*.

If, with the help of a digital computer, all of the possible transmission modifications are optimized according to this method, it is found that some of them have almost identical relative outside diameters. After some preselection only those bicoupled transmission designs have been depicted in fig. 134a which combine a minimum outside diameter with a good efficiency and whose carrier speeds are equal or smaller than the input speeds.

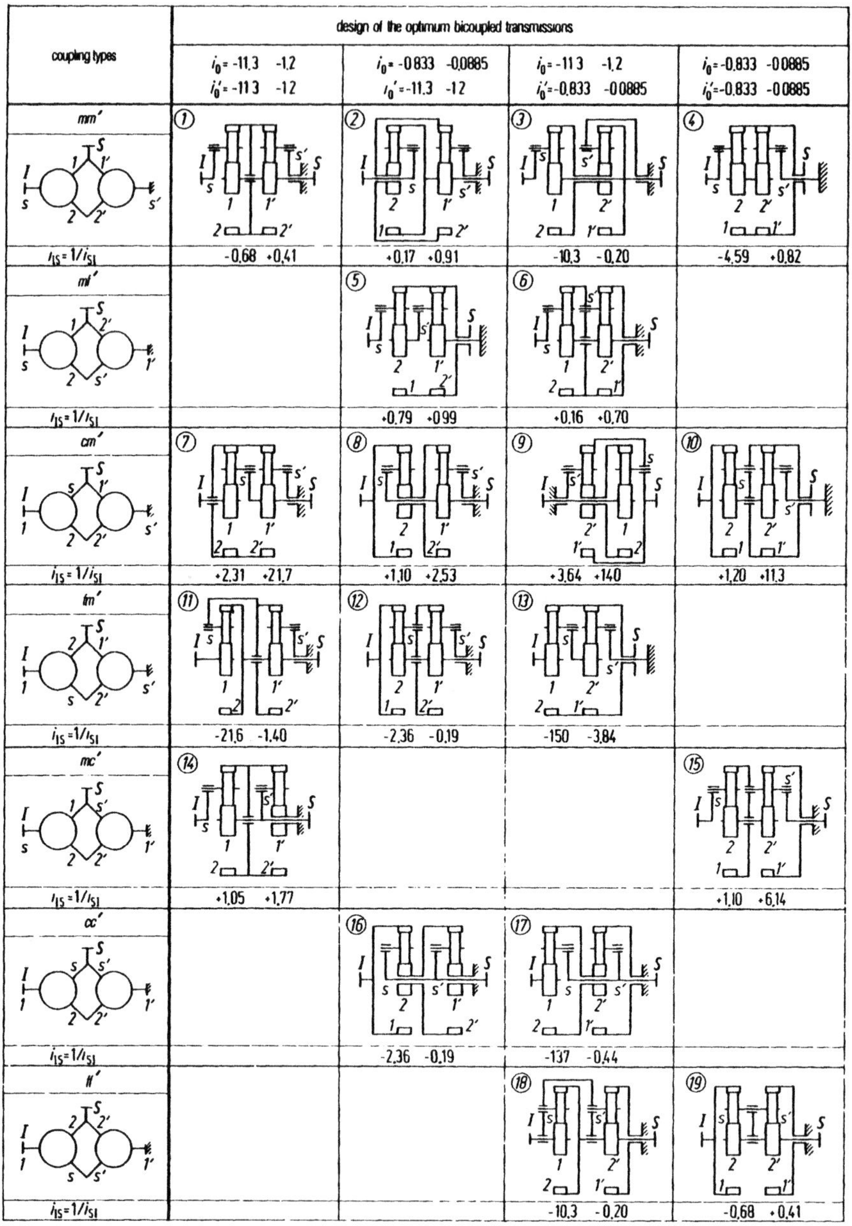

coupling types	design of the optimum bicoupled transmissions			
	$i_0 = -11.3 \quad -1.2$ $i_0' = -11.3 \quad -1.2$	$i_0 = -0.833 \quad -0.0885$ $i_0' = -11.3 \quad -1.2$	$i_0 = -11.3 \quad -1.2$ $i_0' = -0.833 \quad -0.0885$	$i_0 = -0.833 \quad -0.0885$ $i_0' = -0.833 \quad -0.0885$
mm' $i_{1S} = 1/i_{S1}$	① $-0.68 \quad +0.41$	② $+0.17 \quad +0.91$	③ $-10.3 \quad -0.20$	④ $-4.59 \quad +0.82$
ml' $i_{1S} = 1/i_{S1}$		⑤ $+0.79 \quad +0.99$	⑥ $+0.16 \quad +0.70$	
cm' $i_{1S} = 1/i_{S1}$	⑦ $+2.31 \quad +21.7$	⑧ $+1.10 \quad +2.53$	⑨ $+3.64 \quad +140$	⑩ $+1.20 \quad +11.3$
tm' $i_{1S} = 1/i_{S1}$	⑪ $-21.6 \quad -1.40$	⑫ $-2.36 \quad -0.19$	⑬ $-150 \quad -3.84$	
mc' $i_{1S} = 1/i_{S1}$	⑭ $+1.05 \quad +1.77$			⑮ $+1.10 \quad +6.14$
cc' $i_{1S} = 1/i_{S1}$		⑯ $-2.36 \quad -0.19$	⑰ $-137 \quad -0.44$	
tt' $i_{1S} = 1/i_{S1}$			⑱ $-10.3 \quad -0.20$	⑲ $-0.68 \quad +0.41$

Fig. 134a. Schematic representation of the chosen bicoupled transmissions.

It is evident from fig. 134a that the complexity of the various coupling cases may vary substantially. Therefore, the transmission which has the smallest outside diameter does not necessarily constitute the most economical solution.

The ratios of the outside diameters of the transmissions shown in fig. 134a, with respect to the fictitious diameter d^*, that is, relative to an equal input torque $T_I = T^*$, and their efficiencies, are plotted in fig. 134b as functions of their overall speed-ratios i_{IS}.

The outside diameters of the same transmissions relative to an equal input torque $T_S = T^*$, acting on the connected coupling shaft S, and their efficiencies, are plotted in fig. 134c as functions of the speed-ratios i_{SI}. As compared to fig. 134b, where I is the input shaft, the diameter minima of identical transmissions, in this case, lie at reciprocal abscissa values.

For comparison, the outside diameters relative to an identical input torque T^*, and the efficiencies of the simple negative-ratio transmissions according to fig. 32, are plotted in fig. 134b for the six possible two-shaft speed-ratios.

To determine the overall efficiencies of the optimum transmissions shown in fig. 134, the basic efficiencies of their component gear trains have been calculated according to section 6 to account for the influence of the number of teeth, assuming $\mu_t = 0.07$ and $z_{min} = 17$.

Figs. 134d and 134e finally show the basic speed-ratios $i_{o'}$ of the auxiliary component transmissions of the optimum bicoupled transmissions designated by circled numbers in figs. 134b and 134c respectively. These must be known before the basic speed-ratios of the main gear trains can be determined from the speed-ratios i_{IS} or i_{SI}, with the aid of worksheets 1 and 2.

An optimum bicoupled transmission for a given speed-ratio i can now be designed as follows:

1. From fig. 134b for $i = i_{IS}$, and an input shaft I, and fig. 134c for $i = i_{SI}$, and an input shaft S, determine all bicoupled and simple epicyclic transmissions which cover the given speed-ratio i. Select the transmission which is best suited for the intended purpose with regard to the relative outside diameter D/d^*, the efficiency η_{IS} or η_{SI}, and the design complexity as evident from fig. 134a.

2. For the selected design, determine the basic speed-ratio $i_{o'}$ of the auxiliary gear train from fig. 134d or 134e.

3. Use $i_{o'}$ to find $i_{c'f'} = k_{cf}$ from the proper equation given in worksheet 1. With k_{cf} and i_{IS} or i_{SI} known, the basic speed-ratio of the main gear train i_o can be calculated with the help of worksheet 2.

4. Proceeding from the monoshaft I, determine the torques as described in section 35d. Design the gear train by using the standard methods which are described in most books covering machine design, [33], [53].

Proof: Before the final lay out, determine the diameter ratio $d_x/d_y = $

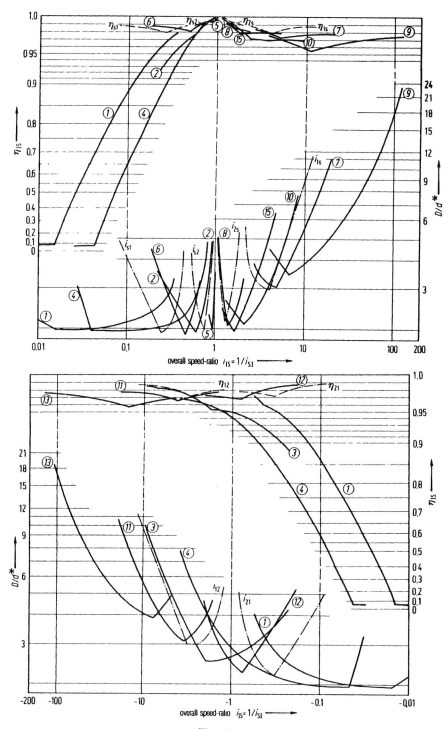

Fig. 134b.

$\sqrt[3]{T_x/T_y}$ of the two component gear trains from the torques of their smallest gears x and y. These torques have been determined in step 4. With d_x/d_y determine the diameter ratio of the two annular gears. For the smallest transmission diameter, the latter becomes equal to one. If substantial deviations from this value are found, repeat the calculations beginning with step 3 and a slightly adjusted value $i_{o'}$.

If, instead of the monoshaft II, the connected coupling shaft S is fixed, then the optimum bicoupled transmission which has been found from fig. 134b for a given speed-ratio $i = i_{IS}$, and which transmits the torque $T^* = T_I$, simultaneously is the optimum series-coupled transmission with respect to its outside diameter and efficiency. The latter transmits the torque $T^* = T_I$ and has the speed-ratio $i_{III} = 1 - i_{IS}$. However, its carrier speeds are different and, if need be, should be checked separately. Therefore, the optimum series-coupled transmission with the speed-ratio $i = i_{III}$ should also be considered as early as step 1 of the described procedure. It can be found in fig. 134b at the speed-ratio $i_{IS} = 1 - i_{III} = 1 - i$. The efficiency η_{III} of the optimum series-coupled transmission can be found from worksheet 5 when η_{IS} as obtained from fig. 134b, or η_{SI} as obtained from fig. 134c, is substituted into the appropriate equations listed in boxes 7, 9, or 11.

k) Design of Precise Speed-Ratios

A given speed-ratio, which cannot be realized with sufficient accuracy by a simple epicyclic transmission, usually can be implemented precisely with a bicoupled transmission. Should a bicoupled transmission be unsuitable because gears with an excessively large number of teeth would be required, then a symmetrical compound transmission normally does satisfy the given conditions. Transmissions of this type will be discussed later in section 38.

To design a bicoupled transmission with a precise speed-ratio i_{IS}, we begin with eq. (96) and define all speed-ratios as integer fractions. According to section 9, the numerators and denominators of these fractions represent the numbers of teeth, or products of numbers of teeth, of the gears of

---

Fig. 134b-e. Optimum bicoupled planetary transmissions with minimal outside diameters which have the same input torques and the same root stresses at all gears, shown as functions of the overall speed-ratios i_{IS} or i_{SI} according to Schnetz [37]. Of the 72 possible transmissions (9 coupling cases · 4 variations · 2 positions of the input shaft), those with the higher efficiencies and the relative carrier speeds n_s, $n_{s'} \leqq n_{in}$ have been selected. Transmissions nos. 4, 13, 14, and 15 with n_s, $n_{s'} > n_{in}$ are exceptions.

←

Fig. 134b. The smallest relative outside diameters and the efficiencies for transmissions whose monoshaft I is the input shaft. The encircled numbers refer to the gear trains shown in fig. 134a. The broken lines refer to a simple negative-ratio gear train according to fig. 32.

the transmission. Thus the equation

$$i_{IS} - 1 = i_{mf}(i_{f'c'} - 1) \, ,$$

can be written as

$$\frac{g - h}{h} = \frac{q}{t}\left(\frac{x}{y} - 1\right).$$

If $i_{IS} = g/h$ is given now in form of an integer fraction and if we further assume a convenient speed-ratio $i_{mf} = q/t$ for the main gear train, then the speed-ratio of the auxiliary gear train becomes:

$$i_{f'c'} = \frac{x}{y} = \frac{(g - h)t + hq}{hq} \tag{99}$$

When the fixed monoshaft m' is also the carrier shaft of the auxiliary transmission, then $i_{f'c'} = (i_{o'})_1$ and eq. (99) represents its basic speed-ratio. The subscript 1 of the basic speed-ratio merely indicates the consecutive number of this solution.

With $i_{f'c'} = i_{1's'} = 1 - i_{o'}$ we find a second version of the auxiliary gear train which has the basic speed-ratio

$$(i_{o'})_2 = \frac{hq - (g - h)t - hq}{hq} = -\frac{(g - h)t}{hq} \, , \tag{100}$$

and with $i_{f'c'} = i_{s'1'} = \dfrac{1}{1 - i_{o'}}$ a third version

$$(i_{o'})_3 = \frac{(g - h)t + hq - hq}{(g - h)t + hq} = \frac{(g - h)t}{(g - h)t + hq} \, . \tag{101}$$

According to section 9, the numerators and denominators of the fractions in eqs. (99) to (101) represent the number of teeth, or the products of numbers of teeth, of the respective basic transmissions. Therefore, their largest prime factors must not exceed the maximum allowable number of teeth for the largest gears. Before we can search for suitable speed-ratios, we must make sure that the given values of h and $(g - h)$ do not contain excessively

$$\longrightarrow$$

Fig. 134c. The smallest relative outside diameters and the efficiencies for transmissions whose connected shaft S is the input shaft. The encircled numbers refer to the gear trains shown in fig. 134a.

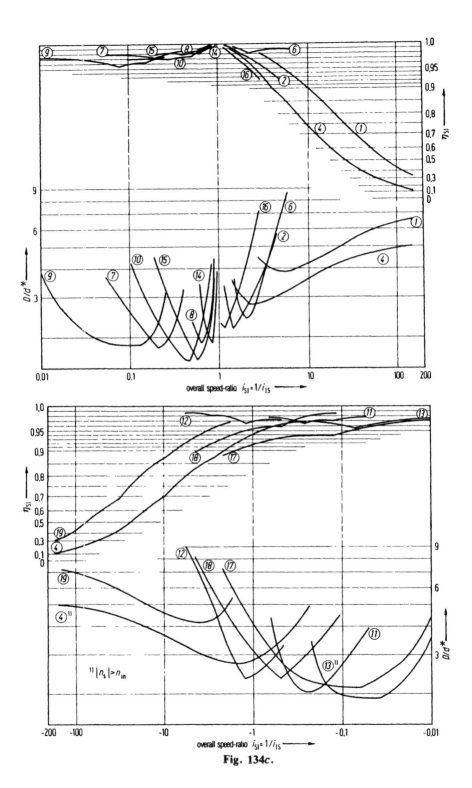

Fig. 134c.

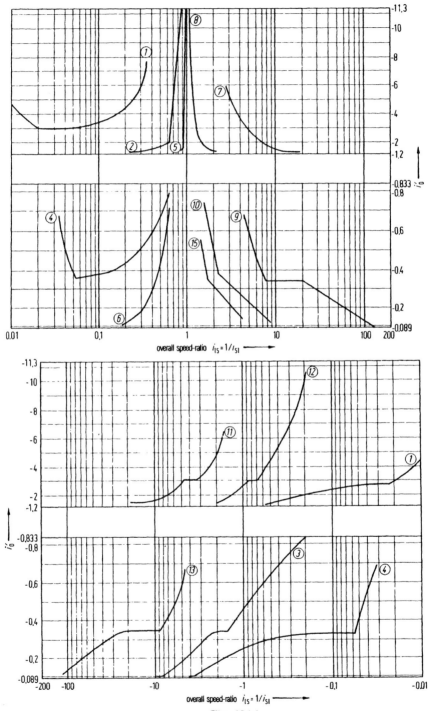

Fig. 134d.

large prime factors. If the prime factors of h are too large, auxiliary gear trains whose carrier shafts are identical with shafts m' and c' and whose speed-ratios are thus given by eq. (99) and (100) respectively, cannot be built. Likewise, if the prime factors of the difference $(g - h)$ are to large, auxiliary gear trains with speed-ratios according to eqs. (100) and (101) are not possible when their carrier shafts are identical with the shafts c' and f' respectively.

With the equations remaining after this test, the basic speed-ratio $i_{o'}$ can be calculated for an assumed q/t or vice versa. Further transmissions can be designed which are based on the reciprocal value of the given speed-ratio. In this case the values of g and h are exchanged and we must investigate whether the new value of h can still be represented by a product of prime numbers.

First example, medium speed-ratio: A coaxial transmission with an accurate speed-ratio $i = 938/176$ is to be designed. Thus, we have either

$$\alpha) \quad i_{IS} = \frac{938}{716} = \frac{g}{h} \quad \text{or} \quad \beta) \quad i_{IS} = \frac{716}{938} = \frac{g}{h} \ .$$

Only case α) shall, subsequently, be investigated. A check of the values

$$(g - h) = 938 - 716 = 222 = 2 \cdot 3 \cdot 37 \ , \quad \text{and}$$

$$h = 716 = 4 \cdot 179 \ ,$$

shows that the prime factors of the difference $(g - h)$ are sufficiently small while h contains the rather large factor *179* which cannot be further reduced. Therefore, the only choice left is an auxiliary gear train according to eq. (101) which has a speed-ratio $i_{f'c'} \triangleq i_{s'1'}$ indicating that the carrier shaft is identical with the free coupling shaft f'. However, eq. (101) yields suitable auxiliary gear trains only when its denominator does not contain excessively large prime numbers, which represent numbers of teeth as explained earlier. Therefore, we start by expanding the denominator into its prime factors. Its values q and t have been arbitrarily chosen so that practical numbers of teeth are obtained.

For $g = 938$, $h = 716$, and with $q = 2$, $t = -2, -3, -4$, the denominator of eq. (101) becomes

a) $-2 \cdot 222 + 716 \cdot 2 = 988 = 4 \cdot 13 \cdot 19$,

b) $-3 \cdot 222 + 716 \cdot 2 = 766 = 2 \cdot 383$,

c) $-4 \cdot 222 + 716 \cdot 2 = 544 = 32 \cdot 17$.

←——————————————————————

Fig. 134d. Basic speed-ratio $i_{o'}$ of the optimum transmissions with input at I according to fig. 134*b*.

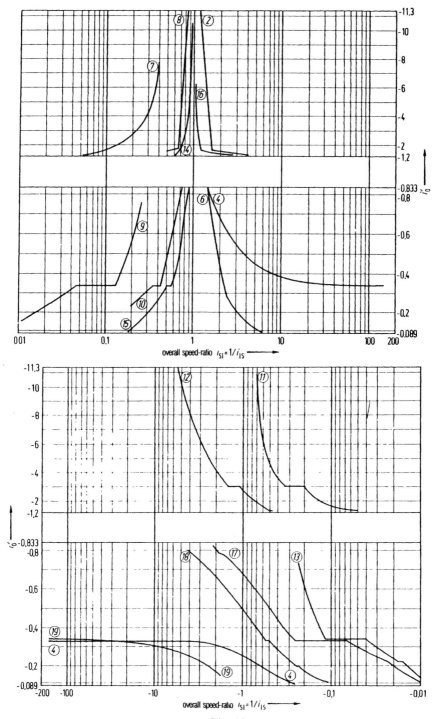

Fig. 134e.

It is obvious that the choice b) is not practical because of the large prime factor 383. Choice a) on the other hand results in a negative-ratio transmission of the type shown in fig. 34, however, with reversed indices *1* and *2*. According to eq. (101), its basic speed-ratio is

$$(i_{0'})_a = \frac{(g-h) \cdot t}{(g-h) \cdot t + h \cdot q} = \frac{-444}{988} \cdot \frac{2}{2} = \frac{24}{-76} \cdot \frac{37}{26}$$

$$= \left(-\frac{z_{p1'}}{z_{1'}}\right)\left(-\frac{z_{2'}}{z_{p2'}}\right),$$

where the negative sign is affixed to the number of teeth of the annular gear. To meet the requirement of equal center distances, it has been tried to match, as closely as possible, the sum of the numbers of teeth of the two meshing pairs of gears by combining the appropriate prime factors in the numerator and denominator and expanding the fraction. A complete adjustment, however, must be achieved by choosing appropriate values for the diametral pitch, the helix angles and/or addendum modifications.

Assumption c) leads to a simple negative-ratio transmission of the type shown in fig. 32.

$$(i_{0'})_c = \frac{-888}{544} = \frac{-111}{68}, \quad \text{where} \quad z_{1'} = 68 \quad \text{and} \quad z_{2'} = -111.$$

Depending on the position of the carrier shaft, the three possible main gear trains become:

$$i_{mf} = \frac{q}{t} = \frac{2}{-4} = i_0, \quad \text{thus} \quad (i_0)_{c1} = -\frac{1}{2},$$

$$i_{mf} = i_{1s}, \quad \text{thus} \quad (i_0)_{c2} = 1 - \frac{2}{-4} = \frac{3}{2} \quad \text{and}$$

$$i_{mf} = i_{s1}, \quad \text{thus} \quad (i_0)_{c3} = 1 - \frac{-4}{2} = 3.$$

To obtain a simple design, we choose the transmission which has the basic speed-ratio $(i_0)_{c2} = 3/2$ and whose carrier is connected to the free coupling shaft. The symbol of this bicoupled transmission is shown in fig. 135/1a. The positions of all of its shafts are determined by the choice of eq. (101)

Fig. 134e. Basic speed-ratio $i_{0'}$ of the optimum transmissions with input at S according to fig. 134c.

and the main gear train $c2$. Fig. 135/1b is a true-to-scale schematic of the chosen auxiliary gear train which also shows the number of teeth for each gear. To obtain a planet p with an integer number of teeth, it is necessary to correct the tooth profiles.

Fig. 135/1c shows a simple positive-ratio transmission of the type described in fig. 19 which constitutes a possible main gear configuaration. If we choose the same number of teeth for its gears 2 and p_2 as for the gears $1'$ and p' of the auxiliary gear train, then we obtain the rather simple reduced bicoupled transmission of fig. 135/1d. According to our calculation the basic speed-ratio of the main gear train then becomes:

$$(i_o)_{c2} = \frac{3}{2} = \left(-\frac{z_{p1}}{z_1}\right)\left(-\frac{z_2}{z_{p2}}\right) = \frac{z_{p1} \cdot 68}{z_1 \cdot 22} . \tag{102}$$

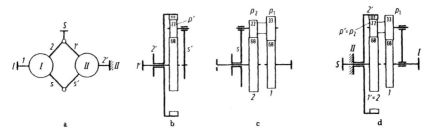

Fig. 135/1. Example for the synthesis of a constrained bicoupled transmission which accurately realizes a given speed-ratio $i_{lS} = 938/716$: *a*, symbolic representation of the bicoupled transmission; *b*, schematic representation of the chosen auxiliary component transmission *II; c*, schematic representation of the main component transmission *I* which has been chosen so that one of its stages has the same number of teeth as the auxiliary component transmission *b; d*, schematic representation of the resulting reduced bicoupled transmission.

To obtain the required equal center distances in both stages of the basic transmission with gears of the same diametral pitch, the sum of the number of teeth in stage *1* must equal the sum of the number of teeth in stage *2*, so that

$$z_1 + z_{p1} = 68 + 22 = 90 . \tag{103}$$

Thus we can solve eq. (102) for z_{p1}, which yields

$$\frac{z_{p1}}{z_1} = \frac{3}{2} \cdot \frac{22}{68} \quad \text{that is} \quad z_{p1} = z_1 \cdot \frac{66}{136}$$

and then obtain

$$z_1 = \frac{90}{1 + 66/136} = 60.59$$

and

$$z_{p1} = 29.41 , \qquad \text{from eq. (103)}.$$

Because these values are not integer numbers as required, we must choose slightly different gear sizes. The pair $z_1 = 68$ and $z_{p1} = 33$ accurately satisfies eq. (102) but not eq. (103). Consequently, the center distances in both gear stages can be made equal only when, after all, gears with different diametral pitch, helix angle, and addendum modifications are used.

Second example: If the speed-ratio is given by two prime numbers, for example, $i = 983/761$, then the following two possibilities exist:

$$\alpha) \quad i_{IS} = 983/761 = g/h , \qquad \text{and}$$

$$\beta) \quad i_{IS} = 761/983 = g/h .$$

Only case α) is subsequently investigated. The expansion into prime factors of

$$(g - h) = 983 - 761 = 222 = 2 \cdot 3 \cdot 37$$

yields a largest value of *37* which represents a sufficiently small number of teeth. Since the prime factors g and h themselves are too large, a bicoupled transmission with the required speed-ratio can only have an auxiliary gear train according to eq. (101), but not according to eqs. (99) or (100). This auxiliary gear train has a speed-ratio $i_{f'c'} = i_{s'1'}$ and its carrier is mounted on the free coupling shaft.

TABLE 9. POSSIBLE COMPONENT GEAR TRAINS FOR SPEED-RATIO $i_{IS} = 983/761$

EXAMPLE	q	t	i_{mf}	DENOMINATOR EQ. (101)	$i_{o'}$
a	6	-2	-3	$4122 = 2 \cdot 3^2 \cdot 229$	prime factor too large
b	6	-3	-2	$3900 = 2^2 \cdot 3 \cdot 5^2 \cdot 13$	$-\dfrac{111}{650} = -\dfrac{37 \cdot 3}{50 \cdot 13}$
c	6	-10	$-6/10$	$2346 = 2 \cdot 3 \cdot 17 \cdot 23$	$-\dfrac{370}{391} = -\dfrac{10 \cdot 37}{17 \cdot 23}$
d	6	-16	$-6/16$	$1014 = 2 \cdot 3 \cdot 13^2$	$-\dfrac{592}{169} = -\dfrac{37 \cdot 16}{13 \cdot 13}$
e	6	-21	$-6/21$	$-96 = -2^5 \cdot 3$	$\dfrac{777}{16} = \dfrac{37 \cdot 21}{8 \cdot 2}$

For the main gear train which has the speed-ratio $i_{mf} = q/t$, we choose $q = 6$ because it cancels the factors 2 and 3 in the numerator of eq. (101). To prevent $i_{o'}$ from becoming too large, negative values are chosen for t as can be seen in table 9. Generally, the values of q and t should not be larger than the acceptable number of teeth and they should always be chosen so that the speed-ratio q/t falls within the possible speed-ratio ranges of the simple positive or negative-ratio transmissions.

Example d in table 9 yields a transmission whose gears have very small numbers of teeth and which, consequently, has a small diameter. The following three transmissions are suitable for its main gear train:

1. $i_{mf} = i_o$, $(i_o)_a = -6/16$.

2. $i_{mf} = i_{1s}$, $(i_o)_b = 1 + \dfrac{6}{16} = \dfrac{22}{16} = \dfrac{13}{16} \cdot \dfrac{22}{13}$

$$= \left(-\frac{z_{2'}}{z_{p2'}}\right)\left(-\frac{z_{p1'}}{z_{1'}}\right) .$$

3. $i_{mf} = i_{s1}$, $(i_o)_c = 1 + \dfrac{16}{6} = \dfrac{22}{6}$.

If designed according to fig. 32, the first of these options has the advantage of the greatest simplicity. As can be seen from fig. 135/2a, the second option has the advantage that its carrier and the carrier of the auxiliary gear train are mounted on the free coupling shaft. Therefore both carriers can be combined into a single unit. If, in addition, its basic speed-ratio is expanded by a factor 13, a positive-ratio transmission as shown in fig. 19 is obtained and the central gears z_2 and $z_{1'}$ of the connected coupling shaft, as well as

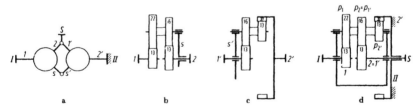

Fig. 135/2. Example for the design of a constrained bicoupled transmission with the accurate speed-ratio $i_{1S} = 983/761$. The chosen numbers of teeth are as shown in b–d: a, symbolic representation of the chosen transmission; b, schematic representation of the main component gear train; c, schematic representation of the auxiliary component gear train; d, overall transmission designed as a reduced bicoupled gear train.

their meshing planets z_{p2} and z_{p1}, can be designed to have equal numbers of teeth. As will be explained later in section 37, these two gear pairs then can be combined into a single gear pair so that we obtain a reduced bicoupled transmission. Figs. 135/2b and 135/2c show the schematic of the two component gear trains and the number of teeth for each gear. Fig. 135/2d finally shows a schematic of the compound transmission for the required speed-ratio. In an actual design study, further transmission modifications based on different q and t values, also for the reciprocal speed-ratio of possibility β, should be systematically explored since, as example 1 illustrates, there is always a chance that an even simpler solution can be found.

Third example: High speed-ratio with a prime factor of 39,989. Desired is a transmission with a speed-ratio $i = 39,989$ and thus $g = 39,989$, $h = 1$ and $(g - h) = 39,988 = 2 \cdot 2 \cdot 13 \cdot 769$. In this case we cannot use auxiliary gear trains whose basic speed-ratios are given by eqs. (100) and (101) since the prime factor 769 is too large to be taken as a reasonable number of teeth. However, it should be possible to find a suitable auxiliary gear train whose basic speed-ratio is given by eq. (99). The speed-ratio q/t of the main gear train should be of the same order of magnitude as the overall speed-ratio i_{IS}, that is, $q \approx (g - h)$, $t = 1$ so that the basic speed-ratio $i_{o'}$ of the auxiliary gear train does not become too large. According to section 22, gear trains of this type can be built as positive-ratio gear trains with $i_{s1} = z_l^2$ and $i_{s2} = -(z_l^2 - 1)$. Therefore, we choose $t = 1$ and $q = z_l^2$ or $q = -(z_l^2 - 1)$, where q is of the same order of magnitude as $(g - h)$. Thus:

$$i_{o'} = \frac{39,988 + z_l^2}{z_l^2} \quad \text{or} \quad i_{o'} = \frac{39,988 - (z_l^2 - 1)}{-(z_l^2 - 1)}$$

where z_1 stands for the number of teeth of gear 1 in the main gear train. As discussed in section 22, the number of teeth of all other gears of the gear train depends on the choice of z_1.

TABLE 10. SUITABLE BICOUPLED TRANSMISSIONS FOR SPEED-RATIO $i_{IS} = 39,989$

EXAMPLE	z_1	$q = -(z_l^2 - 1)$	$39,988 - (z_l^2 - 1)$	$i_{o'}$
a	142	$-20,163 = -3 \cdot 11 \cdot 13 \cdot 47$	$19,825 = 5^2 \cdot 13 \cdot 61$	$-\dfrac{25}{47} \cdot \dfrac{61}{33}$
b	157	$-24,648 = -2^3 \cdot 3 \cdot 13 \cdot 79$	$15,340 = 2^2 \cdot 5 \cdot 13 \cdot 59$	$-\dfrac{25}{79} \cdot \dfrac{59}{30}$
c	79	$-6,240 = -2^5 \cdot 3 \cdot 5 \cdot 13$	$33,748 = 2^2 \cdot 11 \cdot 13 \cdot 59$	$-\dfrac{22}{16} \cdot \dfrac{59}{15}$
d	53	$-2,808 = -2^3 \cdot 3^3 \cdot 13$	$37,180 = 2^2 \cdot 5 \cdot 11 \cdot 13^2$	$-\dfrac{55}{18} \cdot \dfrac{65}{15}$

A suitable auxiliary gear train whose basic speed-ratio is determined by the first of the two equations given above could not be found despite attempts with about *20* different values of z_1^2. Together with a number of values containing excessively large prime factors, however, the second equation yielded the suitable, kinematically-equivalent solutions compiled in table 10.

Obviously, transmission *d* represents the best solution. Its central gear *1* has the smallest number of teeth z_1 and thus transmission *d* has the smallest main gear train with the highest revolving-carrier efficiency, even though the latter is still rather low. The number of teeth of the auxiliary gear train leads to a two-stage negative-ratio gear train. Fig. 135/3 shows the resulting bicoupled transmission. Since only a large number, but not all possible values of z_1 have been checked, it should not be expected that it represents the optimum solution. With the assumption that $\eta_o = \eta_{o'} = 0.98$, the overall efficiency of the transmission *d* becomes $\eta_{1S} = 0.017$.

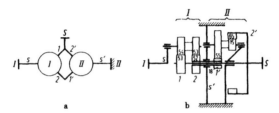

Fig. 135/3. Constrained bicoupled transmission with the precise speed-ratio $i_{1S} = 39,989$ (prime number) representing solution *d* of table 10 with $\eta_{1S} = 0.017$.

A higher efficiency can be obtained when a more costly solution is chosen, that is, when a main gear train with a standard negative speed-ratio of $i_{mf} = 1.5$ to 10 is combined with an auxiliary gear train which, because of the required high speed-ratio, is designed as a multistage conventional transmission.

Section 36. Variable Bicoupled Transmissions

a) Applications, Nomenclature, Design

Variable bicoupled transmissions are special types of constrained bicoupled transmissions. They are characterized by the fact that over a given range, they have an infinitely variable speed-ratio, see section 32f and fig. 113d. Therefore, they are frequently used between a prime mover which can operate only at a constant speed and a machine which requires a variable input speed, such as a boiler feed pump, a cooling tower, etc., or where in open or closed loop control systems infinitely variable speed-ratios are needed.

The continuous change of the speed-ratios between the shafts I and S of a variable bicoupled transmission is usually effected by an auxiliary component transmission which has an infinitely variable speed-ratio $i_{f'c'}$, that is, by a "variable-speed drive." Variable-speed drives usually are belt or link-belt drives, traction drives, or hydrostatic transmissions which consist of a variable displacement hydraulic pump and a hydraulic motor (e.g., axial piston type units). Infinitely variable revolving main drive trains are rarely used. Subsequently, only the most frequently encountered variable bicoupled transmissions will be discussed whose auxiliary component transmission consists of a variable-speed drive.

Usually these auxiliary component transmissions cannot be integrated with the main component transmission into compact units which are comparable with the simple bicoupled transmissions. Frequently, even an intermediate drive train is inserted between the main and auxiliary sections of the transmission. Nevertheless, by maintaining a given speed-ratio between two shafts of the main component transmission, this infinitely variable drive train clearly functions as an auxiliary component transmission. If, therefore, as shown in fig. 136, a variable-speed drive is connected to the main component transmission through one or more intermediate drive trains, then this compound speed drive is considered as the auxiliary section of the transmission. As will be explained in more detail in section 36d, intermediate drive trains are usually needed to match the inherent speed-ratio limits of the component transmissions to the desired overall speed-ratio limits. Likewise, they must frequently be used where spatial conditions do not allow a coaxial arrangement of all component transmissions.

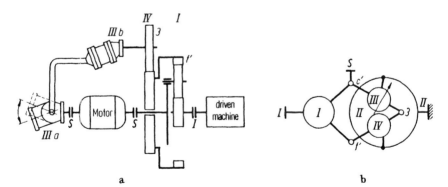

a   b

Fig. 136. Variable bicoupled transmission [38] with a hydrostatic variable component transmission II which consists of the variable displacement pump $IIIa$, the constant displacement motor $IIIb$, and the intermediate gear train IV: a, schematic representation of the transmission; b, symbolic representation.

The mentioned and most frequently used types of auxiliary component transmissions are conventional transmissions. Therefore, we can consider them as basic transmissions whose carrier shaft (which corresponds to the transmission housing) is the fixed monoshaft II (see fig. 136b), while shaft c' corresponds arbitrarily to shaft $2'$ and shaft f' to shaft $1'$, or vice versa. With these designations, their speeds, torques, and efficiencies can be determined by the same equations and mathematical methods which have been used in section 35 to analyze the constrained bicoupled transmissions, that is, the equations for the simple two-shaft transmissions and worksheet 1 can be applied when the analogy of section 33 is utilized.

b) Adjustment Ranges of the Variable Bicoupled Transmission and the Auxiliary Component Transmission

The adjustment range of a variable-speed drive at a constant input speed is frequently defined as the *ratio* or the *difference* between its maximum and minimum output speeds. As a design characteristic, however, the adjustment range of a variable-speed drive is independent of the positions of its input and output shafts. Therefore, we shall not specify which of the shafts of the variable bicoupled transmission is the input or output shaft, but rather shall describe its adjustment range by means of the two possible extreme speed-ratio limits as follows.

All parameters such as speed- or power-ratios which are valid for one of the speed-ratio limits are denoted by an asterisk, for example, i_{1S}^{*}, $i_{f'c'}^{*}$ while all values which pertain to the opposite speed-ratio limit are characterized by a triangle (e.g., i_{1S}^{Δ}, $i_{f'c'}^{\Delta}$). Therefore, all parameters which carry an asterisk bear constant relationships with each other as in constrained bicoupled transmissions. The same is true for all parameters which are identified by a triangle. However, it is immaterial which of the two limits is denoted by the asterisk and which by the triangle.

By comparison with the speed-ratio step φ of a change-gear, we shall call the quotient of the speed-ratio limits, the adjustment-ratio of the variable-speed drive, specifically

$$\varphi = \frac{i_{1S}^{\Delta}}{i_{1S}^{*}} = \text{adjustment ratio of a bicoupled transmission and} \quad (104)$$

$$\varphi' = \frac{i_{f'c'}^{\Delta}}{i_{f'c'}^{*}} = \text{adjustment ratio of an auxiliary} \quad (105)$$
$$\text{component transmission.}$$

Both speed-ratio limits of a bicoupled transmission, as well as of its variable component transmission, can be positive, or both can be negative. In either case, φ and φ' are positive. If they have opposite signs, the adjustment-ratios φ and φ' become negative which means that one of the two shafts I or S, or f', or c', reverses its sense of rotation when the speed-ratio of the drive is continually changed from one limit to the other.

The adjustment-ratio of a bicoupled transmission as a function of the adjustment-ratio of its auxiliary component transmission or vice versa can be obtained immediately from fig. 131. If the adjustment-ratio of a variable bicoupled transmission is specified by its speed-ratio limits

$$i_{IS}^* = -0.6 \quad \text{and} \quad i_{IS}^\Delta = -1.5 \,,$$

and the speed-ratio of the main component transmission is $i_{mf} = 2$, then the appropriate speed-ratio limits of the auxiliary component transmission can be read from fig. 131, or they may be calculated with eq. (97) to be:

$$i_{f'c'}^* = 0.2 \quad \text{and} \quad i_{f'c'}^\Delta = -0.25 \,.$$

Thus, the adjustment-ratios become

$$\varphi = \frac{i_{IS}^\Delta}{i_{IS}^*} = \frac{-1.5}{-0.6} = 2.5 \quad \text{for the bicoupled transmission and}$$

$$\varphi' = \frac{i_{f'c'}^\Delta}{i_{f'c'}^*} = \frac{-0.25}{0.2} = -1.25 \quad \begin{array}{l} \text{for the auxiliary component} \\ \text{transmission.} \end{array}$$

In this example both speed-ratio limits of the variable bicoupled transmission are negative. Its adjustment-ratio, therefore, becomes positive, that is, $\varphi > 0$, while one of the speed-ratio limits of the auxiliary component transmission and consequently its adjustment-ratio is negative, that is, $\varphi' < 0$. The sign change at $i_{f'c'} = 0$ indicates that for this speed-ratio $n_{f'} = 0$, so that the main component transmission operates as a two-shaft transmission and, according to R 18, changes its total-power shaft, a fact that is corroborated by fig. 131. This leads to the following important guide rule:

The total-power shaft changes its position when the speed-   R 33
ratio of a variable bicoupled transmission, or its auxiliary
component transmission, passes through zero or infinity
within its speed-ratio range, or both speed-ratios do so simul-
taneously within theirs. Consequently, the power-flow in the
bicoupled transmission changes from one to another of three
possible modes: power division, positive circulating power, or
negative circulating power.

The relationship between the two adjustment-ratios φ and φ' can be found with eq. (96)

$$i_{\mathrm{IS}}^{\triangle} = 1 - i_{\mathrm{mf}} + i_{\mathrm{mf}} i_{\mathrm{f'c'}}^{\triangle}$$

and with eqs. (104) and (105)

$$i_{\mathrm{IS}}^{*}\varphi = 1 - i_{\mathrm{mf}} + i_{\mathrm{mf}} i_{\mathrm{f'c'}}^{*} \varphi' \ . \tag{106}$$

If according to eq. (97)

$$i_{\mathrm{f'c'}}^{*} = \frac{i_{\mathrm{IS}}^{*} - 1 + i_{\mathrm{mf}}}{i_{\mathrm{mf}}} ,$$

then
$$i_{\mathrm{IS}}^{*}\varphi = 1 - i_{\mathrm{mf}} + i_{\mathrm{mf}}\varphi' \cdot \frac{i_{\mathrm{IS}}^{*} - 1 + i_{\mathrm{mf}}}{i_{\mathrm{mf}}}$$

or
$$i_{\mathrm{mf}} = 1 + \frac{i_{\mathrm{IS}}^{*}(\varphi - \varphi')}{\varphi' - 1} \ . \tag{107}$$

In case we do not want to refer the speed-ratio i_{mf} of the main component transmission to i_{IS}^{*} but rather want to express it as a function of the other speed-ratio limit $i_{\mathrm{IS}}^{\triangle}$, we can use eq. (104) and thus obtain:

$$i_{\mathrm{mf}} = 1 + \frac{i_{\mathrm{IS}}^{\triangle}(\varphi - \varphi')}{\varphi(\varphi' - 1)} = 1 + \frac{i_{\mathrm{IS}}^{\triangle}(1 - \varphi'/\varphi)}{\varphi' - 1} \ . \tag{108}$$

If we then replace i_{IS}^{*} in eq. (106) with eq. (96) we obtain, accordingly,

$$\frac{1}{i_{\mathrm{mf}}} = \frac{i_{\mathrm{f'c'}}^{*}(\varphi - \varphi')}{1 - \varphi} + 1 \ . \tag{109}$$

If next, eq. (98) is applied to both speed-ratio limits, so that

$$i_{\mathrm{mf}} = \frac{i_{\mathrm{IS}}^{*} - 1}{i_{\mathrm{f'c'}}^{*} - 1} = \frac{i_{\mathrm{IS}}^{\triangle} - 1}{i_{\mathrm{f'c'}}^{\triangle} - 1} ,$$

we finally obtain

$$i_{\mathrm{f'c'}}^{\triangle} = \frac{(i_{\mathrm{IS}}^{\triangle} - 1)(i_{\mathrm{f'c'}}^{*} - 1)}{i_{\mathrm{IS}}^{*} - 1} + 1 \ . \tag{110}$$

Three of the five variables in eqs. (107) to (109) can be either known or arbitrarily chosen. The fourth and fifth which then become dependent variables, however, must subsequently be calculated.

c) Power in Variable-Speed Drives

The amount of power which can be transmitted by a variable bicoupled transmission is limited by the capacity of the variable component transmission. Modern traction drives are capable of transmitting up to 20 kw, belt and link-belt drives up to 75 kw and hydrostatic drives up to 700 kw. If the power levels exceed these limits, Ward-Leonard systems are now used exclusively as variable speed drives. Thus, it is obvious that is is desirable to transmit only as small a fraction of the external power as possible through the variable component transmission. This means that the power-ratio ϵ or ϵ_0 as defined in section 35g, should approach zero as closely as possible.

As in section 35g, we want to limit this consideration to the power-ratios $\epsilon_0 = -P_f/P_I$ of transmissions which we assume to be loss-free since such as assumption will permit us to obtain quickly, and with an acceptable accuracy, a general concept of a variable bicoupled transmission which is independent of its eventual design.

With eq. (93) and worksheet 1 we obtain

$$\epsilon_0 = 1 - \frac{i_{SI}}{i_{cm}} = 1 - \frac{i_{SI}}{\dfrac{1}{1 - i_{mf}}} = 1 - \frac{1 - i_{mf}}{i_{IS}}$$

or
$$i_{mf} = (\epsilon_0 - 1)i_{IS} + 1 . \qquad (111)$$

If we refer ϵ_0 to the speed-ratio limit i_{IS}^*, then we obtain ϵ_0^* by letting eq. (107) equal eq. (111) so that:

$$1 + \frac{i_{IS}^*(\varphi - \varphi')}{\varphi' - 1} = (\epsilon_0^* - 1)i_{IS}^* + 1 ,$$

and
$$\epsilon_0^* = \frac{\varphi - \varphi'}{\varphi' - 1} + 1 = \frac{\varphi - 1}{\varphi' - 1} . \qquad (112)$$

However, if we refer ϵ_0 to the other speed-ratio limit, $i_{IS}^\triangle = i_{IS}^*\varphi$, then we obtain $\epsilon_0^\triangle$ by letting eq. (108) equal eq. (111), that is:

$$1 + \frac{i_{IS}^\triangle(\varphi - \varphi')}{\varphi(\varphi' - 1)} = (\epsilon_0^\triangle - 1)i_{IS}^\triangle + 1 , \qquad \text{and}$$

Fig. 137a–b. Power-ratio ϵ_0 (power in the auxiliary component transmission divided by the external power, neglecting all losses) at specific speed-ratio limits as a function of the speed adjustment ratio of the bicoupled transmission $\varphi = i_{\mathrm{IS}}^{\Delta}/i_{\mathrm{IS}}^{*}$ and the speed adjustment ratio of the auxiliary transmission $\varphi' = i_{\mathrm{f'c'}}^{\Delta}/i_{\mathrm{f'c'}}^{*}$.

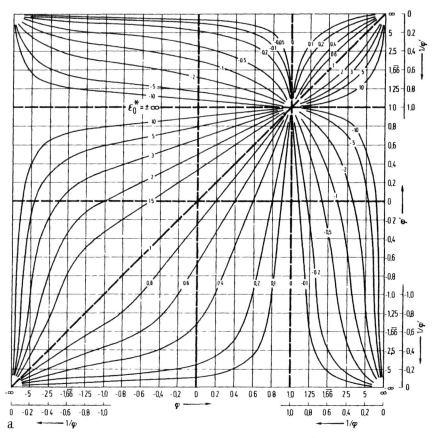

Fig. 137a. ϵ_0^{*} at the speed-ratio limits i_{IS}^{*} and $i_{\mathrm{f'c'}}^{*}$.

$$\epsilon_0^{\Delta} = \frac{\varphi - \varphi'}{\varphi(\varphi' - 1)} + 1 = \frac{\varphi'}{\varphi} \cdot \frac{\varphi - 1}{\varphi' - 1}.$$ (113a)

Reduced to the form

$$\frac{\epsilon_0^{\Delta}}{\epsilon_0^{*}} = \frac{\varphi'}{\varphi},$$ (113b)

Eqs. (113a) and (112) clearly illustrate the relationship between the power-ratios ϵ_0 at the speed-ratio limits and the adjustment-ratios φ and φ' of the

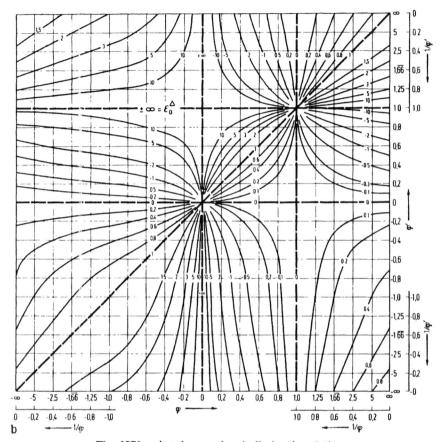

Fig. 137*b*. ϵ_0^Δ at the speed-ratio limits i_{1S}^Δ and $i_{1'c'}^\Delta$.

variable bicoupled and the variable component transmission, respectively. Eqs. (112) and (113a) are plotted in figs. 137a and 137b. With the help of these diagrams, it is easy to find an adjustment-ratio φ' for an auxiliary component transmission which, for a given adjustment-ratio φ of a planned variable bicoupled transmission, yields a sufficiently small power-ratio at either one of the two speed-ratio limits and can be realized with a standard design.

If a variable bicoupled transmission must transmit full power over its entire speed-ratio range, then as a rule both power-ratio values ϵ_0^* and ϵ_0^Δ should lie between 0 and ±1. A comparison between figs. 137a and 137b shows that this condition can be met only if the adjustment-ratio falls into the small common areas of the two diagrams which are not crosshatched in fig. 137c.

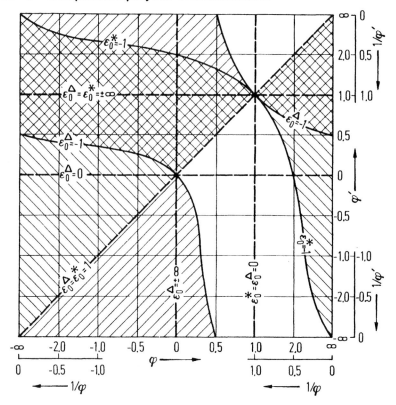

C $\boxed{\quad} \epsilon_0^* > |1|, \boxed{\quad} \epsilon_0^\Delta > |1|, \boxed{\quad}$ both ϵ_0^* and ϵ_0^Δ lie between +1 and -1

Fig. 137c. Synopsis by superimposition of a and b to aid in the design of variable bi-coupled transmissions.

If one of the adjustment-ratios is negative, then it must be verified, for example, with figs. 130 or 131, that for an arbitrary value of i_{IS} within the adjustment range, the associated power-ratio ϵ_0 lies between ϵ_0^* and ϵ_0^Δ. For a negative adjustment-ratio, i_{IS} can pass through zero or infinity while it varies continually between the speed-ratio limits and thus the power-ratio ϵ_0 may lie between ϵ_0^* and ϵ_0^Δ but also outside of these limits.

Some transmissions, such as drives for centrifugal pumps, must transmit full power only at one end of their speed-adjustment range. Thus it is frequently possible to permit a higher power-ratio ϵ_0 at the other end of their speed-adjustment range and thus gain more latitude for their layout and design.

The information contained in figs. 137a to 137c and eq. (113b) is now summarized in the following guide rule:

The power-ratio ϵ_0 in a variable component transmission R 34
becomes smaller the closer φ approaches a value of 1, that is,
the less the variable bicoupled transmission needs to be
adjusted and the more the variable component transmission
can be adjusted.

Fig. 137c specifically shows the following relationships which are important for the practical design of variable bicoupled transmissions:

1. If the adjustment-ratio of a variable *bicoupled transmission* is *negative,* then the power transmitted by the variable component transmission is larger than the external power in at least a part of the speed-adjustment range.

2. The power transmitted by the auxiliary component transmission can be kept smaller than the external power, that is, $|\epsilon_0| < 1$, over the total speed-adjustment range only when the adjustment-ratio φ of the bicoupled transmission is *positive.*

If in this case, the adjustment ratio φ' of the auxiliary transmission is also positive, then for a bicoupled transmission with an adjustment ratio of $\varphi > 2$ or $0 < \varphi < 0.5$, the condition $|\epsilon_0| < 1$ can be met only when the main component transmission is adjusted in the same direction as the auxiliary component transmission. Then both component transmissions have their maximum speed-ratio at the same end of their speed-adjustment ranges. Simultaneously the speed-adjustment range of the bicoupled transmission is smaller than the speed-adjustment range of the auxiliary component transmission; φ is closer to *1* than φ'.

3. If the adjustment-ratio φ of a bicoupled transmission is *positive,* then the condition $|\epsilon_0| < 1$ can be met for a range of $\varphi = 1/3$ to 3 by an auxiliary component transmission with an adjustment-ratio of $\varphi' = -1$, that is, an auxiliary component transmission whose adjustment limits have equal magnitudes but opposite signs. Hydrostatic transmissions with axial piston pumps and motors are suited for this application.

The previously mentioned speed-adjustment range of $\varphi = 1/3$ to 3 can obviously be extended by deviating from $\varphi' = -1$ in either direction, that is, by making the positive and negative speed-adjustment ranges of the auxiliary transmission unequal or, when both are equal, by not fully utilizing one of the ranges.

When we design a power transmission we should investigate (following R 34) whether the required power rating of the variable bicoupled transmission can be obtained by increasing the adjustment-ratio of its auxiliary transmission by coupling two variable component transmissions in series rather than in parallel.

d) Intermediate Transmissions as Means of Shifting Adjustment Limits

If only the speed-ratio limits i_{IS}^* and $i_{IS}^\triangle$ of a planned variable bicoupled transmission are specified, then we can calculate its speed-adjustment ratio φ with eq. (104). Following the considerations outlined in section 36c, we can then choose a suitable speed-adjustment ratio φ' for the variable component transmission from fig. 137 and this choice clearly determines the kinematical behavior of the variable bicoupled transmission. The speed-ratio i_{mf} of the main component transmission can be calculated with eq. (107), and the two speed-ratio limits $i_{f'c'}^*$ and $i_{f'c'}^\triangle$ of the auxiliary component transmission with eq. (97), when first i_{IS}^* and then $i_{IS}^\triangle$ are substituted for i_{IS}.

Suppose we had chosen an adjustment-ratio $\varphi' = 3/0.4 = 7.5$ because we had already decided to use a specific auxiliary component transmission (for example, a variable V-belt or link-belt drive) with the speed-ratio limits $i^* = 0.4$ and $i^\triangle = 3$. Then we can generally find some necessary speed-ratio limits $i_{f'c'}^*$ and $i_{f'c'}^\triangle$ which yield the same adjustment-ratio as the chosen variable-speed drive but are not identical with its speed-ratio limits i^* and $i^\triangle$.

The variable component transmission must then be matched to the main component transmission by one or more intermediate transmissions which simultaneously enable the variable-speed drive to operate at its optimum speed level, where it is capable of transmitting maximum power.

These intermediate transmissions change the speed-ratio of the auxiliary component transmission at both speed-ratio limits by the same factor i. Thus, the adjustment-*ratio*, is *not affected* by the addition of one or more intermediate transmissions with constant speed-ratios.

Intermediate transmissions can be arranged as follows:

1. Intermediate transmissions for matching a variable component transmission as shown in figs. 138a to 138c. To obtain the calculated, required speed-ratio limits $i_{f'c'}^*$ and $i_{f'c'}^\triangle$, even when the available variable component transmission has different speed-ratio limits I^* and $i^\triangle$, an intermediate transmission IV can be added, as shown in fig. 138a, or in fig. 138b. Alternately, two intermediate transmissions IV and V can be arranged as shown

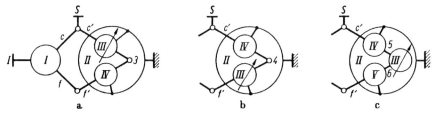

a   b   c

Fig. 138. Possible arrangements of the intermediate transmissions IV and V in the auxiliary component transmission II to match the variable speed drive III to the main component transmission.

in fig. 138c. With the designations given in figs. 138a to 138c, the matching conditions and the resulting intermediate speed-ratios are recorded in table 11.

TABLE 11. MATCHING CONDITIONS

FIG.	MATCHING CONDITIONS	SPEED-RATIOS OF INTERMEDIATE TRANSMISSIONS	EQ.
138a	$i_{f'c'}^* = i_{f'3} i_{3c'}^*$ $i_{f'c'}^\triangle = i_{f'3} i_{3c'}^\triangle$	$i_{f'3} = \dfrac{i_{f'c'}^*}{i_{3c'}^*} = \dfrac{i_{f'c'}^\triangle}{i_{3c'}^\triangle}$	(114)
138b	$i_{f'c'}^* = i_{f'4}^* i_{4c'}$ $i_{f'c'}^\triangle = i_{f'4}^\triangle i_{4c'}$	$i_{4c'} = \dfrac{i_{f'c'}^*}{i_{f'4}^*} = \dfrac{i_{f'c'}^\triangle}{i_{f'4}^\triangle}$	(115)
138c	$i_{f'c'}^* = i_{f'6} i_{65}^* i_{5c'}$ $i_{f'c'}^\triangle = i_{f'6} i_{65}^\triangle i_{5c'}$	$i_{f'6} i_{5c'} = \dfrac{i_{f'c'}^*}{i_{65}^*} = \dfrac{i_{f'c'}^\triangle}{i_{65}^\triangle}$	(116)

The intermediate speed-ratios of all three designs are equal, that is,

$$i_{f'3} = i_{4c'} = i_{f'6} i_{5c'} \,,$$

when the other given speed-ratios remain equal. In the case described by eq. (116) and fig. 138c, both speed-ratio limits of the intermediate transmission *III* are merely multiplied by the product $i_{IV} i_{V}$. Therefore its factors can be freely varied to achieve the optimum speed level of the variable-speed drive.

Of the three possibilities shown in figs. 138a to 138c, that one should be chosen which permits the variable-speed drive to operate at its optimum speed level, that is, we should choose 138a if $n_{c'}$ is the optimum speed of the variable-speed drive, 138b if $n_{f'}$ is its optimum speed, and 138c if neither $n_{c'}$ nor $n_{f'}$ are optimum speeds of the variable-speed drive.

2. Intermediate transmissions for adapting to the speed-ratio of a variable bicoupled transmission as shown in fig. 139. If the adjustment-ratio φ of the variable bicoupled transmission is given and the adjustment-ratio φ' of the variable component transmission is chosen as previously described, then, as a third parameter which can be freely chosen, one of the latter's two speed-ratio limits, for example, $i_{f'c'}^*$, can be used to determine, with eq. (109), a speed-ratio i_{mf} and thus a main component transmission that is matched to the variable component transmission without an additional intermediate transmission. Then the adjustment-ratio φ of the variable bicoupled transmission is indeed acceptable, but the speed-ratio limits i_{IS}^* and $i_{IS}^\triangle$ do not have their desired values $i_{IS, req'd.}^*$ and $i_{IS, req'd.}^\triangle$. Therefore, an additional reduction drive is needed to adapt the bicoupled transmission to the

required speed-ratio limits. This intermediate drive *III* can be connected, either as shown in fig. 139a, where

$$i_{AI} = \frac{i^*_{AS,\text{req'd.}}}{i^*_{IS}} = \frac{i^\Delta_{AS,\text{req'd.}}}{i^\Delta_{IS}} \tag{117}$$

or as shown in fig. 139b, where

$$i_{SB} = \frac{i^*_{IS,\text{req'd.}}}{i^*_{IS}} = \frac{i^\Delta_{IB,\text{req'd.}}}{i^\Delta_{IS}} \tag{118}$$

3. Intermediate transmission as shown in fig. 140 for the simultaneous matching of the variable component transmission and the bicoupled transmission. This arrangement combines both of the possibilities described by figs. 138c and 139b. It is frequently used when, because of space or design reasons, the variable component transmission cannot be directly coupled to the main component transmission and intermediate transmissions must be used to bridge the gap, or when simpler designs, such as shown in figs. 140a and 140b, can be obtained by connecting the input to shaft *A* instead of shaft *S*. Although in this case the coupling shaft *S* is not directly connected, we are able to analyze this transmission like a constrained bicoupled transmission if we place the external connection *A* into the power path between shaft *c'* and the variable component transmission, but not between shaft *f'* and the variable component transmission. This fact must be strictly observed when the shafts are labeled, see fig. 140b. Kinematically, the transmission then behaves as if the coupling shaft *S* were connected through an imaginary duplicate of the intermediate transmission *IV* matching the speed-ratio limits like transmission *III* in fig. 139b. Consequently, the speeds of the connected shafts *A* and (*S*) would be equal. This arrangement is shown explicitly in fig. 140c.

If, in addition to the speed-ratio limits $i^*_{IA,\text{req'd.}}$ and $i^\Delta_{IA,\text{req'd.}}$, and thus the adjustment-ratio

$$\varphi = \frac{i^\Delta_{IA,\text{req'd.}}}{i^*_{IA,\text{req'd.}}} = \frac{i^\Delta_{IS} i_{SA}}{i^*_{IS} i_{SA}} = \frac{i^\Delta_{IS}}{i^*_{IS}} \tag{119}$$

of the bicoupled transmission, the speed-ratio limits i^* and i^Δ, and the adjustment-ratio φ' of its variable component transmission are given, then as a third parameter, together with φ and φ', we can either choose the

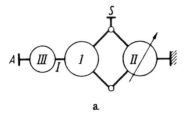

a

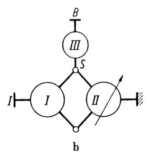

b

Fig. 139. Variable bicoupled transmission with a reduction drive at either of the connected shafts, to adapt the bicoupled transmission to the required speed-ratio limits $i^*_{req'd}$ and $i^\Delta_{req'd}$: *a*, reduction drive at the connected monoshaft *I; b*, reduction drive at the connected coupling shaft *S*.

speed-ratio i_{mf} which determines the main component transmission and calculate first i^*_{IS} with eq. (107) and then the speed-ratio of the intermediate transmission *IV* from:

$$i_{5c'} = \frac{i^*_{IS}}{i^*_{IA,req'd.}} = \frac{i^\Delta_{IS}}{i^\Delta_{IA,req'd.}} , \qquad (120)$$

or we can choose the speed-ratio $i_{5c'}$ of the intermediate transmission *IV* according to the given design requirements and calculate first

$$i^*_{IS} = i^*_{IA,req'd.}.i_{5c'}$$

with eq. (120) and then i_{mf} with eq. (107). The other speed-ratio limit i^Δ_{IS} can be determined with eq. (104). When i^*_{IS} and i^Δ_{IS} are thus known, the speed-

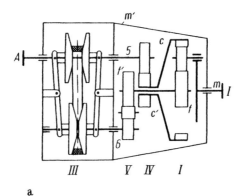

a

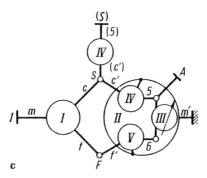

b

c

Fig. 140. PIV-variable bicoupled transmission, type AG [39], with a link-belt drive as the variable component transmission *III* and the two intermediate gear trains *IV* and *V* which have negative and positive speed-ratios, respectively: *A*, input; *I*, output; *S* is not connected: *a*, schematic representation of the transmission; *b*, symbolic representation; *c*, if a hypothetical second gear train *IV* is connected at *S*, the speeds at *A* and (*S*) become equal and the transmission can be treated as a constrained bicoupled transmission.

ratio limits $i_{f'c'}^*$ and $i_{f'c'}^\Delta$ of the auxiliary component transmission can be calculated with eq. (97). The speed-ratio,

$$i_{f'6} = \frac{i_{f'c'}^*}{i_{65}^* i_{5c'}} = \frac{i_{f'c'}^\Delta}{i_{65}^\Delta i_{5c'}} , \qquad (121)$$

of the intermediate transmission V is finally obtained from the matching conditions expressed by eq. (116). In general, it cannot be expected that the first solution thus obtained represents the best possible transmission. Rather, it will be necessary to repeat the calculation with a slightly altered assumption for i_{mf} or $i_{5c'}$ so that the optimum transmission will be the result of an iteration process.

e) Analysis of a Particular Design

A variable bicoupled transmission of the type shown in fig. 140 is to be designed according to the following specifications:

$$\text{input speed:} \quad n_{in} = 16 \; s^{-1}$$

$$\text{output speed:} \quad n_{out} = 0 \text{ to } 15 \; s^{-1}$$

chosen variable speed drive: belt-drive with $i_{max} = 2$ and $i_{min} = 0.5$.

1. Adjustment-ratio of the variable bicoupled transmission: The speed-ratio limits of the bicoupled transmission are $n_{out}/n_{in} = 0$ and 0.9375. Circumventing the decision whether the input should be at shaft I or shaft A, we shall assume temporarily that shaft (S) rather than shaft A is externally connected (fig. 140c) but delay consideration of the intermediate transmission IV until later. Thus we gain the advantage that the transmission can be treated like a constrained bicoupled transmission and the appropriate diagrams can be used directly.

Fig. 131 shows that for $n_S = 0$, that is, for $i_{IS} = \infty$, $i_{mf} = \infty$ also, and thus cannot be realized. For $n_I = 0$, that is, for $i_{IS} = 0$, however, all speed-ratios i_{mf} between $+\infty$ and $-\infty$ are possible. Thus, $n_I = 0$ is possible, and, consequently, the input can only be at S (or A in fig. 140c).

Consequently, the speed-ratio limits of the bicoupled transmission are

$$i_{IA}^* = 0 \quad \text{and} \quad i_{IA}^\Delta = 0.9375 ,$$

where the symbols * and Δ have been assigned arbitrarily and the adjustment ratio becomes:

$$\varphi = \frac{i_{IS}^\Delta}{i_{IS}^*} = \frac{i_{IS}^\Delta \cdot i_{SA}}{i_{IS}^* \cdot i_{SA}} = \frac{i_{IA}^\Delta}{i_{IA}^*} = \frac{0.9375}{0} \rightarrow \infty .$$

2. Adjustment-ratio of the variable component transmission *III*: The given variable component transmission *III* can have two different adjustment-ratios,

$$\varphi_1' = \frac{i_{\text{III,max}}}{i_{\text{III,min}}} = \frac{2}{0.5} = 4$$

or

$$\varphi_2' = \frac{i_{\text{III,min}}}{i_{\text{III,max}}} = \frac{0.5}{2} = 0.25 \ .$$

For $\varphi = \infty$, the point $\varphi' = 4$ lies in the doubly crosshatched area of fig. 137c, which means that the power-ratio $\epsilon_0 > 1$ at both ends of the speed-adjustment range. The point $\varphi' = 0.25$ lies in the simply crosshatched area and thus $\epsilon_0 < 1$ for at least one of the adjustment limits. Therefore we choose:

$$\varphi' = 0.25 = \frac{i^\Delta}{i^*} = \frac{0.5}{2} \ ,$$

and the subscription becomes

$$i^* = i_{65}^* = 2 \ , \qquad i^\Delta = i_{65}^\Delta = 0.5 \ .$$

The order of the subscripts must be *65* rather than *56* since, according to eq. (116), in this form *i* is one of the factors of $i_{f'c'}$ whose subscription in turn is determined by the definition of φ' which is given by eq. (105).

3. Choice of the third parameter: We can now assume one speed-ratio i_{mf} of the main component transmission and determine first i_{1S}^* with eq. (107) and then $i_{5c'}$ with eq. (120) or vice versa. Initially we shall try to work with an arbitrarily chosen value $i_{5c'} = -1$.

4. Determination of the other transmission parameters: With eq. (120), $i_{1S}^* = 0$ and $i_{1S}^\Delta = 0.9375(-1) = -0.9375$. With eq. (108)

$$i_{mf} = 1 + \frac{-0.9375(1 - 0)^{\cdot}}{0.25 - 1} = 2.25 \ .$$

Eq. (107) cannot be used since $i_{1S}^* \cdot \varphi = 0 \cdot \infty$, and thus is indeterminate. With eq. (97) the speed-ratio limits of the auxiliary transmission become:

$$i_{f'c'}^* = \frac{i_{1S}^* - 1}{i_{mf}} + 1 = \frac{0 - 1}{2.25} + 1 = 0.55\overline{5} \ ,$$

and

$$i_{f'c'}^\Delta = \frac{i_{1S}^\Delta - 1}{i_{mf}} + 1 = \frac{-0.9375 - 1}{2.25} + 1 = 0.1389 \ .$$

Check: $$\varphi' = \frac{i_{f'c'}^{\triangle}}{i_{f'c'}^{*}} = \frac{0.1389}{0.555} = 0.25 \ .$$

Intermediate transmission V: The inherent speed-ratio limits of the variable component transmission are adapted to the required values $i_{f'c'}^{*}$ and $i_{f'c'}^{\triangle}$ by the intermediate transmission V whose constant speed-ratio can be found with eq. (121). Thus,

$$i_{f'6} = \frac{i_{f'c'}^{*}}{i_{65}^{*}i_{5c'}} = \frac{0.55\overline{5}}{2(-1)} = -0.277 \ , \qquad \text{and} \ ,$$

contrary to fig. 140a, i_V is negative.

5. Selection of the main component transmission: Depending on whether the carrier is mounted on shaft m, c, or f, three different main component transmissions are possible which correspond to the coupling cases mm', cm', or fm' of fig. 124 where the fixed housing of the auxiliary transmission II is considered as the latter's fixed carrier $s' = m'$. With $i_{mf} = 2.25$, the three speed-ratios are obtained as shown in table 12.

TABLE 12. BASIC SPEED-RATIOS FOR DIFFERENT COUPLING CASES

COUPLING CASE	SUBSCRIPT KEY			BASIC SPEED-RATIO
	m	c	f	
mm'	s	1	2	$i_o = \dfrac{i_{s2}}{i_{s2} - 1} \triangleq \dfrac{i_{mf}}{i_{mf} - 1} = \dfrac{2.25}{1.25} = 1.80$
cm'	1	s	2	$i_o = i_{12} \triangleq i_{mf} = 2.25$
fm'	1	2	s	$i_o = 1 - i_{1s} \triangleq 1 - i_{mf} = 1 - 2.25 = -1.25$

We shall choose the negative-ratio transmission which represents the coupling case fm' since it leads to the simplest possible design as shown in fig. 141. However, its basic speed-ratio is close to -1 and thus requires rather small planets. Therefore, starting with step 3, we repeat the calculation, and choose $i_o = -2$ so that for the coupling case fm', $i_{mf} = i_{1s} =$

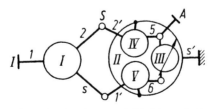

Fig. 141. Shaft designations for the three variable bicoupled transmissions analyzed in the calculations of table 13, where the main component transmission has a negative basic speed-ratio.

$1 - i_o = 3$. Other suitable transmissions can often be found when the signs of the speed-ratio limits are changed, which usually can be done by assuming an appropriate sense of rotation for the input shaft. Then we repeat the calculation steps a third time for a modification of the bicoupled transmission which has the speed-ratio limits $i_{IA}^* = 0$ and $i_{IA}^A = -0.9375$. For convenience the results of the ensuing three calculations are compiled in table 13.

TABLE 13. NUMERICAL EVALUATION OF THREE KINEMATICALLY-EQUIVALENT VARIABLE BICOUPLED TRANSMISSIONS*

VALUES		(EQS.), FIGS. AND WORKSHEETS USED	CALCULATION NO.		
			1	2	3
Given	i_{IA}^*		0	0	0
	i_{IA}^A		0.9375	0.9375	-0.9375
	φ	(104)	∞	∞	-∞
Chosen	φ'	137c	0.25	0.25	0.25
	i_{65}^*		2	2	2
	i_{65}^A		0.5	0.5	0.5
Assumed	i_{mf}		2.25^1	3	0.5
Calculated	i_{IS}^*	(120)	0	0	0
	i_{IS}^A	(108)	-0.9375	-1.5	0.375
	$i_{Sc'}$	(120)	-1^1	-1.6	-0.4
	$i_{f'c'}^*$	(97)	0.555	0.66̄6	-1.0
	$i_{f'c'}^A$	(97)	0.1389	0.16̄6	-0.25
Check	φ'	(105)	0.25	0.25	0.25
	$i_{f'6}$	(121)	-0.27̄7	-0.2083	1.25
Possible main component transmissions $\left\{ \begin{array}{l} mm' \\ cm' \\ fm' \end{array} \right\}$	i_o	1	$\left\{ \begin{array}{l} 1.8 \\ 2.25 \\ -1.25 \end{array} \right.$	1.5 3.0 -2.0	-1.0 0.5 0.5

*Another numerical example of the layout of a variable bicoupled transmission is given in section 6 of [56].
¹ In calculation 1), $i_{Sc'}$ had the chosen value of -1, but i_{mf} was calculated.

f) Efficiency

If the shafts I and S of a variable bicoupled transmission are connected, then its efficiency at a given operating point, for example, at i_{IS}, can be calculated as described in sections 35e and 33e for constrained bicoupled transmissions. Of course, this efficiency changes when the speed-ratio is changed, and, therefore, it is expedient to calculate the efficiency for both ends of the speed-ratio and at least one operating point in between.

If, instead of shaft S, a shaft A of the auxiliary component transmission is connected, then the efficiency at a given operating point can be calculated when the speeds and torques are determined separately as described in section 35. However, even in this case it is simpler to determine the efficiency as if shaft S instead of shaft A in fig. 140 were connected to the outside. The error caused by this assumption becomes very small when the intermediate transmission IV between S and A has a high efficiency. In an unfavorable case, such as $\epsilon_o > 1$, or $\epsilon_o < 0$, it can be expressed by the deviation from the actual friction power-loss $\Delta P_L = \pm \zeta_{IV} P_A$, which represents the loss incurred in the intermediate transmission IV when the total effective power P_A is transmitted from shaft c through transmission IV to shaft A rather than to shaft S directly. When the shaft A rather than the shaft S is connected, then the error in the determination of the efficiency is only affected by the external power but not by the circulating positive or negative futile power. If the transmission operates with power division, the error is even smaller as can be easily proven by means of the power-flow. The error diminishes to zero when $\epsilon = 0.5$.

For the transmissions which have been obtained in the second and third calculation of the previous section 36e, the described approximation yields the following overall efficiencies when the efficiencies of the individual transmission sections are assumed to be:

$$\eta_o = 0.98 \, ; \qquad \eta_{III} = 0.9 \, ; \qquad \eta_{IV} = 0.985 \, ;$$

with $\quad i_V = i_{f'6} < 0 \quad$ as in calculation 2, $\qquad (\eta_V)_2 = 0.985 \, ;$

with $\quad i_V = i_{f'6} > 0 \quad$ as in calculation 3, $\qquad (\eta_V)_3 = 0.97 \, .$

$(\eta_V)_3$ is assumed to be lower than $(\eta_V)_2$ since an additional gear stage is needed to obtain a positive speed-ratio $i_{f'6}$.

The notation and the subscript key are the same as in the previous section 36e and fig. 141.

Transmission 2 with $i_o = -2$. With worksheet 1 we obtain:

$$i_{III}^{\Delta} = i_{mf} i_{m'f'}^{\Delta} = i_{1s} i_{1's'}^{\Delta} = (1 - i_o)(1 - i_o^{\Delta})$$
$$= (1 + 2)(1 - 0.16\overline{6}) = 3 \cdot 0.83\overline{3} = 2.5 \, .$$

Since the series speed-ratio $i_{III} > 0$ the efficiency can be found with box 12 of worksheet 5. Thus

$$\eta_{SI}^{\Delta} = \frac{i_{III}^{\Delta} - 1}{\dfrac{i_{III}^{\Delta}}{\eta_{III}^{\Delta}} - 1} = \frac{2.5 - 1}{\dfrac{2.5}{0.962} - 1} = 0.938 \, ,$$

where $\qquad \eta_{III}^{\Delta} = \eta_{m'f'}^{\Delta} \eta_{fm}^{\Delta} = \eta_{s'1'}^{\Delta} \eta_{s1} \, .$

According to worksheet 5, boxes 11 and 10 and with $i_{o'}^\Delta = i_{f'c'}^\Delta = 0.16\bar{6}$

$$\eta_{s'1'}^\Delta = \frac{i_{o'}^\Delta - 1}{i_{o'}^\Delta \eta_{o'}^\Delta - 1} = \frac{0.16\bar{6} - 1}{0.16\bar{6} \cdot 0.985 \cdot 0.9 \cdot 0.985 - 1} = 0.9753$$

$$\eta_{s1} = \frac{i_o - 1}{\dfrac{i_o}{\eta_o} - 1} = \frac{-2 - 1}{\dfrac{-2}{0.98} - 1} = 0.9866$$

so that

$$\eta_{\text{III}}^\Delta = 0.9753 \cdot 0.9866 = 0.962 \ .$$

With $i_o = 0.5$ and $i_{o'}^\Delta = i_{f'c'}^\Delta = -0.25$, we obtain for transmission No. 3:

$$i_{\text{III}}^\Delta = i_{mf}i_{f'm'}^\Delta = i_{1s}i_{1's'}^\Delta = (1 - i_o)(1 - i_{o'}^\Delta)$$

$$= (1 - 0.5)(1 + 0.25) = 0.625 \ ,$$

$$\eta_{mf} \triangleq \eta_{1s} = \frac{1 - i_o/\eta_o}{1 - i_o} = \frac{1 - 0.5/0.98}{1 - 0.5} = 0.9796 \ ,$$

$$\eta_{f'm'}^\Delta \triangleq \eta_{1's'}^\Delta = \frac{1 - i_{o'}^\Delta \eta_{o'}^\Delta}{1 - i_{o'}^\Delta} = \frac{1 + 0.25 \cdot 0.97 \cdot 0.9 \cdot 0.985}{1 + 0.25}$$

$$= 0.972 \ ,$$

$$\eta_{\text{III}}^\Delta = \eta_{1s}\eta_{1's'}^\Delta = 0.9796 \cdot 0.972 = 0.9522 \ ,$$

$$\eta_{SI}^\Delta = \frac{i_{\text{III}}^\Delta - 1}{i_{\text{III}}^\Delta \eta_{\text{III}}^\Delta - 1} = \frac{0.625 - 1}{0.625 \cdot 0.9522 - 1} = 0.926 \ .$$

A comparison shows that the theoretical efficiency of transmission *2* is slightly higher than that of transmission *3*. The design according to calculation *2*, moreover, saves a reversing gear in transmission *V* which is needed in design *3* in order to obtain a positive speed-ratio $i_V = i_{f'6} > 0$. The higher efficiency of transmission *2*, therefore, is essentially due to the fact that it does not incur the losses of an additional reversing gear stage. Since at the other end i_{AI}^* of the speed-ratio range, the output speed and, consequently, the output power are zero the associated efficiency $\eta^* = 0$. Within the speed-adjustment range, the efficiencies lie between the values calculated at the speed-ratio limits.

Section 37. Reduced Bicoupled Transmissions

a) Derivation from Simple Bicoupled Transmissions

Among the possible coupling cases shown in fig. 124, the coupling cases cc' and ff' may be expected to have an especially simple structure since, as shown in fig. 142, in either of these cases the two carriers are directly coupled and, therefore, can be combined into a single unit requiring only one set of bearings. Consequently, these transmissions have been called single-carrier bicoupled transmissions.

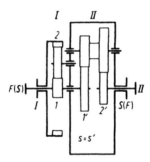

Fig. 142. Single-carrier bicoupled transmission with a common carrier at either the connected coupling shaft S or the free coupling shaft F, according to the structural coupling cases cc' or ff'.

A further simplification is possible if, as shown in fig. 143a, the two central gears of the connected coupling shaft and, likewise, their meshing planets, are designed to have equal sizes. The respectively equal central and planet gears then can also be reduced to a single unit each, as shown in fig. 143b. However, the reduced central gear $1 = 1'$ cannot constitute the free coupling shaft because it would merely idle if it were not externally connected, as can be understood from fig. 143b. Consequently, the bicoupled transmission would become a simple planetary transmission of the type shown in fig. 34. If, therefore, the design complexity of a single-carrier bicoupled transmission is reduced by combining the two identical central gears and their identical meshing planets, then their common carrier must become the free coupling shaft. Transmissions of this type shall subsequently be called "reduced bicoupled transmissions."

b) Functionally-Equivalent Bicoupled Transmissions

The transmissions shown in fig. 143a, and in fig. 143b after the equal gears of fig. 143a have been combined, differ only in one important characteristic, namely, that in fig. 143a the shafts of the equal planets p and $p_{1'}$

are not coupled together. Thus, they can transmit torques or tangential forces, which may have either equal or opposite directions, to the rigidly coupled gears *1* and *1'*. The sum or difference of these torques is formed only at the connected coupling shaft and then transmitted to the outside. In the reduced bicoupled transmission of fig. 143b, however, the two planet gears p and p' are rigidly coupled as a consequence of their combination into a single unit. If, thus, the torques which are transmitted by the gears *2* and $p_{2'}$ act in the same direction, then their sum is formed already at the common planet gear $p = p_{1'}$ and is transmitted from there to the common gear $1 = 1'$. If they have opposite directions their difference is formed at $p_1 = p_{1'}$, transmitted to the central gear $1 = 1'$, and from there to the outside.

When a *sum* of two torques is transmitted from $p = p_{1'}$ to $1 = 1'$, then the rolling-power between these two gears equals the sum of the two rolling-powers, which according to fig. 143a, are transmitted from p to *1* and from $p_{1'}$ to *1'*. In this case the tooth-friction power-loss of the reduced bicoupled transmission equals the sum of the two tooth-friction power-losses incurred by the bicoupled transmission of fig. 143a before its reduction. This means that the tooth-friction losses and, consequently, the efficiency of the bicoupled transmission, do not change when equal gears are combined into single units.

Not considering the two omitted bearings per planet, the spatial arrangement, as well as the function of the other parts of the transmissions shown in figs. 143a and 143b, has remained the same. Therefore, a simple bicoupled transmission and the reduced bicoupled transmission with equal gears derived from it are *functionally equivalent* if a sum of torques or, because the speeds of the coupled shafts are identical, a sum of powers is transmitted to or from the coupling shaft. This case occurs according to section 35 when the constrained bicoupled transmission operates with power division.

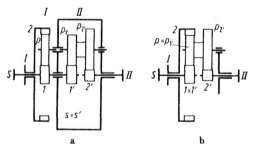

Fig. 143. Formation of a reduced bicoupled transmission: *a*, simple bicoupled transmission according to the coupling case *ff'* which has equal-sized coupled central gears *1* and *1'* and equal-sized meshing planets p and $p_{1'}$; *b*, reduced bicoupled transmission which originated from *a* by reducing the equal sized central gears and planets to one gear each.

If, however, the *difference* between two torques is transmitted from $p = p_1{}'$ to $1 = 1'$, then the rolling-power and, consequently, also the rolling-power loss between these two gears is only equal to the difference between the power-losses which are incurred in the gear meshes $p - 1$ and $p_1{}' - 1'$ of the non-reduced bicoupled transmission of fig. 143a. The efficiency of a reduced bicoupled transmission is then higher than the efficiency of the corresponding, non-reduced, simple bicoupled transmission and, consequently, the two transmissions are not functionally equivalent. According to section 35f, the difference between two torques acts on the connected coupling shaft of *those* constrained bicoupled transmissions which transmit a positive or negative circulating power.

The occurrence of a sum or difference of torques at the connected coupling shaft has been noted only in constrained bicoupled transmissions (sec. 35f). But, this result is also valid when these bicoupled transmissions acquire two degrees of mobility by connecting their fixed shafts, because they have only one static degree of freedom and, consequently, their torques bear a constant ratio relative to each other and are independent of the speeds.

We can conclude that the equations which have been derived for simple bicoupled transmissions, and are compiled in worksheets 4 and 5, can be used directly to determine the efficiency of a constrained bicoupled transmission or a reduced bicoupled transmission with two degrees of mobility, provided they refer to a simple bicoupled transmission which is *functionally equivalent* to the reduced bicoupled transmission. Only those bicoupled transmissions which operate without circulating power can be functionally equivalent to a reduced bicoupled transmission. To operate without circulating power requires that the power-flows in the coupled shafts c and c' must be the same. Their speeds obviously are equal, and their torques also must be equal for this to occur. And if equal under one running condition, such as in a constrained bicoupled transmission, they are equal under all running conditions because of the independence of torques and speeds. Therefore, a fundamental requirement for the functionally-equivalent transmission is that the torques of the shafts c and c' of the component transmissions must have equal signs. This always is realized with dissimilar monoshafts, as can be verified with guide rules 4, 8, and 9. That is, one must be a difference shaft, and the other the summation shaft of its component transmission. Symbolic diagrams c of figs. 114, 115, and 144 show this feature and thus serve to positively identify transmissions which are functionally equivalent to those shown in schematic diagrams a.

Conversely, the speed characteristics of a reduced bicoupled transmission, and its corresponding simple bicoupled transmission, are equal even when both are not functionally equivalent as far as their losses are concerned. We can now formulate the following guide rules:

A reduced bicoupled transmission and its corresponding R 35
functionally-equivalent simple bicoupled transmission have
the same operating characteristics, are represented by the
same transmission symbol, and can be analyzed with the same
equations.

The simple bicoupled transmission which is functionally R 36
equivalent to a reduced bicoupled transmission is character-
ized by the fact that its monoshafts are formed by a summa-
tion shaft and a difference shaft from its component trans-
missions.

c) Analysis of Reduced Bicoupled Transmissions

With the help of the given worksheets and diagrams, the speeds, torques, and efficiencies are determined for the functionally-equivalent constrained bicoupled transmission, or for the functionally-equivalent simple bicoupled transmission with two degrees of mobility rather than for the reduced bicoupled transmission itself (sec. 32d, e).

Therefore, the only new problem consists in finding this functionally-equivalent simple bicoupled transmission when only the reduced bicoupled transmission is known. This is indeed a simple task. First, however, we want to determine how many kinematically-equivalent bicoupled transmissions belong to each reduced bicoupled transmission.

Each reduced bicoupled transmission such as the transmission shown in fig. 144a (described by Wolfrom [2] in 1912), has two degrees of mobility and, consequently, three connected shafts *A*, *B*, and *C*, each of which can be the connected coupling shaft of the corresponding simple bicoupled transmission, and thus can be keyed to the combined central gears. If we assume that shaft *A* is the connected coupling shaft, and draw its central gear together with its meshing planet twice, side by side, connecting the two equal central gears with a common shaft, then the other gears of the reduced bicoupled transmission form two different component transmissions, each of which contains one of the two central gears. These two component gear trains are connected through a common carrier and thus constitute the corresponding bicoupled transmission shown in fig. 144b. Both have a negative speed-ratio and, therefore, according to worksheet 3, their carrier shafts are also their summation shafts. We can now draw the Wolf symbol of this bicoupled transmission which is shown above fig. 144b. Analogously, we can double the central gears of shafts *B* and *C* and obtain the two additional kinematically-equivalent bicoupled transmissions which are shown in figs. 144c and 144d together with their Wolf symbols.

A comparison of the three corresponding bicoupled transmissions shows that we have obtained three different simple epicyclic transmissions *I, II* and *III,* two of which have a negative, one a positive speed-ratio. In other cases all three simple transmissions may have a positive speed-ratio. The three possible combinations of two each of the three simple transmissions constitute the three kinematically-equivalent bicoupled transmissions. Further combinations are not possible.

The monoshafts of only one of the three possible corresponding bicoupled transmissions are formed by a summation shaft and a difference shaft from each of the component transmissions as required by R 32 and R 36. This bicoupled transmission can thus be identified as the functionally-equivalent bicoupled transmission. In our example it is the transmission which, according to fig. 144c, consists of the two component transmissions *I* and *III.* Subsequently, we shall use this transmission as an example for the calculation of the speed-ratios, and the efficiencies of the corresponding reduced bicoupled transmission. These results can be generalized in the following guide rule:

> *Each reduced bicoupled transmission can be expanded into*   R 37
> *three different kinematically-equivalent simple bicoupled*
> *transmissions, but only one of these is functionally equivalent*
> *to it.*

The functionally-equivalent simple bicoupled transmission can be found quickly if, in succession, each of the three central gears of the reduced bicoupled transmissions is assumed to be the connected, reduced coupling gear, and the two gear trains originating from it are then traced to the other connected central gears. We recognize both of these gear trains as planetary transmissions. If their non-reduced central gears are found to be keyed to a summation shaft and a difference shaft, than these gear trains constitute the component gear trains of the functionally-equivalent bicoupled transmission.

If the three possible component gear trains consist of two negative-ratio transmissions and one positive-ratio transmission, then that shaft of the positive-ratio transmission which attains the highest velocity when the carrier becomes locked is always the reduced connected coupling shaft of the functionally-equivalent bicoupled transmission. If all three component transmissions have a positive speed-ratio, it is the shaft which rotates with the median speed.

If one of the three shafts of the reduced bicoupled transmission is locked, we obtain as its corresponding simple bicoupled transmission a series-coupled transmission when the locked shaft is assumed to be a coupling shaft, and *two* real constrained bicoupled transmissions when the two rotating shafts are successively considered as connected coupling shafts.

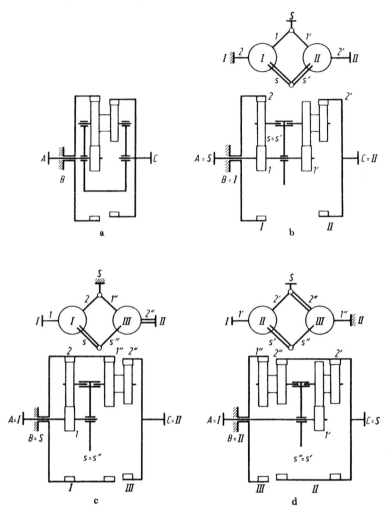

Fig. 144. Reduced bicoupled transmission and its three corresponding simple bicoupled transmissions (in each case S denotes the connected coupling shaft): a, given reduced bicoupled transmission; b–d, corresponding kinematically equivalent simple bicoupled transmissions with their proper designations; b, with the component transmissions I and II; c, functionally equivalent bicoupled transmission with the component transmissions I and III; d, with the component transmissions II and III.

d) Example of an Analysis

The transmission shown in fig. 144a is analyzed using the same assumptions as Wolfrom [2]. With the designations of fig. 144c, the functionally-equivalent bicoupled transmission has the following pitch circle diameters in millimeters: $d_2 = d_{1''} = 400$; $d_p = d_{p1''} = 160$; $d_1 = 80$; $d_{2'} = 384.616$; $d_{p2'} = 144.616$. The efficiency of all gear meshes is assumed to be $\eta_t = 0.975$. For the sake of a comparison, we shall analyze the two kinematically-equivalent transmissions shown in figs. 144b and 144d, as compared with the functionally-equivalent bicoupled transmission of fig. 144c which consists of the component gear trains I and III.

All three component gear trains, I with i_o, η_o, II with $i_{o'}$, $\eta_{o'}$ and III with $i_{o''}$, $\eta_{o''}$, have two successive gear meshes in the direction of the power-flow so that according to the previously made assumptions

$$\eta_o = \eta_{o'} = \eta_{o''} = 0.975 \cdot 0.975 = 0.950626 .$$

(The large number of decimals is carried only for the sake of comparison with the results obtained by Wolfrom and other authors.)

The shafts of the three component gear trains, all of which are represented in the two kinematically-equivalent bicoupled transmissions, have the same designations in both of these transmissions. Thus:

$$i_o = \frac{d_2}{d_1} = \frac{-400}{80} = -5 ,$$

$$i_{o'} = \frac{d_{p1'}}{d_{1'}} \cdot \frac{d_2}{d_{p2'}} = \frac{160}{80} \cdot \frac{-384.616}{144.616} = -5.319 ,$$

$$i_{o''} = \frac{d_{p1''}}{d_{1''}} \cdot \frac{d_{2''}}{d_{p2''}} = \frac{160}{-400} \cdot \frac{-384.616}{144.616} = 1.063827 .$$

If we lock the annular gear B in the transmission of fig. 144a, as Wolfrom did, then, with the aid of worksheets 1 and 5, we can determine the overall speed-ratio and the overall efficiency of the three corresponding constrained bicoupled transmissions as follows:

Transmission according to fig. 144b with the component gear trains I and II:

$$i_{I\,II} = i_{2s}i_{s'2'} = \left(1 - \frac{1}{i_o}\right)\frac{i_{o'}}{i_{o'} - 1} = \left(1 + \frac{1}{5}\right)\frac{-5.319}{-6.319} = 1.0101 ,$$

$$i_{S\,II} = \frac{i_{I\,II}}{i_{I\,II} - 1} = \frac{1.0101}{0.0101} = 100 ,$$

$$\eta_{S\,II} = \frac{i_{I\,II} - 1}{i_{I\,II} - \eta_{III}} = \frac{0.0101}{1.0101 - 0.9837} = 0.3826 ,$$

where

$$\eta_{III} = \eta_{2's'}\eta_{s2} = \frac{i_{o'} - \eta_{o'}}{i_{o'} - 1} \cdot \frac{i_o - 1}{i_o - 1/\eta_o}$$

$$= \frac{-5.319 - 0.9506}{-5.319 - 1} \cdot \frac{-5 - 1}{-5 - 1/0.9506} = 0.9837 .$$

Transmission according to fig. 144c with the component gear trains *I* and *III*:

$$i_{III} = i_{1s}i_{s''2''} = (1 - i_o) \frac{i_{o''}}{i_{o''} - 1} = 6 \cdot \frac{1.0638}{0.0638} = 100 ,$$

$$\eta_{III} = \eta_{1s}\eta_{s''2''} = \frac{i_o\eta_o - 1}{i_o - 1} \cdot \frac{i_{o''} - 1}{i_{o''} - \eta_{o''}} = 0.9588 \cdot 0.5638 = 0.5406 .$$

In this functionally-equivalent transmission, the locked shaft, by coincidence, is the connected coupling shaft. Therefore, its series speed-ratio i_{III} and its series efficiency η_{III} are also the speed-ratio i_{AC} and the overall efficiency η_{AC} of the given Wolfrom transmission.

With only a small error in his complicated analysis, Wolfrom [2] found a value of $\eta = 6/10.97 = 0.5469$ for the overall efficiency, and he added the comment: "We desisted from building a model of this planetary transmission." If his error is corrected, the efficiency is found to equal the previously calculated value of $\eta_{III} = 0.5406$ which is also obtained with the equation given by Poppinga [43].

Transmission according to fig. 144d with the component gear trains *II* and *III*:

$$i_{III} = i_{1's'}i_{s''1''} = (1 - i_{o'}) \frac{1}{1 - i_{o''}} = (1 + 5.319) \frac{1}{1 - 1.0638} = -99 ,$$

$$i_{IS} = 1 - i_{III} = 1 + 99 = 100 ,$$

$$\eta_{IS} = \frac{i_{III}\eta_{III} - 1}{i_{III} - 1} = \frac{-99 \cdot 0.513 - 1}{-100} = 0.518 ,$$

where

$$\eta_{III} = \eta_{1's'}\eta_{s''1''} = \frac{i_{o'}\eta_{o'} - 1}{i_{o'} - 1} \cdot \frac{i_{o''} - 1}{i_{o''}/\eta_{o''} - 1} = 0.958 \cdot 0.536 = 0.513 .$$

A comparison now shows that the speed-ratios of the three corresponding constrained bicoupled transmissions have the same value of $i_{AC} = 100$, so that the three transmissions are kinematically equivalent and, further, that the functionally-equivalent transmission of fig. 144 has the highest efficiency which could be expected according to section 37b. It is also obvious that the latter efficiency represents the efficiency of our reduced bicoupled transmission.

e) Possible Design Configurations of Reduced Bicoupled Transmissions

If the advantage of the reduced bicoupled transmissions, namely, their reduced number of gears, is to be fully utilized, then their component transmissions must contain the smallest possible number of gears. Therefore, it is preferable that negative-ratio gear trains of the type shown in figs. 32 and 34 and positive-ratio gear trains of the type shown in figs. 19, 21 and 22 should be considered. With these we obtain the following transmission types:

1. Not considering any additional gear trains used for power branching the smallest possible number of five gears is obtained only in configurations which consist of one negative-ratio gear train of the type shown in fig. 32 and one negative-ratio gear train of the type shown in fig. 34 or, alternately, one positive-ratio gear train according to figs. 19 or 21 and one negative-ratio gear train according to fig. 34. The two transmission types shown in fig. 145 are practical examples of these combinations. They are drawn to scale for their respective speed-ratio limits assuming that in each case three planets are arranged around the circumference. The Wolfrom transmission, fig. 144, and the Minuteman cover drive [35], fig. 114, which differ only in the choice of their fixed shafts, obviously belong in this group.

2. The next higher number of gears, namely six, results from a combination of two positive-ratio gear trains of the type shown in figs. 19, 21, and

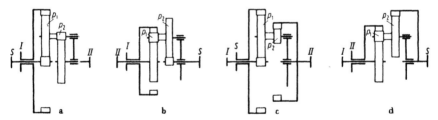

Fig. 145. The two types of reduced bicoupled transmissions with the smallest number of gears. In each case S denotes the reduced connected coupling shaft. For equal root stresses in the geometrically similar pinions, the limits of the subsequently given series speed-ratios $i_{I\,II}$ are determined by the size of the largest three planets that can be distributed around the circumference of the central gear (sec. 20) and by the value ∞ where S changes its position. $A–b$, designs with a single annular gear: a, $p_1 > p_2$, $i_{I\,II} = 1.12 \ldots \infty$; b, $p_1 < p_2$, $i_{I\,II} = 1 \ldots \infty$. $C–d$, designs with two annular gears: c, $p_1 > p_2$, $i_{I\,II} = 25.4 \ldots \infty$; d, $p_1 < p_2$, $i_{I\,II} = 2.46 \ldots \infty$.

22. Fig. 146a shows a transmission which consists of two positive-ratio transmissions of the type described by fig. 19. Fig. 146b shows an actual design of this type; fig. 146c shows the symbolic representation which is valid for both.

The transmission shown in fig. 146b has been described by Gaunitz [36], who stated that it can be used where a high speed-ratio and a good efficiency are required, but he did not give an actual efficiency value. For the gear sizes $z_1 = 18$, $z_{p1} = 90$, $z_2 = z_{2'} = 50$, $z_{p2} = z_{p2'} = 49$, $z_{1'} = 49$, $z_{p1'} = 48$ and under the assumption that the loss factor $\zeta = 0.01$ in each stage, the total speed-ratio and the overall efficiency are obtained as follows:

According to R 36, the shaft which has already been denoted by S is the reduced connected coupling shaft and thus we obtain with worksheets 1 and 5,

$$i_o = \frac{z_{p1}}{z_1} \cdot \frac{z_2}{z_{p2}} = \frac{90 \cdot 50}{18 \cdot 49} = \frac{250}{49} \, ,$$

$$i_{o'} = \frac{z_{p1'}}{z_{1'}} \cdot \frac{z_{2'}}{z_{p2'}} = \frac{48 \cdot 50}{49 \cdot 49} = \frac{2400}{2401} \, ,$$

$$i_{1\,II} = i_{1s} \cdot i_{s'1'} = (1 - i_o) \cdot \frac{1}{1 - i_{o'}} = \left(1 - \frac{250}{49}\right) \frac{1}{1 - \frac{2400}{2401}} = -9489 \, ,$$

$$i_{1S} = 1 - i_{1\,II} = 1 + 9849 = 9850 \, ,$$

$$\eta_o = \eta_{o'} = 0.99 \cdot 0.99 = 0.98 \, .$$

With worksheet 5, boxes 9 and 11 we find that

$$\eta_{1\,II} = \eta_{1s} \eta_{s'1'} = \frac{i_o \eta_o - 1}{i_o - 1} \cdot \frac{i_{o'} - 1}{i_{o'} \eta_{o'} - 1} = 0.975 \cdot 0.0204 = 0.0199 \, ,$$

so that finally with worksheet 5, box 7 the overall efficiency becomes:

$$\eta_{1S} = \frac{i_{1\,II} \eta_{1\,II} - 1}{i_{1\,II} - 1} = \frac{-9849 \cdot 0.0199 - 1}{-9849 - 1} = 0.0200 \, .$$

For a simple positive-ratio transmission (fig. 74) with a similar speed-ratio, $i_{s1} = 9801$, the smallest numbers of teeth are $z_2 = 100$, $z_1 = 99$, $z_{p1} = 98$, and $z_{p2} = 99$ (sec. 9 and worksheet 1), and a loss factor of $\zeta = 0.01$ per stage, leads to a revolving-carrier efficiency of $\eta_{s1} = 0.0051$. With these numbers of teeth, and for equal face widths and gear stresses, the diameter

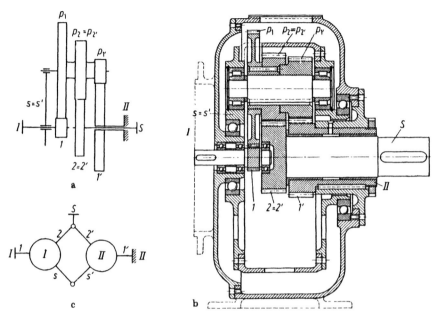

Fig. 146. Reduced bicoupled transmission with six gears which consists of only positive-ratio component gear trains according to fig. 19. S denotes the reduced connected coupling shaft: a, schematic representation of the gear train; b, practical example according to Gaunitz [36]; c, symbolic representation of the functionally equivalent transmission.

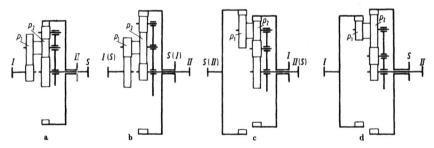

Fig. 147. Reduced bicoupled transmissions with six gears which consist of positive-ratio transmissions according to figs. 22 and 19, or 21. S denotes the reduced connected coupling shaft. Symbols in parentheses designate possible alternate shaft positions: a, design with a single annular gear, $p_1 > p_2$; b, design with a single annular gear, $p_1 < p_2$; c, design with two annular gears, $p_1 > p_2$; d, design with two annular gears, $p_1 < p_2$.

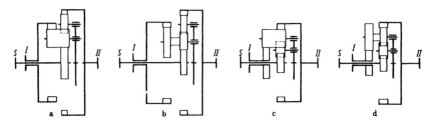

Fig. 148. Reduced bicoupled transmissions with six gears which consist of positive-ratio transmissions according to fig. 22 and negative-ratio transmissions according to fig. 32 or 34. S denotes the reduced connected coupling shaft: *a–b*, designs with two annular gears; *c–d*, designs with a single annular gear.

of the simple planetary gear train is about 20% larger than the gear train of the reduced bicoupled transmission of fig. 146b, and has a lower efficiency!

If two simple epicyclic transmissions (figs. 19 and 22) are combined, bicoupled transmissions (figs. 147a and 147b) are obtained, while gear trains as shown in figs. 21 and 22 combine to form transmissions as shown in figs. 147c and 147d. These types have larger diameters but a shorter length than those shown in fig. 146.

3. Combinations of positive-ratio gear trains, according to fig. 22, and negative-ratio gear trains, according to figs. 32 and 34, also have six gears, provided that the planets with double face width are considered as two gears. Depending upon whether one or two annular gears are used, designs as shown in figs. 148a and 148b, or 148c and 148d, result. Fig. 148c corresponds to the Ravigneaux gear train of fig. 115.

f) Applications

The application of a reduced bicoupled transmission is recommended where a simple planetary transmission could be used, but a reduced bicoupled transmission leads to a smaller transmission diameter or a higher efficiency. It is true that for extremely high speed-ratios, reduced bicoupled transmissions have a higher efficiency than simple positive-ratio transmissions. However, the efficiency is still substantially smaller than for series-coupled transmissions with several stages. Thus, a reduced bicoupled transmission offers an advantage only when a small transmission is more important than a high efficiency, that is, when the required output power is small.

Reduced bicoupled transmissions are most frequently used as change gears, but, as explained in section 40, they are not always connected as real bicoupled transmissions.

Section 38. Symmetrical Compound Transmissions for Precise Speed-Ratios

a) Generation of Precise Speed-Ratios

Using basic transmissions or simple revolving drive trains, the realization of a precise speed-ratio requires unnecessarily large dimensions when this speed-ratio is expressed as an integer fraction whose numerator and/or denominator is either a large prime factor or a product which contains large prime factors (≙ number of teeth). Likewise, if such a speed-ratio cannot be realized with a bicoupled transmission as described in section 35k, then, according to section 32h, it may be generated with the necessary accuracy by a symmetrical compound transmission. With its two constrained component gear trains the latter possesses the required adaptability.

A method for the synthesis of a symmetrical compound transmission whose main gear train has a basic speed-ratio of $i_o = -1$ was described by Willis [1]. However, this method is valid only for a special case where the denominator of the given speed-ratio can be factored into sufficiently small numbers so that only the numerator contains a large prime number. For the general case of an arbitrary speed-ratio, we are able to find a solution as follows:

If eq. (25) is written in general form, for example, with the subscript-key $1 ≙ a, 2 ≙ b, s ≙ c$ (sec. 11d), then with

$$n_c = \frac{i_{ab} n_b - n_a}{i_{ab} - 1} = \frac{i_{ab} i_{b3} n_3 - i_{a3} n_3}{i_{ab} - 1},$$

the required speed-ratio i_{c3} for the transmission shown in fig. 149 becomes:

$$i_{c3} = \frac{n_c}{n_3} = \frac{i_{ab} i_{b3} - i_{a3}}{i_{ab} - 1}. \tag{122}$$

If both the required speed-ratio i_{c3}, as well as the speed-ratio of the main gear train i_{ab}, which must be chosen, are expressed as fractions with integer denominators and numerators so that

$$i_{c3} = \frac{p}{q} \quad \text{and} \quad i_{ab} = \frac{a}{b},$$

then eq. (122) becomes:

$$i_{c3} = \frac{p}{q} = \frac{a i_{b3} - b i_{a3}}{a - b}. \tag{123}$$

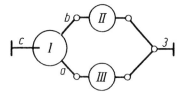

Fig. 149. Symmetrical compound transmission with the main component transmission *I* and the auxiliary component transmissions *II* and *III*.

By trial and error we can now express $q = a - b$ in terms of positive or negative integers which can be factored into sufficiently small prime numbers ($\hat{=}$ minimum number of teeth) and at the same time yield a feasible speed-ratio $i_{ab} = a/b$. Thus, we obtain one or more possible main gear trains with gear sizes which can be practically realized.

To determine suitable auxiliary gear trains *II* and *III* we resolve the numerator p of eq. (123) into the two terms (ai_{b3}) and (bi_{a3}), which then can be factored into sufficiently small prime numbers. From these terms we find the speed-ratios of the two auxiliary gear trains

$$i_{b3} = \frac{(ai_{b3})}{a} \quad ; \quad i_{a3} = \frac{(bi_{a3})}{b} . \tag{124}$$

Speed-ratios with similar values and equal signs are obtained for the auxiliary gear trains when p is decomposed into terms which bear a ratio of approximately

$$\frac{ai_{b3}}{bi_{a3}} \approx \frac{a}{b}$$

relative to each other, where the signs of a and b must be observed. The simplest auxiliary gear trains are obtained when the term (ai_{b3}) has as many prime factors as possible in common with a and the term (bi_{a3}) with b, so that the above expression can be substantially reduced.

To enable a reduction of the two terms (ai_{b3}) and (bi_{a3}), it is required that a and b can be factored into different prime numbers, since, as addends (eq. 123) of a prime number (ai_{b3}) and (bi_{a3}) have no common divisors. On the other hand, a and b must also contain identical prime factors when the speed-ratio i_{ab} is to be reduced to a fraction with a two-digit numerator and denominator which can be realized with a simple transmission. Obviously, these two requirements cannot be reconciled. It becomes clear that speed-ratios which contain large prime numbers necessitate main and auxiliary gear trains with several stages. If the speed-ratio is a ratio of two large prime numbers, that is, if a and b have no common divisor, then we may obtain simple auxiliary gear trains. The main gear, however, will always require several stages.

If, as a result of a trial and error division of the numerator p, a term $(ai_{b3})_0$ has been obtained which contains a sufficiently large common divisor g with a, so that according to eq. (124), i_{b3} becomes a fraction whose numerator and denominator represent feasible numbers of teeth, and if the complementary term $(bi_{a3})_0 = (ai_{b3})_0 - p$ has been determined also, then further pairs of values which contain the same common divisor g of (ai_{b3}) and a can be found. They have the form

$$(ai_{b3}) = (ai_{b3})_0 + gt \qquad (125)$$

$$(bi_{a3}) = (bi_{a3})_0 + gt \qquad (126)$$

where t is an arbitrary positive or negative integer number. By varying the value of t, a term (bi_{a3}) is now sought which has a sufficiently large common divisor h with b so that the speed-ratio i_{a3} of eq. (124) also becomes a reasonably simple fraction whose numerator and denominator represent feasible numbers of teeth. The related term (ai_{b3}) is obtained from eq. (125) using the same factor t.

If we denote the first suitable pair of values which has been found by this method with $(ai_{b3})'$ and $(bi_{a3})'$, then further sets of values with the common divisors g and h can be found from

$$ai_{b3} = (ai_{b3})' + zgh \qquad (127)$$

and $\qquad bi_{a3} = (bi_{a3})' + zgh , \qquad (128)$

where z is an arbitrary positive or negative integer.

These equations indicate that the integer numerators and denominators which represent the speed-ratios of the auxiliary gear trains *II* and *III* become smaller, and thus, fewer stages and smaller gears are required the larger the common divisors g and h become. However, with increasing values of g and h, the number of kinematically-equivalent solutions of eqs. (127) and (128) decreases.

If, at the same time, the required accurate speed-ratio is very large, and the efficiency is not of primary concern, then we should attempt to obtain a positive-ratio gear train with a basic speed-ratio $i_{ab} = i_0$ of approximately 1 (see sec. 22). This would place the large speed-ratio in the main gear train and result in simple auxiliary gear trains. If a good efficiency is important, it is expedient to choose a speed-ratio range of ±10 for a main gear train according to fig. 32, or ±22 for a main gear train according to fig. 35, and accept auxiliary gear trains with large speed-ratios which may necessitate several stages.

If a suitable solution can be found also with $i_{b3} = 1$ or $i_{a3} = 1$, then this auxiliary gear train can be replaced by a coupling, that is, a simpler constrained bicoupled transmission can be used instead of a symmetrical compound transmission.

b) Power-Flow and Efficiency

Depending on the position of the total-power shaft in the main gear train, power division or circulating power (see sec. 35f) can arise in a symmetrical compound transmission as well as in a constrained bicoupled transmission. However, because of the symmetry of the connected shaft *3*, it makes no sense to distinguish between the directions of the circulating power as positive or negative. A corresponding consideration applies to R 28.

If, as indicated in fig. 150, the two auxiliary gear trains are considered as a single multistage transmission, and if it is further assumed that a shaft *S* rather than shaft *3* is connected at *c*, then the symmetrical compound transmission is transformed into a constrained bicoupled transmission which has the same speed-ratios, the same total-power shaft in the main gear train *I*, and the same power-flow, that is, power division or circulating power. With the speed-ratios i_{IS} and $i_{c'f'}$ of this "substitute transmission," its power-ratio ϵ_0, and thus the input power of its auxiliary transmission, can be determined from eq. (94) or fig. 130. This input power also flows through the auxiliary gear train *III*, which ties into the *free* coupling shaft, while the power flowing through the auxiliary gear train *II* differs from it by the amount P_3 of the external power. However, for all practical purposes we can neglect this difference if we only want to know whether or not circulating power exists and how large it is.

Using the same substitute transmission and worksheet 5 we can calculate the efficiency with close approximation. It is more accurate, but also more tedious, to calculate the efficiency from the speed- and torque-ratios as described in the second part of section 36f.

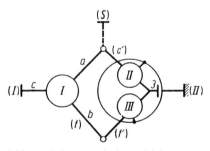

Fig. 150. Constrained bicoupled transmission which serves as a "substitute transmission" to simplify the verification of the power-flow of a symmetrical compound transmission. Shaft *3* of the substitute transmission is assumed to be idling.

c) Design Examples

First example: Around 1823 Péqueur and Pérrelet designed a transmission to represent the cycles of the moon. With a main gear train of $i_{ab} = i_{12} = -1$, this transmission, which has been described by Willis, had an overall gear-ratio of

$$i_{s3} = \frac{\text{avg. moon cycle}}{12 \text{ hours}} = \frac{2{,}551{,}443''}{43{,}200''} = \frac{850{,}481}{14{,}400} .$$

In the historical example shown in fig. 151, $b = -a$. Thus, eq. (123) becomes:

$$\frac{850{,}481}{14{,}400} = \frac{a \cdot i_{b3} - b \cdot i_{a3}}{a - b} = \frac{a \cdot i_{b3} - b \cdot i_{a3}}{7{,}200 + 7{,}200} .$$

Factoring yields $a = -b = 7{,}200 = 2^5 \cdot 3^2 \cdot 5^2$.

The common divisors of the historical example are $g = 9$ and $h = 25 \cdot 32$, and the numerator is resolved into the following addends:

$$p = 850{,}481 = 50{,}481 + 800{,}000 .$$

Thus, the auxiliary gear trains have the following gear-ratios and number of teeth:

$$i_{b3} = \frac{a i_{b3}}{a} = \frac{50{,}481}{7{,}200} = \frac{9 \cdot 71 \cdot 79}{9 \cdot 25 \cdot 32} = \frac{71}{25} \cdot \frac{79}{32} ,$$

$$i_{a3} = \frac{b i_{a3}}{b} = \frac{-800{,}000}{-7{,}200} = \frac{25 \cdot 32 \cdot 1{,}000}{25 \cdot 32 \cdot 9} = \frac{80}{6} \cdot \frac{50}{6} .$$

Besides the historical solution shown in line 1, table 14 lists other solutions which use the same as well as other main gear trains. Because the numerator contains a large prime number, at least one of the auxiliary gear trains must have two stages. This is true even when the gear-ratio of the main gear train already approximates the total required gear-ratio and, consequently, the gear-ratio of the auxiliary gear train is close to 1 as shown in line 5 of table 14. Fig. 152 schematically shows the transmissions specified by lines 3 and 4 of table 14.

Second example: Transmission with a speed-ratio $i = 7{,}108/577$. The denominator is itself a prime number while the numerator contains the prime factor 1,777. The difference $7{,}108 - 577 = 6{,}531 = 3 \cdot 7 \cdot 311$. Therefore, this gear-ratio cannot be realized with a constrained bicoupled transmission according to section 35k. A suitable symmetrical compound

TABLE 14. KINEMATICALLY-EQUIVALENT TRANSMISSIONS REPRESENTING THE CYCLES OF THE MOON

LINE	i_{ab}	a	b	g	h	a_{b3}	bi_{a3}	i_{b3}	i_{a3}	FIG.
1	-1	7,200	$-7,200$	3^2	$2^5 \cdot 5^2$	50,481	$-800,000$	$\dfrac{71 \cdot 79}{25 \cdot 32}$	$\dfrac{80 \cdot 50}{6 \cdot 6}$	151
2	-1	7,200	$-7,200$	$2^5 \cdot 5^2$	3^2	828,800	$-21,681$	$\dfrac{111 \cdot 84}{9 \cdot 9}$	$\dfrac{33 \cdot 73}{32 \cdot 25}$	
3	$-3/2$	8,640	$-5,760$	$3^3 \cdot 5$	2^6	372,465	$-478,016$	$\dfrac{62 \cdot 89}{16 \cdot 8}$	$\dfrac{77 \cdot 97}{10 \cdot 9}$	152a
4	-6.5	12,480	$-1,920$	195	128	847,665	$-2,816$	$\dfrac{69 \cdot 63}{8 \cdot 8}$	$+\dfrac{22}{15}$	152b
5	$+\dfrac{64}{63}$	921,600	907,200	1,024	315	406,016	$-444,465$	$\dfrac{13 \cdot 61}{60 \cdot 30}$	$-\dfrac{17 \cdot 83}{64 \cdot 45}$	

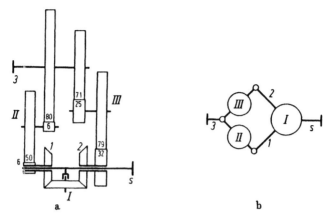

Fig. 151. Transmission by Péqueur and Pérrelet (around 1823) representing the cycles of the moon with the precise speed-ratio $i_{s3} = 850,481/14,400$ [1]: a, schematic representation of the gear train with the chosen number of teeth; b, symbolic representation as a symmetrical compound transmission.

transmission, however, can be found as follows: If, according to eq. (123), the denominator is resolved into two numbers a and b, which can be further resolved into as many small prime factors as possible, e.g.,

$$a = 425 = 5 \cdot 5 \cdot 17 \qquad \text{and} \qquad b = -152 = -2 \cdot 2 \cdot 2 \cdot 19 ,$$

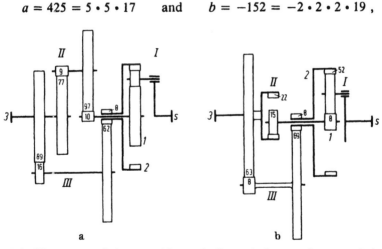

Fig. 152. These transmissions are kinematically equivalent to the transmission of fig. 151 and thus have the same speed-ratio, but their smallest gears have larger numbers of teeth: a, with parallel axes and a negative ratio planetary with $i_o = -3/2$; b, with parallel axes and a negative-ratio planetary with $i_o = -6.5$, which has the smallest maximum number of teeth and the smallest number of gears of all gear trains listed in table 14.

then we obtain a two-stage main gear train with the gear-ratio

$$i_{ab} = \frac{a}{b} = -\frac{17}{19} \cdot \frac{25}{8} = -\frac{68}{19} \cdot \frac{25}{32} .$$

Since, as addends of a prime number, the terms a and b cannot be reduced, the main gear train must have stepped planets. Eq. (123) then becomes

$$\frac{7,108}{577} = \frac{ai_{b3} - bi_{a3}}{425 + 152} .$$

The fastest way to resolve the numerator of this equation into numbers ai_{b3} and ba_{13}, which can be substantially reduced by the factors a and b, consists of trying only those numbers which contain some of the prime factors of the term a, for example, $ai_{b3} = 17 \cdot 300$, $17 \cdot 290$, etc. Then we calculate bi_{a3} and investigate whether it has any common factors with b. For the given example we find a solution with $ai_{b3} = 17 \cdot 170 = 2,890$ and $bi_{a3} = 2,890 - 7,108 = -4,218 = -2 \cdot 3 \cdot 19 \cdot 37$. Thus, the auxiliary gear trains have the gear-ratios:

$$i_{b3} = \frac{2,890}{425} = \frac{34}{5} = \frac{34}{15} \cdot \frac{45}{15} ,$$

$$i_{a3} = \frac{-4,218}{152} = \frac{111}{4} = \frac{74}{16} \cdot \frac{84}{14} .$$

This compound transmission is shown schematically in fig. 153.

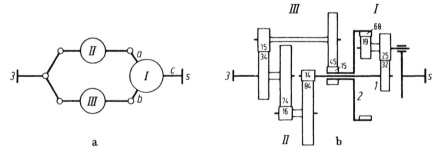

Fig. 153. A possible solution for a symmetrical compound transmission with the precise speed-ratio $i_{s3} = 7,108/577$ where the shafts of the main component transmission $a \triangleq 1$, $b \triangleq 2$, $c \triangleq$ s: a, symbolic representation; b, true-to-scale schematic representation with number of teeth indicated.

Section 39. Four-Shaft Bicoupled Transmissions

Occasionally, compound epicyclic transmissions with more than three connected shafts, and more than one static degree of freedom, are used (see sec. 24b). Transmissions of this type have been described in detail by Wolf [13]. Except for an additional connected "free" coupling shaft, they have the same basic structure as the simple bicoupled transmissions. In the literature they are sometimes called "four-shaft transmissions" as can be readily understood from their symbolic representation which is given in fig. 154. At this point, their combination with infinitely-variable drive trains may be especially mentioned [50, 51].

The speed characteristics of the four-shaft transmissions are completely identical with those of the simple bicoupled transmissions. Because of the analogy which has been introduced earlier, they can be determined with the aid of worksheets 1 and 2, when one of the two, now connected coupling shafts, is denoted by (S), and the speed of the now connected "free" coupling shaft (F) is finally determined as the speed of one of the two component gear trains. For details of the latter procedure see section 34b.

However, the efficiency equations of the simple bicoupled transmissions are not valid for the four-shaft transmissions and, therefore, different symbols are used to denote the latter's shafts. The efficiency of the four-shaft transmissions with two static degrees of freedom depends on a speed-ratio and on the torque-ratio between two arbitrary, connected shafts. Because of this additional parameter, the efficiency equations can no longer be written in a concise and generally applicable form. Rather, the efficiency must be calculated with eq. (53) after the speeds and torques of all input and output shafts have been determined separately. With the symbols used in fig. 154, the torques of the coupling shafts are:

$$T_C = T_c + T_{c'} \quad \text{and} \quad T_D = T_f + T_{f'} . \tag{129}$$

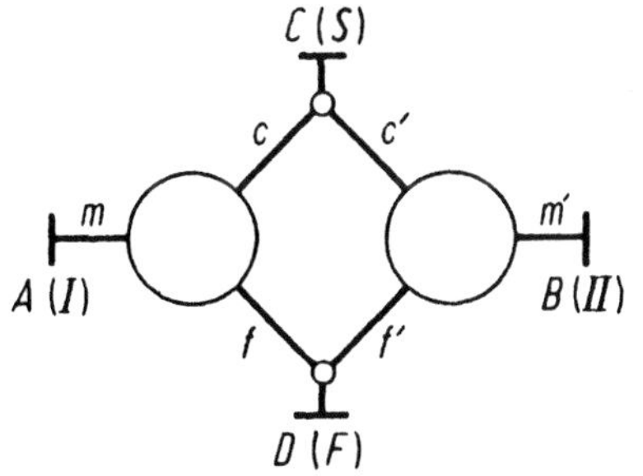

Fig. 154. Four-shaft transmission which originated from the additional connection of the free coupling shaft of a simple bicoupled transmission. The original designations of the latter's connected shafts are shown in parentheses.

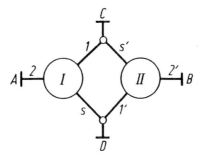

Fig. 155. Four-shaft transmission illustrating the example of sec. 39.

The torques and speeds of the component transmission shafts can be determined as described in sections 11 and 12 and with the help of worksheets 1 and 2.

Example: The transmission of fig. 155 has the same component gear trains as the transmission shown in fig. 125, where $i_o = -2/7$ and $i_{o'} = 42/11$. However, it has an additional connection at shaft D, which in fig. 125 is the free coupling shaft. The given speeds are the same as in the example of section 34b, that is, $n_A = 25s^{-1}$ and $n_B = -12s^{-1}$ where the designations refer to fig. 155. According to section 34b, we find that $n_C = 14.29s^{-1}$ and $n_D = -5.556s^{-1}$. In this case, the external torques of the two coupling shafts are output torques which have the known values: $T_C = 100$ Nm and $T_D = 150$ Nm. T_A, T_B and the overall efficiency must then be calculated.

With the torques T_C and T_D we neither know the torques at shafts $c, c', f,$ and f' nor the external power-flow of the component gear trains. However, with eq. (44) and the simplifying assumption that $\eta_o = \eta_{o'} = 1$, we obtain

$$\frac{T_s}{T_1} = i - 1 = -\frac{9}{7} \quad \text{and} \quad \frac{T_{s'}}{T_{1'}} = i_{o'} - 1 = \frac{31}{11} \,,$$

and with eq. (129)

$$T_D = T_s + T_{1'} = 150\,\text{Nm} \quad \text{and} \quad T_C = T_1 + T_{s'} = 100\,\text{Nm} \,.$$

From these four equations we find the four torques:

$$-T_1\frac{9}{7} + T_{1'} = 150\,\text{Nm} \quad \text{and} \quad T_1 + T_{1'}\frac{31}{11} = 100\,\text{Nm} \,,$$

$$T_1 = -T_{1'}\frac{31}{11} + 100 = (T_{1'} - 150)\frac{7}{9} \,,$$

$$T_{1'}\left(-\frac{31}{11} - \frac{7}{9}\right) = -100 - \frac{150 \cdot 7}{9} \ .$$

Thus,

$$T_{1'} = -\frac{900 + 1,050}{9}\left(-\frac{99}{356}\right) = 60.25 \text{ Nm} \ ,$$

$$T_s = 150 - 60.25 \qquad\qquad = 89.75 \text{ Nm} \ ,$$

$$T_1 = -89.75 \cdot \frac{7}{9} \qquad\qquad = -69.80 \text{ Nm} \ ,$$

$$T_{s'} = 60.25 \cdot \frac{31}{11} \qquad\qquad = 169.80 \text{ Nm} \ .$$

If we now use the appropriate speeds $n_C = n_1 = n_{s'} = -14.29s^{-1}$, and $n_D = n_s = n_{1'} = -5.556s^{-1}$, to calculate the shaft powers, then we recognize shafts 1 and s of the component gear train I as the input and output shafts respectively. Since shaft 2 is a difference shaft of this negative-ratio gear train, its torque has the same negative sign as the torque of the other difference shaft 1, and, because its speed $n_2 = n_A = 25s^{-1}$ is positive, it is an output shaft. According to definition, shafts C and D are also output shafts. Therefore, shaft B must be the sole input shaft which is verified by the fact that $T_{2'} < 0$ (summation shaft) so that with $n_{2'} = n_B < 0$, it follows that $P_{2'} > 0$.

With eq. (43), we can now determine the exponent $r1$ for each component gear train. Thus we find that $r1 = +1$ and $r1' = +1$, and, under consideration of the basic efficiencies, we can now accurately determine the torques of shafts A and B. With the assumptions used in section 34c, namely that $\eta_o = 0.982$ and $\eta_{o'} = 0.977$ and the four equations

$$\frac{T_s}{T_1} = -\frac{2}{7} \cdot 0.982 - 1 = -1.2806 \ ; \qquad \frac{T_{s'}}{T_{1'}} = \frac{42}{11} \cdot 0.977 - 1 = 2.7304 \ ;$$

$$T_D = T_s + T_{1'} = 150 \text{ Nm} \ ; \qquad T_C = T_1 + T_{s'} = 100 \text{ Nm}$$

we find first the four torques:

$$T_{1'} = \frac{150 - 100\dfrac{T_s}{T_1}}{1 - \dfrac{T_s T_{s'}}{T_1 T_{1'}}} = \frac{278.06}{1 + 3.4966} = 61.84 \text{ Nm} \ ,$$

$$T_s = 150 - T_{1'} = 88.16 \, \text{Nm} \ ,$$

$$T_1 = \frac{T_s}{-1.2806} = -68.84 \, \text{Nm} \ ,$$

$$T_{s'} = 100 - T_1 = 168.84 \, \text{Nm} \ .$$

From the equilibrium conditions for each component gear train we then obtain:

$$T_A = T_2 = -T_1 - T_s = -19.32 \, \text{Nm}$$

$$T_B = T_{2'} = -T_{1'} - T_{s'} = -230.68 \, \text{Nm} \ .$$

Finally, the overall efficiency can be calculated from:

$$\eta = -\frac{P_{out}}{P_{in}} = -\frac{P_A + P_C + P_D}{P_B}$$

$$= -\frac{-19.32 \cdot 25 - 100 \cdot 14.29 - 150 \cdot 5.556}{230.68 \cdot 12}$$

$$= 0.992 \ .$$

Section 40. Planetary Change-Gears with Nonchangeable Couplings

a) Speed Analysis

The speed characteristics of these transmissions which have been defined earlier in section 32k, can be determined by calculation or graphically using the Kutzbach method.

1. Computational speed analysis: To determine the speed characteristics by calculation, it is expedient to start from the symbolic representation of the transmission. Each speed is then analyzed individually and only those component transmissions need be considered which lie in the power-flow and thus participate in the power transmission. The calculation proceeds as previously described, and it is advantageous to begin at a component transmission which has a locked shaft.

Example: A transmission as shown in fig. 156g with its appropriate designations and the basic speed-ratios $i_o = -1.72$, $i_{o'} = -1.667$, $i_{o \cdot} = -4.387$, may be given. If the component gear trains I to III are successively assembled into a symbolic representation of the bicoupled transmission, then we obtain the Wolf symbol shown in fig. 157. Shafts A and B are rigidly connected to the input and output, and shafts C, D, and E to a set of brakes. A direct gear $i_{AB} = 1$ is not provided. With the aid of worksheets 1 and 2 we find the following speed-ratios when shaft C, D, or E is locked:

Shaft D locked: $i_{AB} = i_{1s} = 1 - i_o = 2.72$.

Gear trains *II* and *III* are idling, since $T_{1'} = T_{2''} = T_{2'} = 0$; gear train *I* operates as an epicyclic transmission and transmits the total power.

Shaft E locked: $k_{2s} = i_{1's'} = 1 - i_{o'} = 2.667$

$$i_{AB} = k_{1s} = 1 - i_o(1 - k_{2s})$$
$$= 1 + 1.72(1 - 2.667) = -1.867 .$$

Gear trains *I* and *II* operate as a constrained simple bicoupled transmission; gear train *III* is idling.

Shaft C locked:

$$k_{s'2'} = i_{s''2''} = \frac{i_{o''}}{i_{o''} - 1} = \frac{-4.387}{-4.387 - 1} = 0.814 ,$$

$$k_{s2} = k_{s'1'} = \frac{1}{1 - i_{o'} + i_{o'}/k_{s'2}} = \frac{1}{1 + 1.667 - 1.667/0.814} = 1.615$$

$$i_{AB} = k_{1s} = 1 - i_o + i_o/k_{s2} = 1 + 1.72 - 1.72/1.615 = 1.655$$

The three component gear trains form a higher bicoupled transmission (sec. 32a).

2. Graphical speed analysis: Since the couplings between the component gear trains remain the same for all possible gear ratios, we can draw the speed scales for all shafts of the transmission with the aid of a Kutzbach diagram (see sec. 17). For this purpose, we can start from any arbitrary operating condition of the transmission. However, it is expedient to choose one which is characterized by a locked shaft. As described in section 17, it is then possible to find the speed of each shaft, even if it is idling, at the point of intersection between the appropriate speed scale and the speed line.

Example: A 3-speed "Diwabus" automotive transmission [31], with its three component gear trains *I*, *II* and *III*, is shown symbolically in fig. 156a, and as a true-to-scale schematic in fig. 156b. With the aid of three brakes, it is possible to selectively lock one of the shafts C, D, or E. For the purpose of constructing a Kutzbach diagram, it may be assumed that shaft C, that

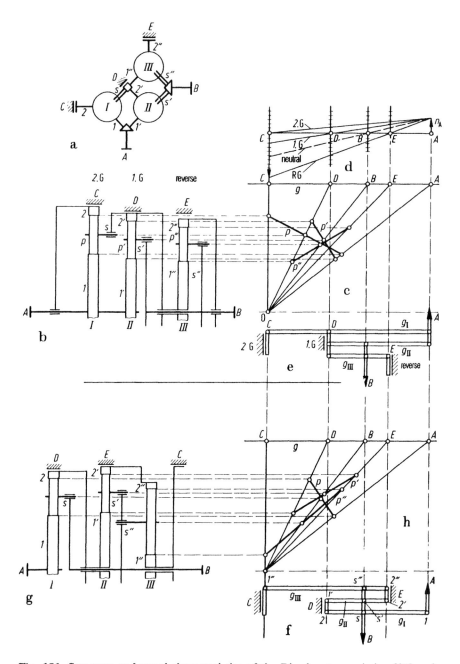

Fig. 156. Structure and speed characteristics of the Diwabus transmission [31] and one of its kinematically equivalent modifications: *a*, symbolic representation of the Diwabus transmission; *b*, schematic representation of the arrangement of the Diwabus transmission; *c*, Kutzbach plan of this gear train; *d*, speed nomograph of

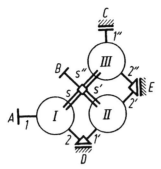

Fig. 157. Symbolic representation of the gear train shown in fig. 156g.

is, shaft 2 of the component gear train *I*, is locked while the input speed has the arbitrary value $n_1 = n_{1'} = n_A$. According to fig. 156c, the points representing the shaft speeds on line *g* are obtained in the following sequence:

assume: points, *0, C* and *A* with the vectors $\overline{C0}$ and $\overline{A0}$;

construct: p; $\overline{0D}$ with *D*; p'; $\overline{0B}$ with *B*; p''; $\overline{0E}$ with *E*.

The speed scales can now be erected as shown in fig. 156d. If n_A is a constant input speed and the speeds n_C, n_D, and n_E are selectively set to zero by locking the appropriate brakes, then three speed lines can be drawn through the foot points of the associated speed scales and intersect the speed scale *B* of the output shaft at two positive speeds and one negative speed. These points of intersection represent the first and second gear of the vehicle, and the reverse gear respectively. For a stopped vehicle the output speed n_B is zero, and we can draw the dashed vector which characterizes the idling drive train. Since it intersects the speed scales *C, D* and *E* at non-zero speed values, the three corresponding shafts rotate. Consequently, their brakes must be released.

A comparative calculation with the basic speed-ratios $i_0 = i_{0^.} = -1.667$ and $i_{0'} = -1.72$, yields the following overall speed-ratios: First gear, brake *D* locked; component gear train *II* operates as a two-shaft transmission with the speed-ratio:

$$i_{AB} = i_{1's'} = 1 - i_{0'} = 1 + 1.72 = 2.72 \; .$$

---

the Diwabus transmission; *e*, analogous connected linkage representing the Diwabus transmission with an individual link for each of the component transmissions *I*, *II*, and *III*; *f*, linkage system which is kinematically equivalent to *e*; *g*, gear train which is analogous to the linkage *f* and kinematically equivalent to the Diwabus transmission; *h*, Kutzbach plan for the gear train *g*, which likewise determines the speed nomograph *d*.

Second gear, brake C locked; component gear trains I and II operate as a constrained bicoupled transmission with the speed-ratios

$$i_{1s} = k_{1'2'} = 1 - i_o = 2.667$$

and

$$i_{AB} = k_{1's'} = \frac{1 - i_o}{1 - i_{o'}/k_{1'2'}} = \frac{1 + 1.72}{1 + 1.72/2.667} = 1.654 .$$

Reverse gear, brake E locked; component gear trains II and III operate as a constrained bicoupled transmission with the speed-ratios:

$$i_{1''s''} = k_{2's'} = 1 - i_{o''} = 2.667$$

and

$$i_{AB} = k_{1's'} = 1 - i_{o'}(1 - k_{2's'}) = 1 + 1.72(-1.667) = -1.867 .$$

As expected, these three speed-ratio values agree with those found by the graphical analysis of fig. 156d. However, they are also equal to the three speed-ratios of the different transmission configurations shown in figs. 156g and 157, which we determined in the previous example. This correspondence will be further discussed in the following section.

b) Synthesis

The synthesis of a planetary change-gear with rigid couplings, that is, the determination of the basic speed-ratios of the component gear trains for a given overall speed-ratio, can be made simple and lucid by a procedure which has been published by Helfer [31]. According to this procedure, the speed vectors along the line g of the Kutzbach diagram can be resolved into the speed vectors of the component gear trains, and the coupling between two shafts can be represented by a line which connects their speed points. Fig. 156e shows a schematic representation of the couplings for the transmission of fig. 156b which have been obtained by this method. The straight lines g_I, g_{II} and g_{III} represent the speed lines of the component gear trains I, II, and III which are constructed separately by the Kutzbach method by omitting the other two gear trains (see figs. 156b and 156c). In this representation, those shafts of the overall transmission which are rigidly coupled, are linked by vertical lines. According to R 23, the intermediate points on lines g_I, g_{II}, and g_{III} identify the summation shafts of the component gear trains. The brakes of the three gear stages are indicated by vertical line segments which are backed by cross-hatching.

Helfer considers the lines g_I, g_{II}, and g_{III} as links which are connected with each other, or with the input and output, by joints located at the speed points A, B, etc. If one of the brakes is locked, then the corresponding joint, or the corresponding group of linked joints, becomes fixed. This constrains the linkage system, and the ratio of the velocities v of the joints A and B, which represent the input and output shafts, becomes equal to the speed-ratio $i_{AB} = n_A/n_B = v_A/v_B$.

The synthesis of compound planetary transmissions can be substantially simplified when the steps of this procedure are reversed. Since in this approach the speed-ratios are normally known, we can begin by drawing the speed lines, and then obtain the speed points where the speed lines intersect g as shown in fig. 156d. However, since the design of the desired transmission is still unspecified, line g can be arbitrarily resolved into three "component transmissions," as shown in fig. 156f, rather than as shown in fig. 156e. Despite their different designs, however, these two transmissions are kinematically equivalent, since both are characterized by the same line g in the Kutzbach diagram. Thus their speed-ratios, which have been determined in the previous example, must also be equal.

The theoretically possible number of kinematically-equivalent transmissions is obtained as follows: For t component gear trains, line g has $(t + 2)$ speed points, any three of which can determine a component transmission type. Without repetition, $(t + 2)$ elements (speed points) can be combined in Z groups of three elements, where

$$Z = \binom{t + 2}{3} = \frac{(t + 2)(t + 2 - 1)(t + 2 - 2)}{1 \cdot 2 \cdot 3}$$

$$= \frac{(t + 2)!}{3!\,(t + 2 - 3)!}.$$

These Z different component transmission types can be arranged in

$$K = \binom{Z}{t} = \frac{Z!}{t!\,(Z - t)!}$$

different transmission combinations which contain t component transmissions each. From these we must subtract q useless combinations which do not simultaneously contain all of the $(t + 2)$ speed points.

Depending on the number of speeds, that is, the number t of the component gear trains, we obtain the theoretically possible number $(K - q)$ of kinematically-equivalent transmission combinations listed in table 15.

Each of these combinations is characterized solely by its lines g_I, g_{II}, etc., and each of these lines can represent three transmission types, namely, a negative-ratio transmission whose carrier shaft is represented by the intermediate speed point, or a positive-ratio epicyclic transmission whose carrier shaft corresponds to the left-hand speed point, or a positive-ratio transmission where it corresponds to the right-hand speed point. Thus, for each of the $(K - q)$ transmission combinations we obtain another $V = 3^t$ possible variations so that, as evident from table 15, the theoretically possible total number

$$G = (K - q) V$$

of kinematically-equivalent transmission types becomes extremely large, considering the fact that each of these can be further varied by changing the positions of the component transmissions relative to each other.

TABLE 15. THEORETICALLY POSSIBLE NUMBER OF
KINEMATICALLY-EQUIVALENT TRANSMISSIONS

t	2	3	4	5
Z	4	10	20	35
K	6	120	4,845	324,632
q		20	1,245	103,236
$K - q$	6	100	3,600	221,396
V	9	27	81	243
G	54	2,700	291,600	$53.8 \cdot 10^6$

A substantial number of these transmissions cannot be realized since some of their rotating parts would penetrate each other. From the others, a design configuration can be chosen, which combines a simple construction with a high efficiency, satisfies all given design conditions and has a small overall diameter, so that the carrier and planet speeds remain low. Transmissions with a small number of gears are obtained when only simple negative-ratio gear trains of the type depicted in fig. 32 are used, or, alternately, two component gear trains are combined to form a reduced bi-coupled transmission. To reduce the number of bearings the carriers of all component gear trains are coupled together so that they can be constructed and supported as a single building block. Such a design, however, is only possible when the transmission contains no couplings between an internal and an external central gear which would have to penetrate the carrier structure. Considerations of this type substantially limit the number of combinations which must be investigated.

The basic speed-ratios of the Z component transmissions can be determined with a speed nomograph as shown in fig. 158a when the gear ratios $i_{1.}$, $i_{2.}$, etc., are known and the rules of similar sections are applied. To simplify matters we may assume that the input speed $n_A = 1$ and the distance $\overline{AB} = 1$, and can then mark the distances $B1. = 1/i_{1.}$; $B2. = 1/i_{2.}$, etc., which represent the associated output speeds $n_{1.}$, $n_{2.}$, etc., on the speed scale B. The lines which pass through their end points and the point $n_A = 1$,

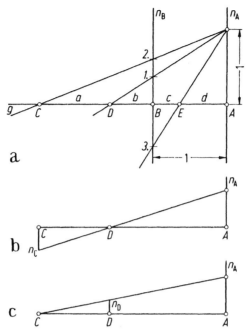

Fig. 158. Determination of the basic speed-ratios of the component gear trains from the nomograph of a compound revolving drive train: *a*, with the input at *A* and the output at *B*, the sections *a* to *d* of the line *g* are determined by the lines passing through the input speed $n_A = 1$ and the associated output speeds *1., 2., 3.,* on n_B; *b*, for a negative-ratio component gear train *CDA*, $i_o = n_A/n_C = -(b + c + d)/a$; *c*, for a positive-ratio component gear train *CDA*, $i_o = n_A/n_D = (a + b + c + d)/a$.

divide line *g* into the sections *a, b, c* ... , where:

$$-\frac{d}{c} = \frac{n_A}{n_{B3.}} = i_{3.} \; ; \qquad \frac{a + b + c + d}{a + b} = \frac{n_A}{n_{B2.}} = i_{2.} \; ;$$

$$\frac{b + c + d}{b} = \frac{n_A}{n_{B1.}} = i_{1.} \; .$$

With these, and the additional equation $c + d = 1$, the length of all sections which represent the basic speed-ratios of the component gear trains can be determined. If, for example, a negative-ratio component gear train with the shafts *C, D, A* is desired, whose shaft *D* corresponds to the carrier shaft, as shown in fig. 158b, the basic speed-ratio $i_o = n_A/n_C$ is obtained from

$$i_o = \frac{n_A}{n_C} = -\frac{\overline{DA}}{\overline{DC}} = -\frac{b + c + d}{a} \; .$$

If, instead, a positive-ratio gear train is sought whose shaft C corresponds to the carrier shaft, then, as shown in fig. 158c, the basic speed-ratio is obtained from

$$i_o = \frac{\overline{CA}}{\overline{CD}} = \frac{a + b + c + d}{a} .$$

By using this method we avoid the less accurate graphical determination of the basic speed-ratio by means of a Kutzbach diagram.

c) Additional Examples Illustrating Transmission Synthesis

A well-designed planetary change-gear with rigid couplings, and the three gear-ratios $i_1 = 2.72$; $i_2 = 1.65$; $i_3 = -1.87$ is to be created. Since two of these gear-ratios are positive and one is negative, the change gear can be treated as described in fig. 158a and, consequently, we can find the sections of line g from the equations given above:

$$i_3 = -\frac{i - c}{c} = -1.87$$

$$i_1 = \frac{b + 1}{b} = 2.72$$

$$i_2 = \frac{a + b + 1}{a + b} = \frac{a + 1.581}{a + 0.581} = 1.65$$

$$c = \frac{1}{2.87} = 0.348$$

$$d = 1 - c = 1 - 0.348 = 0.652$$

$$b = \frac{1}{i_1 - 1} = \frac{1}{1.72} = 0.581$$

$$a = \frac{1.65 \cdot 0.581 - 1.581}{1 - 1.65} = 0.957 .$$

To obtain a short transmission we want to limit ourselves to simple negative-ratio component transmissions as shown in figs. 32 and 33 whose carriers are always represented by the intermediate points on lines g_I, g_{II}, etc. The resulting $Z = 10$ possible component gear train combinations are listed in table 16.

TABLE 16.   THE TEN POSSIBLE NEGATIVE-RATIO COMPONENT GEAR TRAINS
FOR THE SYNTHESIS OF THE DESIRED PLANETARY CHANGE-GEAR
WITH RIGID COUPLINGS

NO.	COMBINED SHAFTS	i_0			$1/i_0$
1	CDB	$-\dfrac{a}{b}$	$= -\dfrac{0.957}{0.581}$	$= -1.647$	-0.607
2	CDE	$-\dfrac{a}{b+c}$	$= -\dfrac{0.957}{0.929}$	$= -1.030$	-0.971
3	CDA	$-\dfrac{a}{b+c+d}$	$= -\dfrac{0.957}{1.581}$	$= -0.605$	-1.652
4	CBE	$-\dfrac{a+b}{c}$	$= -\dfrac{1.538}{0.348}$	$= -4.42$	-0.226
5	CBA	$-\dfrac{a+b}{c+d}$	$= -\dfrac{1.538}{1}$	$= -1.538$	-0.650
6	CEA	$-\dfrac{a+b+c}{d}$	$= -\dfrac{1.886}{0.652}$	$= -2.893$	-0.346
7	DBE	$-\dfrac{b}{c}$	$= -\dfrac{0.581}{0.348}$	$= -1.67$	-0.599
8	DBA	$-\dfrac{b}{c+d}$	$= -\dfrac{0.581}{1}$	$= -0.581$	-1.721
9	DEA	$-\dfrac{b+c}{d}$	$= -\dfrac{0.929}{0.652}$	$= -1.425$	-0.702
10	BEA	$-\dfrac{c}{d}$	$= -\dfrac{0.348}{0.652}$	$= -0.534$	-1.873

Of these, the component gear train no. *2* cannot be used since its basic speed-ratio lies outside of the speed-ratio limits of the simple negative-ratio gear trains according to figs. 32 and 33. If we further eliminate gear train no. *9* because of its relatively small planets, then $\binom{8}{3}$ = 56 combinations

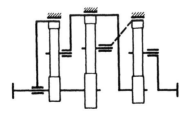

Fig. 159. Gear train which is kinematically equivalent to the Diwabus transmission as shown in fig. 156*b* but which cannot be realized because some of its parts would interfere with each other.

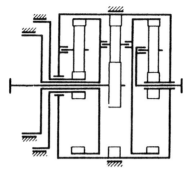

Fig. 160. Gear train which is kinematically equivalent to the Diwabus transmission as shown in fig. 156*b* but which has a very complicated structure.

remain. Ten of these can be eliminated since they do not contain all of the shafts *A* through *D*; for example, a combination of the three component gear trains nos. *1, 4,* and *7* does not contain the shaft *A*. Some of the remaining forty-six combinations cannot be realized since their rotating parts would penetrate each other, as illustrated by the combination shown in fig. 159. Several other combinations, for example, the transmission

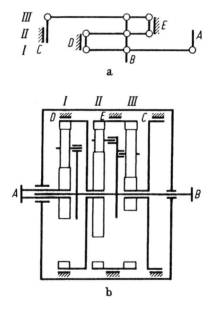

Fig. 161. Gear train which is kinematically equivalent to the Diwabus transmission as shown in fig. 156*b* and which has a simple structure: *a*, kinematically equivalent linkage system; *b*, schematic representation.

shown in fig. 160, lead to very complicated designs when the input and output shaft must lie on opposite sides of the housing. Therefore they are likewise eliminated. Finally about ten transmissions of simple design remain, including the kinematically-equivalent transmissions shown in figs. 156b and 156g, and the combination shown in fig. 161. From these, the most suitable combination must be chosen after a thorough design analysis has been performed.

Section 41. Planetary Change-Gears with Changeable Couplings

Including its reciprocal speed-ratios, a given simple planetary transmission can be operated with either of two basic speed-ratios, or any of four different revolving-carrier speed-ratios which correspond to the six rows of worksheet 1. Based on a simple planetary transmission, therefore, a change-gear transmission with up to four positive and two negative speed-ratios could be constructed when the appropriate coupling and brake mechanisms are provided, see section 29 and R 25.

Consequently, two planetary gear trains with given basic speed-ratios can be connected in $6 \times 6 = 36$ different ways at two of their external shafts as shown in fig. 162. Thus, including their reciprocal speed-ratios, they can realize thirty-six different series speed-ratios $i_{I\,II}$. In turn, each of these thirty-six series-coupled transmissions can operate with three different speed-ratios when its three shafts $I, II,$ or S are locked, one at a time. Thus, including the reciprocal values, a simple bicoupled gear train can realize

$$G_2 = 6 \cdot 6 \cdot 3 = 108$$

different gear-ratios, where the subscript 2 indicates the number of coupled gear trains.

No new speed-ratios are obtained when shafts F and S are interchanged since G_2 already contains the reciprocal values of the speed-ratios. If, however-

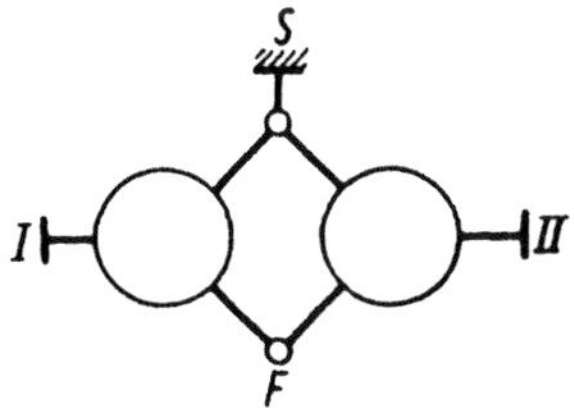

Fig. 162. Bicoupled transmission reduced to two series-connected simple revolving drives by locking shaft S.

296 / Compound Epicyclic Transmissions

ever, the input and output are connected at S and F, respectively, while shaft I or II is locked, each component gear train yields a further $G_1 = 6$ gear-ratios, since, in this case, only the component gear train with the locked shaft operates, while the other merely idles.

Considering all possible coupling changes, including the exchange of the input and output shaft, we thus find that

$$G = 2G_1 + G_2 = 12 + 108 = 120$$

gear-ratios can be realized with a simple bicoupled transmission. Only two of these, however, can be chosen arbitrarily. The other 118 must be calculated with the aid of worksheets 1 and 2.

In 36 of these 120 cases the bicoupled transmission operates as a series-coupled transmission, in 72 as a constrained bicoupled transmission, while in 12 cases, only one of the gear trains transmits power, either in a fixed or rotating carrier mode. According to R 25, which is analogously valid for constrained bicoupled transmissions, 2/3 or 80 of the possible transmissions have a positive speed-ratio, while 1/3 or 40 have a negative speed-ratio.

Depending on how the basic speed-ratios of the two component gear trains are chosen, the gear-ratio steps of the bicoupled transmission will be spaced more or less closely and regularly. For a small number of gear-ratios, for example, 5 to 10, it is usually possible to come reasonably close to some desired values. The necessary expense for clutches and brakes can be kept within reasonable limits when the bicoupled transmission is designed for only relatively few changeable gear-ratios and the two component transmissions are skillfully arranged. The Hydra-Matic transmission shown in fig. 163 may serve as a practical example. In this transmission, the gears are changed without changing the couplings between the two component gear trains, and only the input shaft is alternately coupled to either the outer central gear of the first or the inner central gear of the second component gear train.

In the first and second forward gears the power flows through only one of the component gear trains, but reverse gear is a constrained bicoupled transmission with both components utilized. In the third forward gear which is the direct gear, both clutches are engaged so that—disregarding the slip in the hydrodynamic coupling—both component gear trains rotate together as a rigid block like a coupling. Since this four-speed transmission has a direct gear, and two further gear-ratios can be freely chosen, only one of the four gear-ratios must be selected from the many possible combinations so that its value represents the desired gear stage with sufficient accuracy.

Planetary change-gears of this type can become expensive and complicated if they contain more than two component gear trains. Therefore, they

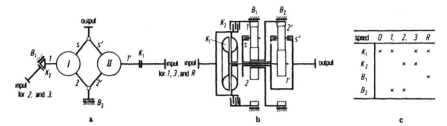

Fig. 163. Hydra-Matic automotive transmission [40] with two clutches K_1 and K_2 and two brakes B_1 and B_2. Depending on the engagement of the clutches, the input torque acts on either shaft 1 or $1'$ or on both shafts simultaneously: a, symbolic representation; b, schematic representation of the gear train; c, table showing the active (x) clutches and brakes for each gear stage.

have been rarely used, although they offer a large variety of speed-ratios and their many possible modifications have not yet been exhaustively investigated. Economical transmission designs which consist of as many as three or four component gear trains are entirely feasible when only one of their couplings is changeable as, for example, in the Hydra-Matic transmission. This coupling change, then, has an effect similar to a two-speed auxiliary gear train which is connected in series with a change-gear. A cost estimate, therefore, should compare the expense of the mechanisms required for a coupling change with the cost of an additional auxiliary gear train.

Section 42. Higher Bicoupled Transmissions

Higher bicoupled transmissions which can be built with an arbitrary number of degrees of mobility, are expediently divided, according to the method used in their analysis, into transmissions with one or two degrees of mobility, and transmissions with more than two degrees of mobility (see also sec. 321 and figs. 111n to 111q). As yet, only a few constrained transmission types have found practical applications, while the operating characteristics of transmissions with three or more degrees of mobility have rarely been investigated.

a) Constrained Higher Bicoupled Transmissions and Higher Bicoupled Transmissions with Two Degrees of Mobility

The analogy which has been described in section 33 for simple bicoupled transmissions is likewise valid for these rarely used transmissions which also have three rotating shafts and two degrees of freedom. If the connected shafts are denoted by *I*, *II*, and *S,* worksheets 1 to 3, as well as worksheets 4 and 5, are valid without restrictions for the analogous subscripts. Proof of this fact can be obtained from the considerations described in sections 11d and 15g, when it is assumed that the transmission with the unknown internal structure, which is contained in the housing, is now a higher bicoupled transmission with two degrees of mobility, as shown in figs. 164 to 166.

The determination of $i_{\text{I II}}$, $\eta_{\text{I II}}$, and $\eta_{\text{II I}}$ which must all be known before the mentioned worksheets can be used, causes some additional difficulties when higher, rather than simple, bicoupled transmissions must be analyzed. However, this determination can be facilitated by an expedient choice of the

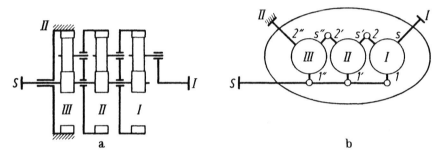

a   b

Fig. 164. Constrained higher bicoupled transmission according to fig. 111n: *a,* schematic representation of the Type HG heavy-duty transmission of the Desch K. G. [42]; *b,* symbolic representation of the gear train which operates as a series-coupled transmission when the shaft *S* is locked.

shaft designations which consist of labeling one or two of the possibly existing monoshafts with *I* or *II* respectively and the connected coupling shaft with *S*. If the transmission has more than one connected coupling shaft, then that one which generates the simplest and clearest power-flow when locked, is denoted by *S*. The determination of $i_{\text{I II}}$, $\eta_{\text{I II}}$ and $\eta_{\text{II I}}$ will then be simplified as much as possible.

In figs. 164 to 166, the higher bicoupled transmissions with two degrees of mobility (figs. 111n, 111o, 111q) are shown together with these suggested designations. To determine $i_{\text{I II}}$, the connected coupling shaft *S* must be fixed. To illustrate how this lowers the degree of complexity of a higher bicoupled transmission and simplifies the determination of $i_{\text{I II}}$, figs. 165 and 166 show the coupling shaft *S* disconnected and its two component transmission shafts individually locked. The described procedure will now be illustrated by three practical examples.

First example: In fig. 164, the speed-ratio $i_{\text{I II}}$ can be found as the series speed-ratio of a three stage series-coupled transmission. This procedure is analogous to the one described for simple bicoupled transmissions. If we assume that the basic speed-ratios and the basic efficiencies of the three component transmissions are $i_o = i_{o'} = i_{o''} = -3$ and $\eta_o = \eta_{o'} = \eta_{o''} = 0.985$, then we obtain, with worksheets 1 and 5,

$$i_{1 \text{II}} = i_{s2} i_{s'2'} i_{s''2''} = i_{s2}^3 = \left(\frac{i_o}{i_o - 1} \right)^3 = \left(\frac{-3}{-4} \right)^3 = 0.422 \ ,$$

$$\eta_{1 \text{II}} = \eta_{s2}^3 = \left(\frac{i_o - 1}{i_o - 1/\eta_o} \right)^3 = \left(\frac{-3 - 1}{-3 - 1/0.985} \right)^3 = 0.9887 \ .$$

The gear-ratio and the efficiency of the constrained version of the transmission with a locked shaft II (fig. 164) become, according to worksheets 1 and 5:

$$i_{SI} = \frac{1}{1 - i_{1 \text{II}}} = \frac{1}{1 - 0.422} = 1.73 \ ,$$

and

$$\eta_{SI} = \frac{i_{1 \text{II}} - 1}{i_{1 \text{II}} \eta_{1 \text{II}} - 1} = \frac{0.422 - 1}{0.422 \cdot 0.9887 - 1} = 0.9918 \ .$$

To realize the same gear-ratio with a simple planetary transmission whose carrier is connected at the output, its basic speed-ratio would have to be:

$$i_o = 1 - i_{1s} = 1 - 1.73 = -0.73$$

as can be verified with box 3 of worksheet 1. The form of this equation and the negative fractional basic speed-ratio indicate that the input gear 1 is the annular gear of a gear train as shown in fig. 32, while the inner central gear 2 is locked. According to worksheet 5, box 7, the efficiency of such a transmission which is kinematically equivalent to the higher bicoupled transmission of fig. 164 becomes:

$$\eta_{1s} = \frac{i_o \eta_o - 1}{i_o - 1} = \frac{-0.73 \cdot 0.985 - 1}{-0.73 - 1} = 0.9937 \ .$$

A comparison of the loss factors

$$\zeta_{SI} = 1 - 0.9918 = 0.0082 \ ,$$

and

$$\zeta_{1s} = 1 - 0.9937 = 0.0063 \ ,$$

of the two kinematically-equivalent transmissions, shows that the tooth-friction losses in the higher bicoupled transmission are approximately 30% higher than the friction losses in the corresponding simple planetary transmission. Nevertheless, the higher bicoupled transmission merits consideration when large powers must be transmitted since the heat which is generated by friction is distributed over several component transmissions and, consequently, is easier to remove. Component gear train *I* in fig. 164a must be designed to carry the full output torque while the component gear trains *II* and *III* can be built progressively lighter because the torques decrease with increasing speed.

Second example: If the connected coupling shaft is locked, the higher bicoupled transmission of fig. 111o is transformed into a simple two-shaft transmission and a constrained bicoupled transmission connected in series. This simplification can be clearly recognized in fig. 165 where, instead of the locked coupling shaft *S*, the two locked components of this coupling shaft are depicted to facilitate the following description. If the basic speed-ratios and the basic efficiencies of the three component gear trains are known, then the speed-ratio

$$i_{1\,II} = i_{1A} i_{A\,II}$$

is found as the product of the speed-ratios of the component gear train *I*, and the constrained bicoupled transmission which consists of component gear trains *II* and *III*. The speed-ratio of gear train *I* can be obtained with the aid of worksheet 1, and the speed-ratio of the constrained bicoupled transmission by reference to section 35c.

With a locked coupling shaft *S* the series efficiency,

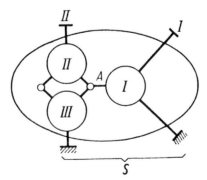

Fig. 165. Higher bicoupled transmission with two degrees of mobility according to fig. 111o; the connected coupling shaft is resolved into its components, which are shown individually locked.

$$\eta_{I\,II} = \eta_{IA}\eta_{A\,II}\,, \quad \text{or} \quad \eta_{I\,II} = \eta_{II\,A}\eta_{AI}\,,$$

can likewise be found as the product of the efficiencies of gear train I and the constrained bicoupled transmission $(II + III)$. See section 35e and worksheet 5.

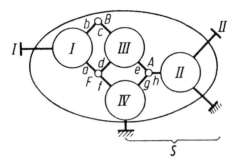

Fig. 166. Higher bicoupled transmission with two degrees of mobility according to fig. 111p; the connected coupling shaft S is resolved into its components, which are shown individually locked.

With these initial values, all gear-ratios and efficiencies for two-shaft operation can be calculated without restriction using worksheets 1 and 5, and all speed-ratios and efficiencies for three-shaft operation using worksheets 2 and 4, provided the previously given subscript key $I \triangleq 1$, $II \triangleq 2$ and $S \triangleq s$ is observed. When thus the speeds of all external shafts have been found, the speeds of the free coupling shafts can be readily determined from the speeds of shafts S and I of component gear train I and worksheet 2.

Third example: If the connected coupling shaft S of the higher bicoupled transmission of fig. 111p is locked and disconnected, the simpler transmission shown in fig. 166 is obtained. To find $i_{I\,II}$ when the basic speed-ratios and the positions of the coupled shafts of all component gear trains are known, we determine in succession with the aid of worksheets 1 or 2 for:

gear train II: $i_{A\,II} = i_{h\,II'}\,,$

gear train IV: $i_{FA} = i_{fg'}\,,$

gear train III: k_{dc} from $k_{de} = i_{fg}\,,$

gear train I: k_{IF} from $k_{ab} = k_{dc}\,.$

The series speed-ratio then becomes:

$$i_{I\,II} = k_{IF}\,i_{FA}\,i_{A\,II}\,.$$

With this speed-ratio $i_{1\,II}$, all the external gear-ratios and speed characteristics can be readily determined when worksheets 1 and 2 are used with the analogous subscripts.

However, in this case the efficiencies $\eta_{1\,II}$ and $\eta_{II\,I}$ can be determined only with the aid of an accurate calculation of the torque-ratio T_I / T_{II} and the previously obtained series speed-ratio. Then:

$$\eta_{II\,I} = -i_{1\,II} \cdot \frac{T_I}{T_{II}} \quad \text{and} \quad \eta_{I\,II} = -\frac{1}{i_{1\,II}} \cdot \frac{T_{II}}{T_I} .$$

A first approximation, which neglects all friction losses, yields the signs of all shaft torques when the power, torque and speed of shaft I are arbitrarily defined as

$$P_1 = 1 , \quad T_1 = 1 , \quad n_1 = 1 \quad \text{respectively} .$$

Consequently,

$$P_1 = T_1 n_1 = -P_{II} = -T_{II} n_{II} .$$

With eqs. (42), (44) and (45) and $\eta_0 = 1$ we obtain in succession for

gear train I: T_a and T_b ,

gear train III: T_d and T_e from $T_c = -T_b$,

gear train IV: T_g from $T_f = -T_a - T_d$,

gear train II: T_{II} from $T_h = -T_e - T_g$.

Check: $T_{II} = -T_I i_{1\,II}$.

With the known torques and the previously determined speed-ratios we can now determine the exponent $r1$ for each component gear train from eq. (43). It holds true under the condition that $P_1 = +1$, that is, when shaft I is the input shaft.

We then repeat the torque calculation for this power-flow $I \rightarrow II$ with the correct basic efficiency and the just determined exponent $r1$ for each component gear train. Thus, we obtain the accurate value of T_{II} and consequently are able to calculate the efficiency $\eta_{1\,II}$ from the appropriate equation given earlier in this example.

For the reverse power-flow $II \rightarrow I$ we repeat the torque calculation a second time. However, in this calculation the exponent $r1$ for each component gear train assumes the opposite sign. Thus we obtain a higher torque T_{II} than before which is then used to calculate the efficiency $\eta_{II\,I}$ as previously described.

With the two efficiencies $\eta_{1\,II}$ and $\eta_{II\,I}$, we can readily determine the efficiency of the bicoupled transmission for any operating condition from worksheets 4 or 5.

b) Higher Bicoupled Transmissions with More Than Two Degrees of Mobility

If, as shown in figs. 111n to 111p, two shafts of one component gear train are coupled with two shafts of a second and third component gear train, higher bicoupled transmissions with, at most, two degrees of mobility, are generated. However, if these two shafts of a component gear train are coupled with one shaft each of a second and third component gear train, higher bicoupled transmissions with a cyclical arrangement (fig. 111q) arise which can have any number of degrees of mobility larger than 2. However, their number of degrees of mobility equals, at most, their number of component gear trains. Since transmissions of this type are not practically used, they will not be further investigated.

IV. DESIGN HINTS

Section 43. Geometric Conditions for Assembly of Planetary Transmissions

The design of a new transmission usually begins with the determination of the basic speed-ratio and the selection of the most suitable gear train as explained in earlier sections. The calculation of the gear parameters and the arrangement of the gears relative to each other can then proceed according to the well-known methods established for conventional gear trains. To determine the number of teeth for the gear train with the minimum overall diameter, *before* the gearing is actually designed for the given torques and the available materials, the following procedure can be employed.

a) Correlation between Number of Teeth and Gear Diameter

In reverted planetary transmissions, the center distance between the meshing gears must be selected in such a way that the two central gears can be arranged coaxially.

In this respect, transmission designs with meshing planets (figs. 22, 23, 36, and 37) cause the least problems. Their center distances initially need only be approximated and the gear sets can then be designed to meet cost and manufacturing requirements. Finally, the center distances can be adjusted to the correct values by "jackknifing" the drive train to a greater or lesser degree.

However, in transmissions with simple stepped planets of the types shown, (figs. 19, 21, 24, 34, and 39), the two series-coupled gear stages must, from the very beginning, be designed to have the same center distances. In general, the overall basic speed-ratio is given and must be split between the two stages:

$$i_o = i_{1_{p1}} i_{p_2 2} = \left(- \frac{z_{p1}}{z_1} \right) \left(- \frac{z_2}{z_{p2}} \right) . \tag{130}$$

In standard involute gear trains without addendum modification which have the diametral pitches $P_1 + P_2$ the center distances are

$$C_{1_{p1}} = \frac{1}{2} \cdot \frac{1}{P_1} (z_{p1} + z_1) , \quad \text{and} \quad C_{2_{p2}} = \frac{1}{2} \cdot \frac{1}{P_2} (z_2 + z_{p2}) ,$$

where the subscripts of the gears indicate the respective stages.

The two stages attain identical center distances when the ratio of their diametral pitches, that is,

$$\frac{P_2}{P_1} = \frac{m_1}{m_2} = \left| \frac{z_2 + z_{p2}}{z_1 + z_{p1}} \right|. \tag{131}$$

As discussed in section 9, the number of teeth of an annular gear must always be preceded by a negative sign. Therefore, external gear stages are characterized by a sum, annular gear stages by a difference, of numbers of teeth.

Different partitions of the basic speed-ratio as defined by eq. (130) result in different transmission diameters. Thus, gear trains, according to figs. 19 and 35, assume a minimum diameter when the pinions of both stages simultaneously have the smallest admissible number of teeth $z_{min.}$, and both gear meshes are subject to the same root-bending stresses σ_b. As a rule, this optimum partition of the basic speed-ratio can be realized only with fractional numbers of teeth. If, however, a small deviation from the desired basic speed-ratio is tolerable, then the smallest transmission can be found by searching for the theoretical optimum partition of i_o with the aid of a fast converging iteration. During this calculation the gears are permitted to assume fractional numbers of teeth. The final partition of i_o, however, is obtained by rounding off all numbers of teeth to their nearest integer numbers. If the resulting deviation from the desired basic speed-ratio is too large, a closer approximation can usually be obtained by slightly adjusting the numbers of teeth of some or all of the gears, which does not substantially increase the size of the gear train.

The root-bending stresses of the two stages can be approximately equalized as follows: It is assumed that the chordal thickness t_{cb} of the teeth is proportional only to the module m and that, likewise, the face width b of the teeth bears a constant ratio to the module, that is,

$$\frac{t_{cb}}{m} = P \cdot t_{cb} = \text{const.}, \quad \text{and} \quad \frac{b}{m} = Pb = \text{const.}$$

Then the root-bending stress σ_b caused by the bending moment M_b

$$\sigma_b = \frac{M_b}{I/c} = \frac{M_b}{b t_{cb}^2/6}$$

becomes

$$\sigma_b \sim \frac{M_b}{m^3}.$$

Since also the dedendum of standard gears bears a constant ratio to the module, the root-bending moment can be expressed in terms of the tangential force F acting at the pitch point of the gear train. Thus:

$$M_b \sim Fm$$

and consequently

$$\sigma_b \sim \frac{F}{m^2}$$

or with

$$T = F \cdot d/2 , \qquad \text{that is,} \qquad F \sim \frac{T}{d} ,$$

$$\sigma_b \sim \frac{T}{dm^2} \qquad \text{where}$$

T is the torque transmitted by the gears. Thus, in gear systems, the module m is a suitable characteristic length. Since, according to the Theory of Models (Similitude Analysis) [45, 46], similar external forces produce similar stresses in similar cross-sections of similar rigid bodies, the following relationship exists between the root stresses of two similar *spur* gears a and b:

$$\frac{\sigma_a}{\sigma_b} = \frac{F_a}{F_b}\left(\frac{m_b}{m_a}\right)^2 ,$$

or expressed in terms of the torques acting at the pitch circles

$$\frac{\sigma_a}{\sigma_b} = \frac{T_a d_b}{T_b d_a}\left(\frac{m_b}{m_a}\right)^2 .$$

Since $P = z/d$, that is, $m = d/z$, $d = mz$ and thus,

$$\frac{\sigma_a}{\sigma_b} = \frac{T_a m_b^3 z_b}{T_b m_a^3 z_a} .$$

If this equation is applied to a stepped planet whose root stresses are assumed to be equal so that $\sigma_{b_{p1}}/\sigma_{b_{p2}} = 1$, and whose gears transmit torques of equal magnitude, then we obtain with eq. (131)

$$\frac{\sigma_{b_{p1}}}{\sigma_{b_{p2}}} = 1 = \frac{z_{p2}}{z_{p1}}\left(\frac{m_{p2}}{m_{p1}}\right)^3 = \left|\frac{z_{p2}}{z_{p1}}\left(\frac{z_1 + z_{p1}}{z_2 + z_{p2}}\right)^3\right| , \qquad (132)$$

where, according to section 9, the number of teeth of an annular gear must be preceded by a negative sign. If we now introduce the condition $z_1 = z_{p2} = z_{min}$, for which transmissions according to figs. 19 and 35 assume a minimum diameter, this becomes

$$z_{min} + z_{p1} = |z_2 + z_{min}| \sqrt[3]{\frac{z_{p1}}{z_{p2}}} \, ,$$

or, after division by z_{min},

$$\frac{z_{p1}}{z_{min}} = \left|\frac{z_2}{z_{min}} + 1\right| \sqrt[3]{\frac{z_{p1}}{z_{min}}} - 1 \, .$$

With $z_{p1}/z_{min} = Z$ and $z_2/z_{p2} = i_0/Z$ [eq. (130)] this equation can be rewritten in the form $Z = f(Z)$, that is,

$$Z = \left|\frac{i_0}{Z} + 1\right| \sqrt[3]{Z} - 1 \, . \tag{133}$$

Since $Z = f(Z)$ represents a real number, and eq. (133) maps real numbers back into real numbers, it can be solved by the iterative method of successive approximations [54, Art. 70], that is, we may let Z assume an arbitrary value Z_0 and then use it to calculate a new value Z_1, that is,

$$Z_1 = \left|\frac{i_0}{Z_0} + 1\right| \sqrt[3]{Z_0} - 1 \, .$$

The initially assumed value Z_0 can be adjusted after each step until the error becomes sufficiently small, for example,

$$\left(\frac{Z_1 - Z_0}{Z_1}\right)_n \leqq 0.001 \, .$$

In this case the calculation converges rapidly when the mean value of Z_0 and Z_1 is chosen as the initial value of the next step of the iteration, that is, generally when

$$(Z_0)_n = \left(\frac{Z_0 + Z_1}{2}\right)_{n-1} \, .$$

With the best approximation Z, we can then calculate the numbers of teeth

$$z_1 = z_{min} , \qquad z_{p2} = z_{min} ,$$

$$z_{p1} = Z \cdot z_{min} , \qquad z_2 = \frac{i_o z_{min}}{Z}$$

of the smallest gear train whose planets have a stress ratio $\sigma_{b_{p1}}/\sigma_{b_{p2}} \approx 1$. If z_{p1} and z_2 are fractional numbers, they must be rounded off to their nearest integer numbers. Inserted into eq. (130) the chosen integers are then used to check the resulting basic speed-ratio i_o. If found necessary, the deviation from the desired basic speed-ratio can be minimized by changing some or all the numbers of teeth. In cases where the given basic speed-ratio must be *accurately* realized, the numerators and denominators of eq. (130) can be rounded off to only those integer numbers which contain the same prime factors as the given i_o, plus whatever arbitrary prime factors are needed to expand the fractions. The gear train can then be designed following the well-known conventional methods where the center distances of the two stages are equalized by addendum modifications, or possibly by using helical gearing.

Finally it must be checked whether the obtained center distance is sufficiently large to accommodate the required bearings. If not, that is, if the center distance and, consequently, the gear diameters and the overall transmission diameter, must be enlarged, it is usually better to choose a different partition of the basic speed-ratio i_o. For this purpose we shift the speed-ratio of the stage which requires the larger bearings and thus the larger center distance, closer to 1. This tends to equalize the required minimum center distances of the bearings.

Example: Determination of the numbers of teeth of a negative-ratio transmission of the type shown in fig. 167. The desired basic speed-ratio is

$$i_o = -\frac{190}{13} = -\frac{2 \cdot 5 \cdot 19}{13} = -14.615$$

and the minimum number of teeth $z_{min} = 20$.

The previously described procedure yields the following results:

α) Theoretical numbers of teeth as obtained by a series of successive approximations,

$$\text{with} \quad Z = 3.639$$

$$z_1 = 20 , \qquad z_{p2} = 20$$

$$z_{p1} = 72.780 , \qquad z_2 = -80.324 ,$$

and with eq. (130)

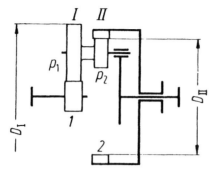

Fig. 167. Example for the determination of the number of teeth leading to a revolving drive train with a minimum outside diameter.

$$i_o = \frac{72.780}{20} \cdot \frac{-80.324}{20} = -14.615 \ .$$

β) *Approximated integer numbers of teeth which lead to a minimum transmission diameter* but result in a slightly smaller basic speed-ratio i_o,

$$z_1 = 20 \ , \qquad z_{p2} = 20$$

$$z_{p1} = 73 \ , \qquad z_2 = -80$$

$$i_o = \frac{73}{20} \cdot \frac{-80}{20} = -14.600 \ .$$

The deviation between the possible speed-ratio and the required speed-ratio is

$$\frac{i_o}{i_{o\,req'd.}} = \frac{14.600}{14.615} = 0.999 \ .$$

γ) *Number of teeth required to realize the exact speed-ratio.* For this case we obtain eight different combinations of numbers of teeth which are close to the optimum solution found in section α. They are compiled in table 17. Because i_o contains the prime factor 13, one of the pinions must have 2 · 13 = 26 teeth; z_{min} = 20 is chosen for the other. Because of the prime factor 19, the number of teeth of one of the gears must become 4 · 19 = 76 or 5 · 19 = 95. All other gear sizes are manipulated by expansion until they approach the optimum solution of section α as closely as possible.

δ) *Comparison of the overall diameters of the considered transmissions.* The overall transmission diameters are expressed in terms of the outside diameters D_I and D_{II} of the two component gear trains of fig. 167. D_I and

TABLE 17. POSSIBLE GEAR SIZES FOR THE NEGATIVE-RATIO TRANSMISSION OF FIG. 167

ROW*	$i_o = \dfrac{z_{p1}}{z_1} \cdot \dfrac{z_2}{z_{p2}}$	$\dfrac{\sigma_{bp1}}{\sigma_{bp2}}$	$\dfrac{i_o}{i_{o\,req'd}}$	$\dfrac{D_1}{m_o}$	$\dfrac{D_{11}}{m_o}$	REMARKS
1	$-\dfrac{190}{13} = \dfrac{72.78}{20} \cdot \dfrac{-80.32}{20}$	0.991	1	165.7	123.7	theoretical
2	$-\dfrac{190}{13} \approx \dfrac{73}{20} \cdot \dfrac{-80}{20}$	1.02	0.999	166	124	{ smallest transmission with an approximated speed-ratio
3	$= \dfrac{76}{26} \cdot \dfrac{-100}{20}$	0.545	1	200†	143	
4	$= \dfrac{95}{26} \cdot \dfrac{-80}{20}$	1.73	1	198	148	
5	$= \dfrac{95}{20} \cdot \dfrac{-80}{26}$	2.64	1	210	170	
6	$-\dfrac{2\cdot5\cdot19}{13} \cdot \dfrac{2\cdot20}{2\cdot20} = \dfrac{76}{20} \cdot \dfrac{-100}{26}$	0.747	1	189†	143	
7	$= \dfrac{100}{20} \cdot \dfrac{-76}{26}$	3.59	1	220	182	
8	$= \dfrac{80}{20} \cdot \dfrac{-95}{26}$	0.989	1	181†	138	{ smallest accurate transmission with $D_1/D_{opt.} = 1.09$
9	$= \dfrac{80}{26} \cdot \dfrac{-95}{20}$	0.706	1	191†	138	
10	$= \dfrac{100}{26} \cdot \dfrac{-76}{20}$	2.28	1	207	157	

* *Row 1*: theoretically smallest transmission according to section α. *Row 2*: smallest possible transmission with an approximate speed-ratio according to section β. *Rows 3 to 10*: combination of gear sizes for an exact realization of the basic speed-ratio according to section γ. These combinations are selected for their closeness to the theoretically smallest transmission diameter. The smallest possible exact transmission is obtained for the combination of gear sizes shown in row 8. $i_{o\,req'd} = -190/13$.

† D_1/m_o over-dimensioned because $\sigma_{bp1}/\sigma_{bp2} < 1$.

D_{II} are referred to the module m_o of a pinion with twenty teeth whose root-bending stress as caused by the torque T_1 of shaft I, is equal to that of gear I itself. If the gear I has just twenty spur teeth as assumed in sections α and β, then $m_o = m_1$. Otherwise, for $T_o/T_1 = 1$, and $\sigma_{b0}/\sigma_{b1} = 1$, eq. (132) yields

$$\frac{m_1}{m_0} \approx \sqrt[3]{\frac{20}{z_1}} \ .$$

This somewhat cumbersome reference to m_o enables us to compare the sizes of several transmissions whose stages I operate with the same bending stresses when the applied torques are equal, or, in other words, when the operating conditions are the same.

When the stress ratio of a considered transmission is $\sigma_{b_{p1}}/\sigma_{b_{p2}} = 1$, both of its stages I and II operate at their design stress level. $\sigma_{b_{p1}}/\sigma_{b_{p2}} > 1$ indicates that stage II is overdesigned and $\sigma_{b_{p1}}/\sigma_{b_{p2}} < 1$ that it is under-designed and thus overstressed. To reduce the stress level in the latter case to the design value, its module, according to eq. (132), must be increased by a factor of $\sqrt[3]{\sigma_{b_{p2}}/\sigma_{b_{p1}}}$. According to eq. (131), and for a constant number of teeth, this then increases the module of stage I and all the gear sizes of both stages so that necessarily now stage I becomes over-dimensioned. If the size of this transmission is subsequently compared to the size of other considered transmissions, then this comparison must be on the basis of its corrected larger stage diameter.

A review of the results listed in table 17 shows that the smallest practical transmission, whose speed-ratio deviates by only about 0.1% from the required speed-ratio, has nearly the same diameter as the theoretically smallest transmission whose gear sizes are shown in row 1. Row 8 represents the smallest possible transmission which accurately realizes the given speed-ratio. However, all of the listed gear train modifications can only be realized with two opposing planets rather than three planets which are uniformly spaced around the circumference of the carrier. This fact will be further discussed in the following section 43b.

The free choice of the numbers of teeth is most severely restricted in the most frequently used design which has only simple planets and is shown in fig. 32. For standard gearing where the pitch circles and the rolling circles coincide we have

$$|d_2| = d_1 + 2d_p \qquad \text{or} \qquad m|z_2| = mz_1 + 2mz_p \ ,$$

so that

$$|z_2| - z_1 = 2z_p \ .$$

Since z_p can only be an integer number, the difference $(|z_2| - z_1)$ must be an even number and, consequently, both gears simultaneously must either have an even or odd number of teeth. However, a transmission of this type can also be built for an odd difference term when the center distances between the planets and the central gears are equalized by using gears with addendum modification.

The number of teeth of the planet gear is obtained by rounding off the value,

$$z_p \approx \frac{|z_2| - z_1}{2},$$

to the next lower integer number. To eliminate backlash between the gears, the teeth of the inner central gear and the planets must have positive addendum modifications, that is, the planets must have long addendum teeth, and the annular gear short addendum teeth. It is also possible to round up to the next higher integer number, but then the addendum modifications must be reversed. However, for the same torques, this results in a somewhat smaller diametral pitch and, consequently, a somewhat larger transmission diameter.

These considerations show that a planetary transmission may have any basic speed-ratio which can be expressed as an integer fraction—a result which is fully valid also for planetary transmissions with bevel gears.

b) Distribution of Several Planets around a Central Gear Circumference

A planetary transmission which contains only one simple or stepped planet, or only one pair of meshing planets, can always be assembled without further conditions. However, with the installation of the first planet, the angles between the two central gears and the axis of the planet on the arm are determined. A second planet can only be installed in certain locations between the two central gears. The possible index angles δ_s between the first and all further planet axes depend on the number of teeth of the gears as has been described by F. Hill [44] and other authors. However, if the smallest index angle $\delta_{s\,min.}$ between two possible assembly locations of the planet axes is known then we can mark any integer multiple of $\delta_{s\,min.}$ on the carrier as a further possible index angle.

The minimum index angle $\delta_{s\,min.}$ can be found by reasoning as follows: A planetary transmission with only a single planet may occupy an arbitrary initial position. If we now lock gear 1 and turn gear 2 by one tooth, the carrier moves through an angle,

$$(\delta_s)_{1\,fixed} = \delta_2 \cdot \frac{n_s}{n_2} = \frac{360}{z_2} \cdot i_{s2} = \frac{360}{z_2} \cdot \frac{i_o}{i_o - 1}, \tag{134}$$

into a new location which, likewise, is a possible assembly position. Since we have turned gear 2 by exactly one tooth while gear 1 is still in its initial position, we could theoretically install a new planet at the initial location of the first planet. Therefore, $(\delta_s)_{1\,\text{fixed}}$ represents a possible assembly angle and, consequently, all integer multiples $g \cdot (\delta_s)_{1\,\text{fixed}}$, where g may be any positive integer number, represent other possible index angles.

The same reasoning can be repeated with a locked gear 2 where gear 1 is moved ahead by one tooth. The carrier consequently turns by an angle

$$(\delta_s)_{2\,\text{fixed}} = \delta_1 \cdot \frac{n_s}{n_1} = \frac{360}{z_1} \cdot i_{s1} = \frac{360}{z_1} \cdot \frac{1}{1 - i_o} \qquad (135)$$

whose multiples $g(\delta_s)_{2\,\text{fixed}}$ are again possible index angles.

α) *Minimum spacing angle in gear trains with simple planets.* According to section 9, the basic speed-ratio of the transmission types shown in figs. 22, 32, 36, 38, 40, and 41, which either have a single, or a single pair, of simple meshing planets, is determined by

$$i_o = + \left| \frac{z_2}{z_1} \right| \quad \text{for positive-ratio transmissions, and}$$

$$i_o = - \left| \frac{z_2}{z_1} \right| \quad \text{for negative-ratio transmissions.}$$

If we substitute these values into eqs. (134) and (135), we obtain for *negative-ratio* transmissions:

$$(\delta_s)_{1\,\text{fixed}} = \frac{360(-|z_2/z_1|)}{z_2(-|z_2/z_1| - 1)} = \frac{360}{|z_2| + |z_1|},$$

$$(\delta_s)_{2\,\text{fixed}} = \frac{360}{z_1(1 + |z_2/z_1|)} = \frac{360}{|z_2| + |z_1|}.$$

Since the index angle δ_s is identical for both cases, it represents the minimum index angle $\delta_{s\,\text{min}}$ for transmissions with *simple* planets. For *positive-ratio* gear trains of the above mentioned types we find analogously

$$(\delta_s)_{1\,\text{fixed}} = (\delta_s)_{2\,\text{fixed}} = \delta_{s\,\text{min}} = \frac{360}{|z_2| - |z_1|}.$$

The number of possible assembly positions for the planets of these transmission types are thus given by the sum of the number of teeth ($|z_2| + |z_1|$) when the gear train has a negative speed-ratio and by the difference of the numbers of teeth ($|z_2| - |z_1|$) when the gear train has a positive speed-ratio.

β) *Minimum spacing angle in transmissions with stepped planets.* According to section 9, the basic speed-ratio of transmissions with a single stepped planet as shown, in figs. 19, 21, 26, 27, 34, 35, 37, and 42 is determined by

$$i_0 = + \left| \frac{z_{p1}}{z_1} \cdot \frac{z_2}{z_{p2}} \right| \quad \text{for positive-ratio gear trains, and}$$

$$i_0 = - \left| \frac{z_{p1}}{z_1} \cdot \frac{z_2}{z_{p2}} \right| \quad \text{for negative-ratio gear trains.}$$

If we substitute these numbers of teeth into eqs. (134) and (135) then we obtain for negative-ratio gear trains

$$(\delta_s)_{1 \text{ fixed}} = \frac{360}{z_2} \cdot \frac{- \left| \frac{z_{p1}}{z_1} \cdot \frac{z_2}{z_{p2}} \right|}{- \left| \frac{z_{p1}}{z_1} \cdot \frac{z_2}{z_{p2}} \right| - 1} = \frac{360 |z_{p1}|}{|z_{p1} z_2| + |z_1 z_{p2}|}, \quad (136)$$

$$(\delta_s)_{2 \text{ fixed}} = \frac{360}{z_1} \cdot \frac{1}{1 + \left| \frac{z_{p1}}{z_1} \cdot \frac{z_2}{z_{p2}} \right|} = \frac{360 |z_{p2}|}{|z_{p1} z_2| + |z_1 z_{p2}|}, \quad (137)$$

and for positive-ratio gear trains

$$(\delta_s)_{1 \text{ fixed}} = \frac{360}{z_2} \cdot \frac{\left| \frac{z_{p1}}{z_1} \cdot \frac{z_2}{z_{p2}} \right|}{\left| \frac{z_{p1}}{z_1} \cdot \frac{z_2}{z_{p2}} \right| - 1} = \frac{360 |z_{p1}|}{|z_{p1} z_2| - |z_1 z_{p2}|}, \quad (138)$$

$$(\delta_s)_{2 \text{ fixed}} = \frac{360}{z_1} \cdot \frac{1}{1 - \left| \frac{z_{p1}}{z_1} \cdot \frac{z_2}{z_{p2}} \right|} = \frac{360 |z_{p2}|}{|z_1 z_{p2}| - |z_{p1} z_2|}. \quad (139)$$

This shows that for gear trains with stepped planets the two index angles $(\delta_s)_{1 \text{ fixed}}$ and $(\delta_s)_{2 \text{ fixed}}$ are not equal. If we develop the resulting carrier spacings and place them side by side as shown in fig. 168, then we observe that $\delta_{s \, min}$ becomes even smaller than the smaller of the two calculated spac-

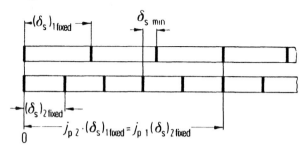

Fig. 168. Development of the circumference of the carrier showing the possible index angles.

ings. In the example of fig. 168, we can generate $\delta_{s\,min}$ if we turn gear 2 by two teeth relative to the fixed gear 1 so that the arm turns by $2\ (\delta_s)_{1\,fixed}$, and then lock gear 2 and turn gear 1 back by three teeth, which causes the arm to move back by $3\ (\delta_s)_{2\,fixed}$. Thus, we obtain

$$\delta_{s\,min} = 2(\delta_s)_{1\,fixed} - 3(\delta_s)_{2\,fixed}\ .$$

The teeth of the stepped planet which are brought into mesh by these two motions do not have the same angular position relative to each other as the teeth which were in mesh before these two motions took place.

Since, during this procedure, both central gears have been turned by an integer number of teeth, a new planet could be installed at the initial location of the first planet, provided its two stepped gears have precisely the same position relative to each other as those of the first planet.

According to eqs. (136) and (137), or (138) and (139), the reduced ratio of the index angles δ_s for a given gear train becomes

$$\frac{(\delta_s)_{1\,fixed}}{(\delta_s)_{2\,fixed}} = \frac{z_{p1}}{z_{p2}} = \frac{tj_{p1}}{tj_{p2}} = \frac{j_{p1}}{j_{p2}}\ , \tag{140}$$

where t is the largest common divisor of the two numbers of teeth of the stepped planets and j_{p1} and j_{p2} are simply the integer residues of the numerator and denominator. As fig. 168 and eq. (140) show, the index angles

$$j_{p2}(\delta_s)_{1\,fixed} = j_{p1}(\delta_s)_{2\,fixed}\ ,$$

are equal, and determine the angle at which a stepped planet can be inserted between the central gears, with the same teeth as were at the initial position fitting the meshes at the new position.

For negative-ratio transmissions, the smallest index angle $\delta_{s\,min}$ as shown in fig. 168, is obtained from eqs. (136) and (137). Thus,

$$\delta_{s\,min} = x(\delta_s)_{1\,fixed} - y(\delta_s)_{2\,fixed}$$

$$= \frac{360}{|z_{p1}z_2| + |z_1 z_{p2}|} \cdot t(xj_{p1} - yj_{p2}) \, ,$$

provided x and y are chosen in such a way that the expression in parentheses has a minimum value. Since the latter contains only integer numbers, its smallest nontrivial value is 1. For this minimum value and, thus, without prior determination of x and y, we obtain

$$\delta_{s\,min} = \frac{360t}{|z_{p1}z_2| + |z_1 z_{p2}|} \, , \qquad (141)$$

for negative-ratio trains, and analogously with eqs. (138) and (139)

$$\delta_{s\,min} = \frac{360t}{|z_{p1}z_2| - |z_1 z_{p2}|} \, , \qquad (142)$$

for positive-ratio gear trains.

γ) *Even spacing of planets around a central gear circumference.* Usually, an even spacing of the planets around the circumference of the carrier is desired since the gear train is then automatically balanced and the radial forces which act on the central gears cancel each other, provided, of course, the planets are manufactured with sufficient accuracy. However, an even spacing of q planets around the circumference of the carrier is possible only when

$$\frac{360}{\delta_{s\,min}q} = g \, , \qquad (143)$$

where g is an integer number. This condition is valid for all types of planetary gear trains but requires that two gears of each simple stepped planet of a particular gear train have precisely the same position relative to each other.

However, disregarding the condition expressed by eq. (143), stepped planets can be equally spaced when we drop the condition that the relative positions of the two gears of each planet must be exactly the same, and instead are willing to bear the extra cost of either cutting each individual stepped planet to fit a specific location, or of making the two gears of each stepped planet adjustable. In the latter case, the required angular displacement between the two gears relative to a reference planet at the zero location can be easily found when each of the planet assemblies is rolled on its associated central gears from the zero position to its final assembly location. There they must be locked together in the exact angular position which each gear reached at this location. That is, relative to the reference planet, both gears must be turned against each other by the difference between their rolling angles. Each further planet will normally have a different angular displacement between its two gears after it has been rolled from the zero position to its intended assembly location.

It is expedient to mark the gearing of transmissions with stepped planets during fabrication so that the planets can later be assembled in their correct angular positions without further trials.

Table 18 lists the equations for $\delta_{s\,min}$ and the conditions under which an equal spacing of the planets is obtained. Transmissions with *two or more* meshing *stepped* planets as shown in fig. 23 are not considered in this table because they are of no practical importance. The table is valid, independently of possible dedendum modifications or helix angles, since the index angle δ_s is only a function of the number of teeth.

TABLE 18. SYNOPSIS OF EQUATIONS DESCRIBING PLANET SPACING
ON THE CARRIER*

TRANSMISSION TYPE		POSITIVE-RATIO TRANSMISSION	NEGATIVE-RATIO TRANSMISSION
With simple planets or meshing planet pairs		according to fig. 22	according to figs. 32, 33, 36, 38, 40, 41
	$\delta_{s\,min}$ [°]	$\dfrac{360}{\|z_2\| - \|z_1\|}$	$\dfrac{360}{\|z_2\| + \|z_1\|}$
	even spacing when	$\dfrac{\|z_2\| - \|z_1\|}{q} = g$	$\dfrac{\|z_2\| + \|z_1\|}{q} = g$
With one stepped planet or with additional simple meshing planets, per planetary gear set		according to figs. 19, 20, 21, 26, 27	according to figs. 34, 35, 37, 42
	$\delta_{s\,min}$ [°]	$\dfrac{360t}{\|z_{p1}z_2\| - \|z_1z_{p2}\|}$	$\dfrac{360t}{\|z_{p1}z_2\| + \|z_1z_{p2}\|}$
	even spacing when	$\dfrac{\|z_{p1}z_2\| - \|z_1z_{p2}\|}{qt} = g$	$\dfrac{\|z_{p1}z_2\| - \|z_1z_{p2}\|}{qt} = g$

*δ_s, smallest possible index angle relative to the carrier; g, arbitrary integer number; t, largest common divisor of z_{p1} and z_{p2} of the stepped planets; q, number of planets or meshing planet pairs spaced around the circumference of the carrier.

δ) *Uneven spacing of planets around a central gear circumference.* If the numbers of teeth are such that eq. (143) or table 18 do not indicate an even spacing of the planets on the carrier, then we can choose other index angles which must be integer multiples of $\delta_{s\,min}$, but need not be equal to each other.

If eq. (143) indicates that two planets can be spaced 180° apart, but an equal spacing is not possible for a larger even number of planets, then it is always possible to arrange these planets in pairs which can be placed opposite to each other. In this special case the transmission has a symmetrical carrier which is balanced so that all radial forces acting on the central gears cancel out, although the planet spacing is not equal.

ε) *Hints for compound planetary transmissions.* If, in a compound planetary transmission, the carriers of two adjacent component gear trains are rigidly coupled and thus can be combined to form a common structure, this common carrier will assume the simplest form when both component gear trains have the same number q of planets and the same planet spacing δ_s. If the transmission cannot be designed with gear sizes which, according to table 18, allow the same equal planet spacing in both component gear trains, then a skillfully chosen unequal spacing, as described in the previous section, may yet make it possible to achieve a suitable distribution of the two planet sets in a common carrier.

In reduced bicoupled transmissions the combined planets of the two component gear trains must definitely have the same spacing. To investigate the planet distribution in these transmissions it is sufficient to look at two of the three possible component gear trains or the component gear trains of any one of the three kinematically-equivalent bicoupled transmissions, described in section 37 and illustrated in fig. 144 for the Wolfrom transmission. If it is not possible to achieve the same even spacing for both component gear trains through an appropriate choice of the numbers of teeth, then the following alternative possibilities can be examined:

1. If the reduced bicoupled transmission contains a component gear train with simple planets which can be equally spaced according to table 18, then the identical spacing can be realized for the other component gear train with stepped planets by individually adjusting the gears of the stepped planets as described in section 43b, γ.

2. The same procedure can be followed when all component gear trains contain stepped planets. In the worst case, each of the three-stage planets requires a different adjustment between any two of its three gears.

3. If, in a component gear train with simple planets, only an unequal spacing, as described in section 43b, δ, can be achieved, then the stepped planets of the other component gear train can be arranged with the same unequal spacing when the gears of the planets are again individually adjusted.

Section 44. Centering the Shafts of a Planetary Transmission

Planetary transmissions offer the possibility of power-branching by splitting the torque of a central gear between the gear meshes of several planets spaced around the circumference of the carrier. Therefore the gearing need be designed for only that fraction of the total torque which is transmitted by a single planet. This fact explains why planetary and star transmissions build smaller than other types of transmissions which carry the same load. An even distribution of the load between all planets, however, requires a high precision during fabrication and assembly and/or a sufficient elasticity of all power transmitting components.

a) Self-Centering of Central Gears

If three planets are arranged around the circumference of the carrier, one of the central gears can be installed without radial bearings since under load its support at the three pitch points is statically determinate. From a statics point of view, dimensional deviations in the gears and the carrier merely displace the radially unsupported central gear from its central position, but essentially do not affect the symmetry of the load distribution.

This statically determinate support exists whenever *one* of the two central gears, *or* the carrier, is free to move radially while the other two members are supported in bearings. If two of the three members, for example, both central gears, remain free to move radially, then the gear train obtains two excess degrees of freedom, as an investigation by Bastert [48] has shown. In this case, it becomes possible to freely move the outer central gear in both the vertical and horizontal direction as far as the backlash between the teeth allows. If the gear train is distorted by external torques, then any arbitrary motion of this type positively displaces the inner central gear.

The following considerations will make it clear why, in actual designs, it is expedient to leave the lightest gear unsupported. During operation, when continually varying teeth with different profile deviations mesh, and the planet axes change their positions relative to the central gear because of deviations in the planet spacing on the carrier, the free gear must continually adjust to these deviations by varying radial displacements. Obviously the required acceleration of the free gear increases directly as the square of its displacement velocity while the necessary displacement force F_a is directly proportional to its mass and its acceleration.

If we assume that all shafts are rigidly supported, and the gears mesh without backlash, then the gear train forms a statically indeterminate system. In this case dimensional deviations cause elastic deformations in the gear train which give rise to elastic reaction forces F_{el} and additional stresses. If now suddenly one of the three shafts is released, the system becomes statically determinate and the elastic force F_{el} accelerates the freed member in the direction of its static equilibrium position. If the elastic reaction forces F_{el} which then represent the largest *available* acceleration forces, are compared with the acceleration forces F_a which are *required* during normal operation to move the free gear between its continually changing static equilibrium positions, then we make the following observations:

When the free gear has a small mass and rotates with a sufficiently low speed, then $F_a \ll F_{el}$. Therefore, the free gear can continually move between its changing equilibrium positions, and no appreciable reaction forces F_{el}, that is, deformations and additional dynamic stresses, arise.

However, when the free gear has a large mass and rotates with a sufficiently high speed, so that, $F_a \gg F_{el}$, the elastic reaction forces are not nearly large enough to effect the theoretically desired displacement of the free gear. In the limiting case, its inertia causes the axis of the free gear to remain in a median position as if it were rigidly supported. All dynamic deformations, stresses, and forces caused by dimensional deviations then have the same value as if the free gear were supported without backlash in this median position.

This consideration not only shows that the lightest gear, usually the sun gear, should be freed from axial constraints and lightened as far as possible, for example, by perforating the gear blank, but also shows that those proposed balancing devices which incorporate large rotating masses, must become ineffective at higher speeds.

b) Displacement of a Central Gear Caused by Dimensional Deviations

In an analytical investigation of the displacement of the central gears as caused by deviations in the tooth profiles and the planet spacing on the carrier, Bastert [48] showed that the involutes can be replaced by their tangents at the momentary contact points since the normal backlash between the meshing teeth allows only small displacements. With this simplification the displacements of the gears can be clearly illustrated by graphical methods.

First we will consider a perfectly accurate negative-ratio transmission as shown in fig. 169 which has the three planets 3, 4, and 5. Its carrier and annular gear 2 are locked and the sun gear 1 is—radially—supported only at its momentary contact points with the planets. The torques T_1, T_2, and T_s shall act on the gears in the directions indicated by the arrows and thus determine the working flanks of the teeth.

If now dimensional deviations are introduced which shift the working tooth flank of *only* the planet 3 in the direction E_{13}, then the sun gear moves away from its previous position by sliding along the flanks of the stationary planets 4 and 5 in a direction perpendicular to the momentary lines of action E_{14} and E_{15}. This causes the sun gear to turn about its instantaneous center of rotation, the velocity pole $P_{1/45}$, which is identical with the point of intersection between the two lines of action E_{14} and E_{15}. Consequently, its center 0, and the foot A, of the perpendicular to the line of action E_{13} also turn

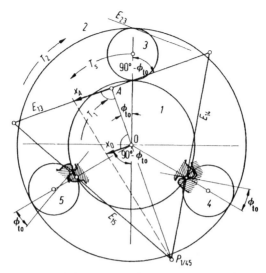

Fig. 169. Displacement x_o of the center of the sun gear *1* resulting from a shift x_A (of this gear) along its line of action E_{13} as caused by dimensional deviations in the meshing teeth: $x_o/x_A = 2/3$.

about the velocity pole $P_{1/45}$. Since it is the perpendicular median of an equilateral triangle, this normal line simultaneously represents the radius vector through the center *0*, whose displacement we set out to investigate, and the radius vector through the point *A*. Its angular displacement equals the dimensional deviation in the direction of the line of action as caused by a deviation from the theoretical circular pitch, as long as the involutes can be represented by their tangents through the pitch point. Thus we can conclude:

1. The center of the sun gear is displaced parallel to that line of action to which the dimensional deviation is referenced, and in the same direction as the contact point.

2. The ratio between the displacements of the points *0* and *A* equals the constant ratio between their distances from their common velocity pole *P* which can be found from the geometry of the equilateral triangle as:

$$\frac{x_o}{x_A} = \frac{\overline{P0}}{\overline{PA}} = \frac{2}{3} .$$

3. This ratio is independent of the pressure angle Φ and any possible gear modifications because it has been derived solely from the geometry of the equilateral triangle.

4. These arguments are valid only for planetary transmissions with three

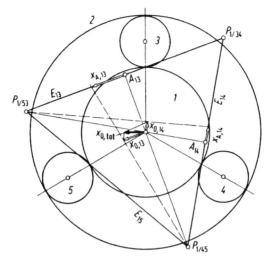

Fig. 170. Total center displacement $x_{0,tot}$ as obtained by vectorial addition of the center displacements $x_{0,13}$ and $x_{0,14}$.

simple planets which are equally spaced around the circumference of the central gears.

If several dimensional deviations occur which cause the point A to shift along the same line of action, then the individual displacements can be added algebraically. However, if some or all of the planets have dimensional deviations, then the displacements along each of the lines of action E_{13}, E_{14}, E_{15} must be determined *individually* as previously described, disregarding momentarily all other deviations. The total displacement of the center of the sun gear is then obtained by vectorially adding the individual displacements x_{13}, x_{14}, and x_{15} as shown in fig. 170. As illustrated in the geometrically similar fig. 171, the displacement of the center of a free annular gear due to dimensional deviations can be determined analogously and contingent on the same limitations. However, when the planets are fixed, the momentary centers of rotation $P_{1/45}$ and $P_{2/45}$ of the sun gear, or the annular gear, are offset against each other by an angle $2\,\Phi_{to}$ as shown in fig. 172.

c) Effect of Some Dimensional Deviations on Displacement of a Central Gear

To find the displacement x_A of the point A along the line of action of *one* particular planet mesh, and the resulting displacement *component* x_0 of the center of the free central gear, we shall subsequently assume that the transmission is dimensionally perfect, except for the one specific deviation whose influence we want to investigate.

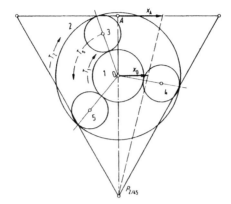

Fig. 171. Displacement x_0 of the center of the annular gear *2* resulting from a shift x_A along the line of action E_{23} when the inner central gear *1* is rigidly supported.

α) *An "effective" normal pitch error* $e_{P,\text{eff}}$ in each of the four tooth flanks which participate in the transmission of a force from one central gear to the other through a single planet, causes a displacement of the point A. The individual displacements, however, are identical in sign and value. The "effective" normal pitch error $e_{P,\text{eff}}$ is composed of the actual normal pitch error e_P, the profile form error e_F, the base circle error e_b, the pressure angle error e_Φ, and the deviation in tooth thickness e_t. Longitudinal (refers to the "length" of the teeth, that is, the face width) form errors such as longitudinal form waviness, also influence the value of $e_{P,\text{eff}}$. In terms of the effective normal pitch error, the displacement of the axis of the free central gear can then be expressed by:

$$x_0 = \frac{2}{3} x_A = \frac{2}{3} e_{P,\text{eff}} \ .$$

β) *Radial run-out of the supported central gear.* If, as shown in fig. 172, the center of the annular gear *2* is radially displaced from the center of the pitch circle (radial run-out) by a distance x_r, and if we further assume that this displacement is caused by a rotation around the velocity pole $P_{2/45}$, then the planets *4* and *5* remain at rest. The point A_{23} moves parallel to this displacement along the line of action, by a distance $x_{A23} = 3/2 x_r$. This causes the planet *3* to rotate clockwise so that the point A_{13} also moves by a distance $x_{A13} = 3/2 x_r$ along the line of action E_{13}. Consequently the center of the free central gear *1* is displaced by a distance $x_{0,13} = 2/3 x_{A13} = x_r$ paral-

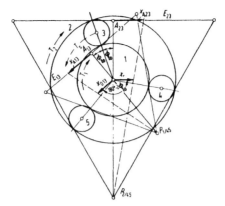

Fig. 172. Displacement $x_{0,13}$ of the free central gear *1* as caused by the radial run-out x_r of the radially supported central gear *2*.

lel to the line of action E_{13}. Since the planets *4* and *5* remain at rest during this displacement, the central gear *1* must turn about its velocity pole $P_{1/45}$. Thus the displacements of both central gears have the same magnitude. However, their directions are turned by an angle $(180 - 2\Phi_{to})°$ relative to each other because the radius vectors in the direction of the two corresponding poles $P_{1/45}$ and $P_{2/45}$ include the angle $2\Phi_{to}$. This consideration shows that a radial run-out of the supported central gear causes a displacement of the free central gear which is equal in magnitude but is offset by an angle of $(180 \pm 2\Phi_{to})°$. The sign of the term $2\Phi_{to}$ depends on which pair of tooth flanks transmits the tangential force, that is, on the direction of the external torques.

γ) *Spacing error on the carrier.* The affected planet is displaced by a distance $x_{s\delta}$ along the circumference:

$$x_0 = \frac{2}{3} x_A = \frac{2}{3} \cdot 2 x_{s\delta} \cos \Phi_{to} = \frac{4}{3} x_{s\delta} \cos \Phi_{to} \ .$$

δ) *Radial deviation on the carrier* causing the center of the affected planet to be radially displaced by a distance x_r. If this radial displacement is considered as a pivoting about the velocity pole $P_{3/12}$ as shown in fig. 173, then small angular displacements of the planet do not displace the free central gear, since the two contact points between the planet and the central gears move along the tangents to the involutes at the contact points and, therefore, vertically to the lines of action E_{13} and E_{23}. Thus,

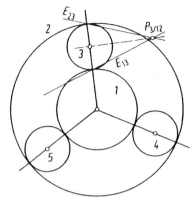

Fig. 173. A radial shift of a planet on the carrier which results from an angular motion around the momentary pole $P_{3/12}$ does not cause a displacement of the axis of the free central gear.

$$x_o = \frac{2}{3} x_A = 0 \ .$$

During its radial displacements the planet turns about its own axis by an angle which is equal to the pivot angle about the velocity pole $P_{3/12}$.

d) Displacement Limits

If the *sun* gear is supported only by three planets, it is free to move radially in all directions until the distance travelled equals the backlash and it

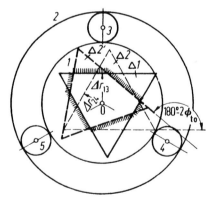

Fig. 174. Displacement limits of the free central gears *1* and *2* when the carrier is rigidly supported. The arbitrarily assumed distances Δr between the sides of the triangles $\Delta 1$ and $\Delta 2$ and the origin of the carrier represent the respective free motions of the inner and outer central gears before meshing play-free with a planet gear.

hits the tooth flanks of a planet gear. The limits of the field where this free motion is possible are circular arcs whose centers are identical with the centers of the planets. Because the backlash between the gears is small, these circular arcs can be replaced by their tangents. With this assumption it is then possible to represent the field of free motion of the sun gear *1* by an equilateral triangle, for example $\Delta 1$ in fig. 174. The center of gravity of this triangle coincides with the center of the carrier only in the hypothetical case that all gears and the carrier are free of dimensional errors and the backlash is the same in all gear meshes. If the *annular* gear *2* instead of the sun gear *1* is free to move radially its field of free motion is represented by the triangle $\Delta 2$ which is rotated relative to the field $\Delta 1$ of the sun gear by an angle of *180°*, as can be verified from fig. 174.

If the carrier of such a planetary gear train is supported without radial play, but both central gears *1* and *2* are free gears, then we can determine a secondary field of motion of the sun gear *1* which arises when the annular gear *2* moves within its field of motion $\Delta 2$, as explained in section 44c, β. Consequently, this secondary field of motion is obtained as the triangle $\Delta 2'$ when the field $\Delta 2$ is rotated about the center *0* of the carrier by an angle $(180 \pm 2\Phi_{to})°$. The remaining play of the sun gear is then determined by the limits of whichever triangle $\Delta 1$ or $\Delta 2'$ is reached first. The resulting hexagonal field is emphasized by cross-hatching in fig. 174. An irregular hexagon of equal size which is, however, rotated backwards by $(180 \pm \Phi_{to})°$ about the center *0* of the carrier, represents the remaining field of free motion of the annular gear *2*.

During operation the two hexagons rotate together with the carrier and represent the somewhat bumpy support of the free central gears in transmissions where only the carrier is rigidly supported.

COMPLETE LIST OF
GUIDE RULES

R 1 The speeds of all parallel shafts of a drive train which rotate in the same direction are designated by the same sign.

2 Torque and speed have the same signs for an input shaft but opposite signs for an output shaft.

3 An input power is always positive; an output power always negative.

4 The two equal torques which act on a free coupling shaft have opposite signs.

5 The degree of freedom F of a compound transmission is given by the sum of the degrees of freedom of its component transmissions minus the number of its constraints and linkages.

6 Conventional transmissions are always constrained transmissions with $F = 1$. However, simple revolving drive trains with three rotating shafts have two degrees of freedom, that is, $F = 2$.

7 Two revolving drive trains are kinematically equivalent if the numerical values of any random pair of two-shaft speed-ratios, drawn from each train's complete set of six i-ratios, coincide.

8 The torque of the summation shaft and the torques of the difference shafts have opposite signs.

9 The torques of the two difference shafts have equal signs.

10 The sign of the exponent $r1$ is always equal to the sign of the rolling-power P_{R1} of shaft 1 and its magnitude is 1.

11 As long as friction losses can be neglected, torque ratios of revolving drive trains are determined solely by their basic speed-ratios. They are independent of the shaft speeds.

12 The external torques always must be matched to those of the revolving drive train. The reverse is impossible, because the torques of the three connected shafts of the transmission bear an invariable ratio to each other.

13 In negative-ratio drives the carrier shaft always is the summation shaft; but in positive-ratio drives the central gear shaft carrying the larger absolute torque is the summation shaft.

14 If the total-power shaft is an input shaft, then both partial-power shafts are output shafts and vice versa.

15 In negative-ratio gear trains, the rolling power always is smaller than the external power.

16 In positive-ratio gear trains, the rolling power can be larger, smaller, or equal to the external power, depending on the existing power-flow mode.

17 If, during a speed change, the shaft speeds pass through the coupling point, then the rolling power-flow changes its direction and the exponent $r1$ changes its sign.

18 The position of the total-power shaft and thus the external power-flow of a revolving drive train changes if the direction of rotation of one or two of its three shafts reverses.

19 At the limit between two different power-flow modes, a three-shaft transmission operates as a two-shaft transmission.

20 Only those positive-ratio, two-shaft or three-shaft transmissions with a basic speed-ratio $\eta_o < i_o < 1/\eta_o$, are capable of self-locking. However, these transmissions do not actually lock unless the carrier shaft is the sole output shaft.

21 The total-power shaft and thus the external power-flow of a revolving drive train can be determined from its basic speed-ratio i_o and an arbitrary speed-ratio k. Its self-locking capability need not be considered.

22 The operating condition with the steepest speed-line is associated with the highest speed of the planet bearings.

23 The speed scale of the summation shaft always lies between the speed scales of the two difference shafts.

24 The speed of the summation shaft always lies between the speeds of the difference shafts.

25 A two-shaft transmission has a negative speed-ratio when the summation shaft is locked, and a positive speed-ratio when one of the difference shafts is at rest.

26 The rolling-power ratio $|P_R/P_{in}|$ is always higher in a positive-ratio gear train than in a kinematically-equivalent negative-ratio gear train.

27 The number of connected shafts of a power transmission must exceed the number designating its kinematic degree of freedom by at least one.

28 If at least one component transmission of a series-coupled compound drive train is self-locking in one of the power-flow directions, then the compound drive train itself is also self-locking in this power-flow direction.

29 All equations derived for simple epicyclic transmissions are equally valid for simple bicoupled transmissions when shafts 1 and 2 are replaced by the analogous monoshafts I and II, the carrier shaft s by the analogous connected coupling shaft S, and the fixed-carrier efficiencies η_{12} and η_{21} by the analogous series-efficiencies $\eta_{I\,II}$ and $\eta_{II\,I}$.

30 For the majority of its possible couplings, series-coupled epicyclic transmissions are not loss-symmetrical, and, therefore, as a rule $\eta_{I\,II} \neq \eta_{II\,I}$.

31 A constrained bicoupled transmission operates with internal power division when the speed-ratio of the auxiliary component transmission is chosen in such a way that the monoshaft m of the main component transmission becomes the latter's total-power shaft. If the total-power shaft of the main component transmission is part of the free coupling shaft, then it operates with positive circulating power; if part of the connected coupling shaft, it operates with negative circulating power.

32 A constrained bicoupled transmission operates with power division when its two monoshafts are a summation shaft and a difference shaft. If both monoshafts are summation shafts or difference shafts, a circulating power flows in the transmission.

33 The total-power shaft changes its position when the speed-ratio of a variable bicoupled transmission, or its auxiliary component transmission, passes through zero or infinity within its speed-ratio range, or both speed-ratios do so simultaneously within theirs. Consequently, the power-flow in the bicoupled transmission changes from one to another of three possible modes: power division, positive circulating power, or negative circulating power.

34 The power-ratio ϵ_0 in a variable component transmission becomes smaller the closer φ approaches a value of 1, that is, the less the variable bicoupled transmission needs to be adjusted, and the more the variable component transmission can be adjusted.

35 A reduced bicoupled transmission and its corresponding functionally-equivalent simple bicoupled transmission have the same operating

characteristics, are represented by the same transmission symbol, and can be analyzed with the same equations.

36 The simple bicoupled transmission which is functionally-equivalent to a reduced bicoupled transmission is characterized by the fact that its monoshafts are formed by a summation shaft and a difference shaft from its component transmissions.

37 Each reduced bicoupled transmission can be expanded into three different kinematically-equivalent simple bicoupled transmissions, but only one of these is functionally equivalent to it.

WORKSHEETS

The following worksheets are directly applicable to all simple epicyclic drive trains. With the analogous subscripts $1 \triangleq I$, $2 \triangleq II$, $s \triangleq S$, so that $i_o \triangleq i_{III} = i_{mf}i_{f'm'}$, they are also valid without restriction for simple and higher bicoupled drive trains. Worksheets 1, 3, and 5 apply to transmissions with one degree of mobility, and 2 and 4 to those with two degrees of mobility.

Worksheet 1. Relationship between the possible two-shaft speed-ratios i of the epicyclic drive trains. Those shafts not represented in the subscripts are fixed.

i	$f(i_0)$	$f(i_{21})$	$f(i_{1s})$	$f(i_{s1})$	$f(i_{2s})$	$f(i_{s2})$
i_0	i_0 ^{1}	$\dfrac{1}{i_{21}}$ ^{2}	$1-i_{1s}$ ^{3}	$1-\dfrac{1}{i_{s1}}$ ^{4}	$\dfrac{1}{1-i_{2s}}$ ^{5}	$\dfrac{i_{s2}}{i_{s2}-1}$ ^{6}
i_{21}	$\dfrac{1}{i_0}$ ^{7}	i_{21} ^{8}	$\dfrac{1}{1-i_{1s}}$ ^{9}	$\dfrac{i_{s1}}{i_{s1}-1}$ ^{10}	$1-i_{2s}$ ^{11}	$1-\dfrac{1}{i_{s2}}$ ^{12}
i_{1s}	$1-i_0$ ^{13}	$1-\dfrac{1}{i_{21}}$ ^{14}	i_{1s} ^{15}	$\dfrac{1}{i_{s1}}$ ^{16}	$\dfrac{i_{2s}}{i_{2s}-1}$ ^{17}	$\dfrac{1}{1-i_{s2}}$ ^{18}
i_{s1}	$\dfrac{1}{1-i_0}$ ^{19}	$\dfrac{i_{21}}{i_{21}-1}$ ^{20}	$\dfrac{1}{i_{1s}}$ ^{21}	i_{s1} ^{22}	$1-\dfrac{1}{i_{2s}}$ ^{23}	$1-i_{s2}$ ^{24}
i_{2s}	$1-\dfrac{1}{i_0}$ ^{25}	$1-i_{21}$ ^{26}	$\dfrac{i_{1s}}{i_{1s}-1}$ ^{27}	$\dfrac{1}{1-i_{s1}}$ ^{28}	i_{2s} ^{29}	$\dfrac{1}{i_{s2}}$ ^{30}
i_{s2}	$\dfrac{i_0}{i_0-1}$ ^{31}	$\dfrac{1}{1-i_{21}}$ ^{32}	$1-\dfrac{1}{i_{1s}}$ ^{33}	$1-i_{s1}$ ^{34}	$\dfrac{1}{i_{2s}}$ ^{35}	i_{s2} ^{36}

Worksheet 2. Relationship between the three-shaft speed-ratios k of the epicyclic drive trains.

k =	$f(k_{12})$	=	$f(k_{21})$	=	$f(k_{1s})$	=	$f(k_{s1})$	=	$f(k_{2s})$	=	$f(k_{s2})$
k_{12}	k_{12} (1)		$\dfrac{1}{k_{21}}$ (2)		$\dfrac{k_{1s}\cdot i_0}{k_{1s}-1+i_0}$ (3)		$\dfrac{i_0}{1-k_{s1}(1-i_0)}$ (4)		$\dfrac{1-i_0(1-k_{2s})}{k_{2s}}$ (5)		$k_{s2}(1-i_0)+i_0$ (6)
k_{21}	$\dfrac{1}{k_{12}}$ (7)		k_{21} (8)		$\dfrac{k_{1s}-1+i_0}{k_{1s}\cdot i_0}$ (9)		$\dfrac{1-k_{s1}(1-i_0)}{i_0}$ (10)		$\dfrac{k_{2s}}{1-i_0(1-k_{2s})}$ (11)		$\dfrac{1}{k_{s2}(1-i_0)+i_0}$ (12)
k_{1s}	$\dfrac{1-i_0}{1-i_0/k_{12}}$ (13)		$\dfrac{1-i_0}{1-i_0\cdot k_{21}}$ (14)		k_{1s} (15)		$\dfrac{1}{k_{s1}}$ (16)		$1-i_0(1-k_{2s})$ (17)		$1-i_0+\dfrac{i_0}{k_{s2}}$ (18)
k_{s1}	$\dfrac{1-i_0/k_{12}}{1-i_0}$ (19)		$\dfrac{1-i_0\cdot k_{21}}{1-i_0}$ (20)		$\dfrac{1}{k_{1s}}$ (21)		k_{s1} (22)		$\dfrac{1}{1-i_0(1-k_{2s})}$ (23)		$\dfrac{1}{1-i_0+i_0/k_{s2}}$ (24)
k_{2s}	$\dfrac{1-i_0}{k_{12}-i_0}$ (25)		$\dfrac{1-i_0}{1/k_{21}-i_0}$ (26)		$\dfrac{k_{1s}-1+i_0}{i_0}$ (27)		$\dfrac{1/k_{s1}-1+i_0}{i_0}$ (28)		k_{2s} (29)		$\dfrac{1}{k_{s2}}$ (30)
k_{s2}	$\dfrac{k_{12}-i_0}{1-i_0}$ (31)		$\dfrac{1/k_{21}-i_0}{1-i_0}$ (32)		$\dfrac{i_0}{k_{1s}-1+i_0}$ (33)		$\dfrac{i_0}{1/k_{s1}-1+i_0}$ (34)		$\dfrac{1}{k_{2s}}$ (35)		k_{s2} (36)

Worksheet 3. Relationship of the range of the basic speed-ratio i_0 (for simple revolving drives) or the series speed-ratio $i_{I\,II}$ (for bicoupled drives) to the ranges of two other two-shaft speed-ratios (i_{1s} and i_{2s} or i_{IS} and i_{IIS}), the position of the summation shaft and the transmission type, for non-self-locking epicyclic drive trains.

i_0	i_{1s}	i_{2s}	Summation Shaft	Type of Drive
<0	>1	>1	Carrier Shaft s	Negative-Ratio Drive
>1	<0	$0\ldots1$	Central Gear Shaft 2	⎫
$0\ldots1$	$0\ldots1$	<0	Central Gear Shaft 1	⎬ Positive-Ratio Drive
1	0	0	no Summation Shaft	⎭

Worksheet 4. The efficiency η of revolving drive trains with three rotating shafts as a function of the basic speed-ratio i_0 (or series ratio $i_{1\,\text{II}}$) the three-shaft speed-ratio k_{12} (or $k_{1\,\text{II}}$) and the direction of power-flow. If, instead of k_{12} (or $k_{1\,\text{II}}$) another three-shaft speed-ratio is known for the same operating condition, then it can be used to determine k_{12} (or $k_{1\,\text{II}}$) from worksheet 2. The associated exponent $r1$ (or $r\text{I}$) which is already considered in these equations and the associated total-power shaft (TPS) are also shown. Caution: $\eta_{21} = \eta_{12} = \eta_0$, but $\eta_{\text{III}} \neq \eta_{1\,\text{II}}$. The latter can be obtained as follows: $\eta_{1\,\text{II}} = \eta_{mf}\eta_{f'm'}$ and $\eta_{\text{III}} = \eta_{m'f'}\eta_{fm}$.

I_0	k_{12}	TPS	Pfl	Power Division — Efficiency η	$r1$	Pfl	Power Summation — Efficiency η	$r1$
$\leqq 0$	$\leqq i_0$	1	$1<\frac{2}{s}$	$\dfrac{k_{12}-I_0+I_0\eta_{12}(1-k_{12})}{k_{12}(1-I_0)}$	$+1$	$\frac{2}{s}>1$	$\dfrac{k_{12}\,\eta_{21}\,(1-I_0)}{\eta_{21}(k_{12}-I_0)+I_0(1-k_{12})}$	-1
	I 0	2	$2<\frac{1}{s}$	$\dfrac{k_{12}-I_0+\eta_{21}(1-k_{12})}{1-I_0}$	-1	$\frac{1}{s}>2$	$\dfrac{\eta_{12}(1-I_0)}{\eta_{12}(k_{12}-I_0)+1-k_{12}}$	$+1$
	0.1	s	$s<\frac{1}{2}$	$\dfrac{(k_{12}-I_0\eta_{12})(1-I_0)}{(k_{12}-I_0)(1-I_0\eta_{12})}$	$+1$	$\frac{1}{2}>s$	$\dfrac{(k_{12}-I_0)(\eta_{21}-I_0)}{(k_{12}\eta_{21}-I_0)(1-I_0)}$	-1
	$\geqq 1$	s	$s<\frac{1}{2}$	$\dfrac{(k_{12}\eta_{21}-I_0)(1-I_0)}{(k_{12}-I_0)(\eta_{21}-I_0)}$	-1	$\frac{1}{2}>s$	$\dfrac{(k_{12}-I_0)(1-I_0\eta_{12})}{(k_{12}-I_0\eta_{12})(1-I_0)}$	$+1$
$0\;1$	$\leqq 0$	s	$s<\frac{1}{2}$	$\dfrac{(k_{12}-I_0\eta_{12})(1-I_0)}{(k_{12}-I_0)(1-I_0\eta_{12})}$	$+1$	$\frac{1}{2}>s$	$\dfrac{(k_{12}-I_0)(\eta_{21}-I_0)}{(k_{12}\eta_{21}-I_0)(1-I_0)}$	-1
	$0\;I_0$	2	$2<\frac{1}{s}$	$\dfrac{k_{12}-I_0+\eta_{21}(1-k_{12})}{1-I_0}$	-1	$\frac{1}{s}>2$	$\dfrac{\eta_{12}(1-I_0)}{\eta_{12}(k_{12}-I_0)+1-k_{12}}$	$+1$
	$I_0..1$	1	$1<\frac{2}{s}$	$\dfrac{k_{12}-I_0+I_0\eta_{12}(1-k_{12})}{k_{12}(1-I_0)}$	$+1$	$\frac{2}{s}>1$	$\dfrac{k_{12}\,\eta_{21}\,(1-I_0)}{\eta_{21}(k_{12}-I_0)+I_0(1-k_{12})}$	-1
	$\geqq 1$	1	$1<\frac{2}{s}$	$\dfrac{\eta_{21}(k_{12}-I_0)+I_0(1-k_{12})}{k_{12}\,\eta_{21}(1-I_0)}$	-1	$\frac{2}{s}>1$	$\dfrac{k_{12}(1-I_0)}{k_{12}-I_0+I_0\eta_{12}(1-k_{12})}$	$+1$
$\geqq 1$	$\leqq 0$	s	$s<\frac{1}{2}$	$\dfrac{(k_{12}\,\eta_{21}-I_0)(1-I_0)}{(k_{12}-I_0)(\eta_{21}-I_0)}$	-1	$\frac{1}{2}>s$	$\dfrac{(k_{12}-I_0)(1-I_0\eta_{12})}{(k_{12}-I_0\eta_{12})(1-I_0)}$	$+1$
	$0\;1$	2	$2<\frac{1}{s}$	$\dfrac{\eta_{12}(k_{12}-I_0)+1-k_{12}}{\eta_{12}(1-I_0)}$	$+1$	$\frac{1}{s}>2$	$\dfrac{1-I_0}{k_{12}-I_0+\eta_{21}(1-k_{12})}$	-1
	$1\;I_0$	2	$2<\frac{1}{s}$	$\dfrac{k_{12}-I_0+\eta_{21}(1-k_{12})}{1-I_0}$	-1	$\frac{1}{s}>2$	$\dfrac{\eta_{12}(1-I_0)}{\eta_{12}(k_{12}-I_0)+1-k_{12}}$	$+1$
	$\geqq I_0$	1	$1<\frac{2}{s}$	$\dfrac{k_{12}-I_0+I_0\eta_{12}(1-k_{12})}{k_{12}(1-I_0)}$	$+1$	$\frac{2}{s}>1$	$\dfrac{k_{12}\,\eta_{21}(1-I_0)}{\eta_{21}(k_{12}-I_0)+I_0(1-k_{12})}$	-1

TPS = Total Power Shaft   Pfl = Power-flow (for transmissions incapable of self-locking)

Worksheet 5. The efficiency η of the two-shaft revolving drive trains as a function of the basic speed-ratio i_0 (or series speed-ratio $i_{I\,II}$) and the direction of the power-flow. The first subscript η represents the input shaft; the second subscript the output shaft; the third shaft is fixed. The exponent $r1$ (or rI) which is already considered in these equations, is also listed. Caution: $\eta_{21} = \eta_{12} = \eta_0$, but $\eta_{II\,I} \neq \eta_{I\,II}$. The latter can be obtained as follows: $\eta_{I\,II} = \eta_{mf}\eta_{f'm'}$ and $\eta_{II\,I} = \eta_{m'f'}\eta_{fm}$.

Group of drives	Negative-Ratio Trains	Positive-Ratio Trains	
Basic Speed-Ratio i_0:	≤ 0	$0\ .1$	≥ 1
Basic Efficiencies — η_{12}	*1* $\quad \eta_{12}$	*2* $\quad \eta_{12}$	*3* $\quad \eta_{12}$
rl	$+1$	$+1$	$+1$
η_{21}	*4* $\quad \eta_{21}$	*5* $\quad \eta_{21}$	*6* $\quad \eta_{21}$
rl	-1	-1	-1
Revolving Carrier Efficiencies — η_{1s}	*7* $\quad \dfrac{i_0\eta_{12}-1}{i_0-1}$	*8* $\quad \dfrac{i_0/\eta_{21}-1}{i_0-1}$	*9* $\quad \dfrac{i_0\eta_{12}-1}{i_0-1}$
rl	$+1$	-1	$+1$
η_{s1}	*10* $\quad \dfrac{i_0-1}{i_0/\eta_{21}-1}$	*11* $\quad \dfrac{i_0-1}{i_0\,\eta_{12}-1}$	*12* $\quad \dfrac{i_0-1}{i_0/\eta_{21}-1}$
rl	-1	$+1$	-1
η_{2s}	*13* $\quad \dfrac{i_0-\eta_{21}}{i_0-1}$	*14* $\quad \dfrac{i_0-\eta_{21}}{i_0-1}$	*15* $\quad \dfrac{i_0-1/\eta_{12}}{i_0-1}$
rl	-1	-1	$+1$
η_{s2}	*16* $\quad \dfrac{i_0-1}{i_0-1/\eta_{12}}$	*17* $\quad \dfrac{i_0-1}{i_0-1/\eta_{12}}$	*18* $\quad \dfrac{i_0-1}{i_0-\eta_{21}}$
rl	$+1$	$+1$	-1

Worksheet 6. Torque Relationships

$$T_2/T_1 = -i_o\eta_o^{r1} \qquad T_s/T_1 = i_o\eta_o^{r1} - 1 \qquad T_s/T_2 = (1/i_o\eta_o^{r1}) - 1$$

$r1$ (or rI) can be obtained directly from worksheet 4 for drive trains with three rotating shafts and from worksheet 5 for drive trains with two rotating shafts, or calculated from eqs. (43) or (83). By definition, $\eta_o^{+1} = \eta_{12} \triangleq \eta_{I\,II}$ and $\eta_o^{-1} = 1/\eta_{21} \triangleq 1/\eta_{II\,I}$. Only for the basic gear trains of simple planetary drives are the efficiencies for reversed power-flows approximately equal, that is, $\eta_{12} \approx \eta_{21}$.

APPENDIX

Generalization of Worksheets 1 and 2

Close scrutiny of worksheet 1 reveals that three general relationships between the i-ratios occur in each row, namely, $i_{ab} = i_{ab}$, $i_{ab} = 1/i_{ba}$ and $i_{ab} = 1 - i_{ac}$. We define these as identical, reciprocal, and complemental relationships, respectively. Worksheet 1 may be constructed from scratch in specific or general form, starting with a knowledge of only these three relationships. An abbreviated one-line general format may be set up to provide for the reproduction of the specific content (but not the same order of entries) of the worksheet as follows:

$$i \quad f(i_{ab}) \quad f(i_{ba}) \quad f(i_{ac}) \qquad f(i_{ca}) \qquad f(i_{bc}) \qquad f(i_{cb})$$

$$i_{ab} \quad i_{ab} \quad 1/i_{ba} \quad 1 - i_{ac} \quad 1 - 1/i_{ca} \quad 1/(1 - i_{bc}) \quad i_{cb}/(i_{cb} - 1)$$

Example of conversion to a specific i, with a $\triangleq$ S, b $\triangleq$ I, c $\triangleq$ II:

$$i_{SI} \quad i_{SI} \quad \frac{1}{i_{IS}} \quad 1 - i_{SII} \quad 1 - \frac{1}{i_{IIS}} \quad \frac{1}{(1 - i_{III})} \quad i_{III}(i_{III} - 1) \ .$$

Six conversion codes are required to reproduce worksheet 1 as a complete table, with either uppercase or lowercase specific subscripts.

Worksheet 2 describes the mutual dependence between any two speed-ratios k of a three-shaft transmission and its basic speed-ratio which appears as a design constant in all expressions except where an identical or reciprocal relationship exists. It can be generalized by replacing the specific subscripts by a, b, c in that order or any other. Thus a generalized worksheet 2 has thirty-six entries. Using all possible conversion codes, six different specific worksheets result from this, one for each of the six i's which are design constants of the drive train. A given derivative worksheet 2 can be reduced to the specific one published in this book by using the appropriate conversion formula expressed in worksheet 1 by columns 1 and 2.

While unnecessary for epicyclic drive train analysis, generalized worksheets 1 and 2 do facilitate synthesis. For synthesis of bicoupled transmissions of the types described in sections 34 to 37, for example, they can be written specifically with the general subscripts m, f, c or m', f', c' used there, in place of a, b, c. If, furthermore, these generalized equations are stored in a computer the synthesis is easily programmable. For instance, only after deciding whether the carrier of the main component transmission should be at m, f, or c would a conversion to $1, 2, s$ in a particular order be made and the actual calculations performed.

SELECTED BIBLIOGRAPHY

1. WILLIS, R.: Principles of Mechanism. London: Parker 1841.
2. WOLFROM, U.: Der Wirkungsgrad von Planetenrädergetrieben. Werkstatts-technik, VI. Jahrg. (1912) 615.
3. KUTZBACH, K.: Mehrgliedrige Radgetriebe und ihre Gesetze. Maschinenbau 6 (1927), H. 22, 1080.
4. ALTMANN: Antriebe von Hebezeugen durch hochübersetzende, raumsparende Stirnradgetriebe. Maschinenbau Heft "Getriebe," Berlin (1928).
5. BRANDENBERGER, H.: Wirkungsgrad und Aufbau einfacher und zusammenge-setzter Umlaufrädergetriebe. Maschinenbau 8 (1929), H. 8 u. 9.
6. ALTMANN: Zwei neue Getriebe für gleichförmige Übersetzung. Maschinenbau "Der Betrieb" 8 (1929) 721.
7. KUTZBACH, K.: Hütte II, 26. Auflage, Berlin: Ernst & Sohn 1931.
8. FORSTER, H. J.: Föttinger-Getriebe in Leistungsverzweigungen. VDI-For-schungsheft Nr. 444, Ausg. B, 20 (1954) Auszug s. ATZ, Jahrg. 58 (1956), Nr. 9, 258.
9. DAHL, A.: Selbstspannende Riementriebe. Konstruktion 6 (1954) H. 8.
10. LAUGHLIN, H. G., HOLOWENKO, A. R., HALL, A. S.: Epicyclic Gear System. Machine Design. 28 (1956), No. 6.
11. LAUGHLIN, H. G., HOLOWENKO, A. R., HALL, A. S.: Design Charts and Equa-tions for Controlled Epicyclic Gear Systems. Machine Design. 28 (1956), No. 29.
12. THOMAS, S. K.: Grundzüge der Verzahnung. München: Hanser 1957, S. 147.
13. WOLF, A.: Die Grundgesetze der Umlaufgetriebe. Braunschweig: Vieweg 1958.
14. OHLENDORF, H.: Verlustleistung und Erwärmung von Stirnrädern. Diss. T. H. München 1958.
15. BRASS, E. A.: Two Stage Planetary Arrangements for the 15:1 Turbo Prop Reduction Gear. ASME Paper Number 60–SA–1 (1960).
16. NEUSSEL, P.: Untersuchung von rückkehrenden Umlaufgetrieben mit und ohne Selbsthemmung unter besonderer Berücksichtigung von Koppelgetrieben. Diss. T. H. Darmstadt 1962.
17. VID-Richtlinie 2127: Getriebetechnische Grundlagen. Düsseldorf: VDI-Verlag, 1962, S. 16
18. SEELIGER, K.: Das einfache Planetengetreibe. Antriebstechnik, 3. Jahrg. (1964), 216.
19. LICHTENAUER, G., ROGG, O., KALLHARDT, K.: Zahnradschaben. München: Carl Hurth, Maschinen- und Zahnradfabrik (1964), S. 105.
20. DIZIOĞLU: Getriebelehre 1. Braunschweig: Vieweg 1965. S. 22.

21. NIEMANN, G.: Maschinenelemente, I, Berlin/Heidelberg/New York: Springer 1965.
22. NEUMANN, R.: Drehzahlberechnung von Umlaufrädergetrieben. Deutsche Textiltechnik 17 (1967), H. 2.
23. NEUMANN, R.: Umlaufrädergetriebe in Textilmaschinen. Deutsche Textiltechnik 15 (1967), H. 5, 6.
24. SUNAGA, T.: Innenverzahnungen mit geringer Zähnezahldifferenz als Planeten-Reduktionsgetreibe. Bulletin of JSME (Japan Society of Mechanical Engineers) (1967).
25. BIRKLE, H. G.: Das Betriebsverhalten der stufenlos einstellbaren Koppelgetriebe. Diss. T. H. Darmstadt 1968.
26. MULLER, H. W., SCHAFER, W. F.: Geometrische Voraussetzungen bei innenverzahnten Getriebestufen kleinster Zähnezahldifferenzen. Forschung im Ingenieurwesen 36 (1970), H. 5, VDI-Verlag.
27. DUDA, MANFRED: Der geometrische Verlustbeiwert und die Verlustunsymmetrie bei geradverzahnten Stirnradgetrieben. Forschung im Ingenieurwesen 37 (1971), H. 1, VDI-Verlag.
28. SCHAFER, W. F.: Einfaches Verfahren zur Ermittlung der Profilüberdeckung ϵ_α bei außen- und innenverzahnten Getriebestufen. Forschung im Ingenieurwesen 36 (1970), H. 5, VDI-Verlag.
29. NIEMANN, G.: Hütte II A, 28. Auflage (1954), Siete 153, Berlin: Ernst & Sohn.
30. KRAUSS, A.: Hütte II A, 28. Auflage (1954), Seite 277, Berlin: Ernst & Sohn.
31. HELFER, FRIEDRICH: Eine Analogie zur Untersuchung von Planetengetrieben, ATZ 69 (1967), Nr. 5, S. 149–152 und Forschung und Konstruktion, Voith, Heidenheim, H. 14.
32. CLARENBACH, J.: Lorenz-Verzahnwerkzeuge. Karlsruhe: Badenia-Verlag 1961.
33. NIEMANN, G.: Maschinenelemente, Band II. Berlin/Heidelberg/New York: Springer 1965.
34. RAVIGNEAUX, P.: 1re addition au brevet d'invention No. 823.757, No. 53.264, demandée le 15 octobre 1943, Paris.
35. GLOVER, J. H.: Efficiency and Speed-Ratio-Formulas for Planetary Systems. Product Engng. 36 (1965) Nr. 19.
36. GAUNITZ, A.: Planetengetriebe mit großer Übersetzung. VDI-Z. 92, Nr. 33. 21 Nov. 1950, S. 956.
37. SCHNETZ, K.: Optimierung zusammengesetzter Planetengetriebe. Fortschrittsberichte der VDI-Zeitschriften, Reihe 1, Nr. 30. Düsseldorf: VDI-Verlag 1971.
38. JARCHOW, FRIEDRICH: Leistungsverzweigung im Getriebe. VDI-Nachrichten Nr. 49, 1967.
39. Werner Reimers KG: Firmendruckschrift 103/3, Bad Homburg.
40. ATZ: Die automatischen Getriebe der amerikanischen "Compact Cars," ATZ 63, H. 8, S. 227–235.
41. FÖRSTER, H. J.: Zur Berechnung des Wirkungsgrades von Planetengetrieben. Konstruktion Jahrg. 21 (1969), Heft 5, S. 169.
42. Messebericht: Werkstatt und Betrieb, 101 (1968), H. 7, S. 301.
43. POPPINGA, REEMT: Stirnrad-Planetengetriebe. Stuttgart: Franckh'sche Verlagshandlung, W. Keller & Co. 1949.

44. HILL, F.: Einbaubedingungen bei Planetengetrieben. Konstruktion 19 (1967), H. 10, S. 393/394.
45. Hütte I. Berlin: Ernst & Sohn 1955.
46. DUBBEL, Taschenbuch für den Maschinenbau. 13 Aufl. Berlin/Heidelberg/New York: Springer 1970.
47. DUDLEY, WINTER: Zahnräder. Berlin/Heidelberg/New York: Springer 1961.
48. BASTERT, C.-CH.: Die Verlagerung der Zentralräder in Planetengetrieben. Forschung im Ingenieurwesen 37 (1971), H. 1, VDI-Verlag.
49. JENSEN, P. W.: Raumbedarf und Wirkungsgrad zusammengesetzter Planetenrädergetriebe mit einstufigem Planetenrad. Konstruktion 21 (1969), H. 5, S. 178–184.
50. SEELIGER, K.: Das Leistungsverhalten von Getriebekombinationen. VDI-Z. 106 (1964), Nr. 6, S. 206–211.
51. GACKSTETTER, G.: Leistungsverzweigung bei der stufenlosen Drehzahlregelung mit vierwelligen Planetengetrieben. VDI-Z. 108 (1966), Nr. 6, S. 210–214.
52. MÜLLER, H. W.: Hydrostatische Umlaufgetriebe. Oelhydraulik und Pneumatik 14 (1970) 11, S. 513–516.
53. DUDLEY, D. W.: Gear Handbook. New York: McGraw-Hill 1962.
54. SCARBOROUGH, J. B.: Numerical Mathematical Analysis. 4th ed. Baltimore: Johns Hopkins Press 1958.
55. MÜLLER, H. W.: Revolving Mechanisms. ASME paper. San Francisco 1972.
56. MULLER, H. W.: Adaptation of Continuously Variable Transmissions to the Characteristic of a Drive Machine by Bicoupled Planetary Transmissions. Journal of Mechanical Design 103 (1981), pp. 41–47.

INDEX

Herbert W. Müller obtained his doctorate at the Technische Hochschule in Dresden in 1961. He is now chairman of machine elements at the Technische Hochshule in Darmstadt. He has published a number of articles in technical journals in Germany, and has presented papers to the American Society of Mechanical Engineers on several occasions.

The manuscript was prepared for publication by Jacqueline A. Nash. The book was designed by Ed Frank. The typeface for the text is Times Roman, designed by Stanley Morison in 1932, and the display face is Helvetica, designed by M. Miedinger in 1957.

Manufactured in the United States of America.